Tips and Techniques for Electronics Experimenters

2nd Edition

Don Tuite and Delton T. Horn

SECOND EDITION
FIRST PRINTING

Library of Congress Cataloging-in-Publication Data

Tuite, Don.
Tips and techniques for electronics experimenters / by Don Tuite and Delton T. Horn. — 2nd ed.
p. cm.
Rev. ed. of: Electronic experimenter's guidebook. 1st ed. 1974.
ISBN 0-8306-9145-6 ISBN 0-8306-3145-3 (pbk.)
1. Electronics—Amateurs' manuals. 2. Electronics—Experiments.
I. Horn, Delton T. II. Tuite, Don. Electronic experimenter's guidebook. III. Title.
TK9965.T755 1989 89-39622
621.381'078—dc 20 CIP

TAB BOOKS Inc. offers software for sale. For information and a catalog, please contact TAB Software Department, Blue Ridge Summit, PA 17294-0850.

Questions regarding the content of this book should be addressed to:

Reader Inquiry Branch
TAB BOOKS Inc.
Blue Ridge Summit, PA 17294-0214

Acquisitions Editor: Roland S. Phelps
Technical Editor: Laura Bader
Production: Katherine Brown

Contents

Introduction **v**

1 Foundation for Electronic Construction **1**
Recommended Tools—Soldering—Layout—Reminders

2 Construction Methods **15**
Breadboarding—Perforated Boards with Push-in Terminals—Perforated Boards with Adhesive-Backed Circuits—True Printed Circuits—Universal PC Boards—Metal Chassis—Wirewrapping—Generally Applicable Techniques—Multi-Strobe Project—12 V DC to 110 V AC Inverter Project—Triac Controller Project—Organ Project

3 Finishing Touches **56**
Cases and Enclosures—Laying Out a Control Panel—Legible Labels—Selecting Knobs, Switches, and Indicators—Making Your Own Meter Scales—Randomizer Project—FET-Set Shortwave Receiver Project

4 Troubleshooting Your Projects **80**
Troubleshooting Equipment—Systematic Troubleshooting—After You Find the Trouble—Capaci-Bridge Project—BJT-FET Transistor Checker Project—Theory of Transistor Checker—Signal Generator Project—Deluxe Logic Probe Project

5 Making Successful Substitutions **118**
Resistors—Capacitors—Coils and Transformers—Diodes—Bipolar Transistors—Field-Effect Transistors—Unijunction Transistors—Integrated Circuits

6 Special Tips for Using ICs **131**
Miniaturization—Installing ICs—Static Electricity

Appendix A Color Codes for Electronic Components and Wiring **137**

Appendix B Electronic Symbols Used in Schematics **145**

Index **150**

Introduction

This book is intended as a kind of bridge. On one side, there's you. You know a little about electronics, but want to know a great deal more. You've read through some books on basic theory, so you know about Ohm's law and you know what the lines and squiggles on a schematic diagram mean, even if you aren't always sure why all of them are there. You might have enjoyed putting together a hi-fi or test equipment kit.

On the other side of the bridge are all the home-built projects and roll-your-own pieces of gear that appear in hobby magazines and books. Beyond that lies the wide world of electronic projects that you can pull together out of ideas in your own head and create from scratch with your own hands.

Somehow, there's a gulf between the place where you stand now and all those fascinating homebrew projects. Unpacking the kit parts from the box with the Benton Harbor postmark is different from facing the clerk at the radio parts store with a handwritten list of parts and no idea of what to say when he tells you that a certain item is indefinitely out of stock! Somehow, too, you have the feeling that if you unwittingly make a mistake, you will wind up with $30 worth of junk parts. "See these cufflinks?" you envision yourself telling your friends, "They used to be $5 transistors!"

What you need is a bridge—a bridge to let you cross from where you are now to a place where you feel confident to undertake a project from a magazine or book, and maybe, ultimately, projects entirely of your own creation. This book is intended to be your bridge.

When you have finished this book, you should know quite a bit about the tools required to make a neat, functional project; the various ways of finishing projects off to give them that *professional* appearance; scientific methods of troubleshooting gear; and finally, some of the considerations involved in making substitutions when the exact component specified isn't available. To help make the main points of the text clearer, there are ten original projects that illustrate the major concepts in each chapter. You might want to build some or all of them or just read through them and look at the diagrams to see how the ideas are embodied in real hardware.

In this second edition, new information has been included, particularly emphasizing integrated circuits and newer construction techniques.

Building your own electronic gear is a rewarding pastime, and this book is intended to be a pleasant introduction to that pastime. So relax, take your time, and enjoy yourself.

1

Foundation for Electronic Construction

The best place to begin is with the tools you will need for building your projects. The tools listed in this chapter are just about all you will need to tackle any of the projects in this book, and in fact, for just about any project you will come across anyplace else.

RECOMMENDED TOOLS

You won't need all of these recommended tools for every project in this book, but if you will be doing much electronics experimenting, you will need each of the tools on the list. There are some other tools like socket punches, tin snips, or heavy-duty soldering irons that you might need from time to time. You're not likely to need them right off, so it's best to wait until you find yourself with a specific project that calls for them before you buy them.

Pliers

You will need *long-nose pliers* for getting into tight spaces, fastening wires to solder lugs, and retrieving dropped parts from crowded corners of chassis. It's handy having two sizes, but if you can only afford one, get the longer variety. Before you buy a pair of long-nose pliers, hold them up to the light and look through the closed jaws. The less daylight you can see between the jaws, the better. However, don't reject a pair that has a ridged gripping surface at the tip and daylight showing the rest of the way down to the hinge (Fig. 1-1). If the jaws meet evenly all over the ridged area, you've found a good pair of pliers.

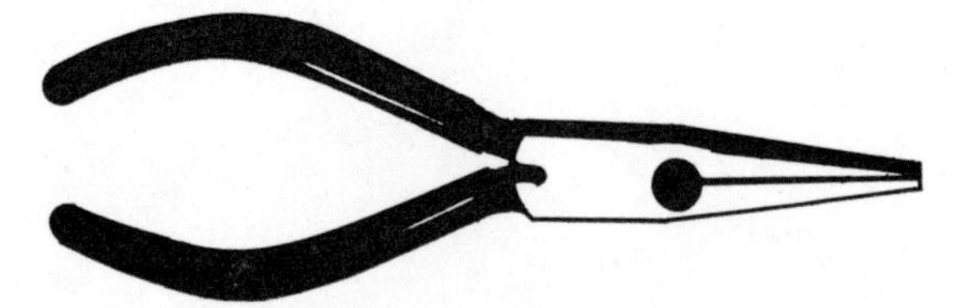

Fig. 1-1. Long-nose pliers are useful for working in tight areas.

Long-nosed pliers are sometimes called ***needle-nose pliers.*** Almost every pair of long-nosed pliers features a cutting area near the base. This cutting area is handy for cutting light wires or snipping off excess lengths of component leads.

However, this long-nosed pliers won't do for all of your needs. Sooner or later you will need to cut a wire or lead in a cramped space. It may be awkward, or even impossible, to get the cutting portion of the long-nosed pliers into the necessary position. Also, if you are cutting a lot of wires, the long-nosed pliers can get to be rather awkward and uncomfortable. For such wire-cutting tasks you need a dedicated tool, in this case ***diagonal side-cutting pliers,*** or ***dikes*** (Fig. 1-2).

Fig. 1-2. Diagonal side-cutting pliers, or dikes, *are intended specifically for wire cutting tasks.*

When you are working on your projects, do yourself a favor and don't try to cut oversize wire with an undersize pair of dikes. Oversize wire is anything over No. 12 (AWG) for most dikes. When you want to cut oversize wire, first nick it with a knife where you want to cut, then bend the wire back and forth a few times at that place. The wire will break cleanly right where you want it to, and you'll avoid dulling your wire cutters.

After you've cut a length of wire, you still have to strip the insulation off. Some hobbyists strip wires with a pair of dikes, an X-acto knife, or even a razor blade. In addition to being awkward, and (if a knife or razor blade is used) potentially dangerous, such make-shift measures really don't do a very good job. It's almost impossible not to nick the wire, which could weaken it and possibly cause it to break, usually at the most inconvenient time. Once again, it is best to use a tool specifically designed for the job at hand.

One of the most popular types of ***wire strippers*** is a very simple type with notched shear-type blades, shown in Fig. 1-3. This device is handy to use and quite inexpensive. It can easily be adjusted for different wire sizes. You may need to use a pair of long-nosed pliers to hold the wire securely with one hand while using the stripper with the other hand.

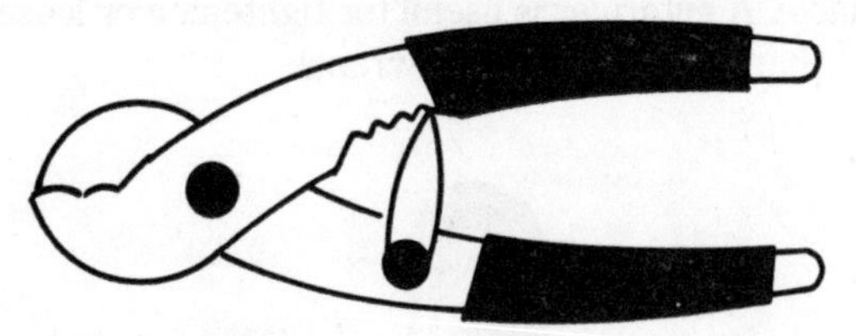

Fig. 1-3. A simple wire stripper.

A variation on this device is a combination tool for wire stripping and crimping solderless terminals. This tool tends to be a bit more expensive, but there are some household uses for solderless connectors, so it isn't a bad investment.

The best tool for wire stripping is a nightmarish-looking pair of pliers with a viselike arrangement in one set of jaws to hold the body of the wire and a set of knife edges in the other jaw with notches in them to fit different sizes of wire. It is naturally more expensive than a simple manual wire stripper, but it is faster, more convenient, and does a better job. This tool never nicks wires, and it is fascinating to watch. Variations are priced in the $7 to $25 range.

Wrenches and Nutdrivers

As a general rule, you should never use a pair of pliers on a nut. Use a wrench, preferably one of the right size, or at least a good adjustable wrench. Like all general rules, however, there are times when this one can be broken. Sometimes it is very handy to have a pair of 10-inch ***vice-grip pliers.*** These pliers will also prove themselves invaluable for holding small work while you are soldering it, and for hundreds of other uses. They are a good investment.

The two sizes of hex-head nuts you will encounter most frequently are the ¼- and ½-inch sizes. The ¼-inch size is found most frequently on nuts for 4-40, 6-32, and 8-32 screws. It is also found on a variety of sheet metal screws. The ½-inch hex nut is used for the bushings of most potentiometers and toggle switches. It is handy to have a nutdriver for ¼-inch hex nuts and an open-end wrench for ½-inch hex nuts.

Again, do not mangle these nuts with some kind of pliers—use a wrench to hold them. Nothing is more certain to spoil your attitude about a project than slipping while trying to tighten a nut with the pliers and scratching the finish on a neatly painted control panel.

From time to time, you will encounter other sizes of hex-head nuts. For these occasions, you will need either a small, adjustable open-end wrench or an assortment of nutdrivers and wrenches in various sizes.

A set of ***nutdrivers*** are not absolutely essential, but they can come in handy in many instances. A nutdriver, as shown in Fig. 1-4, is a lot like a screwdriver, except instead of a blade there is a socket for gripping a nut at the end of the long handle. A nutdriver is useful for tightening or loosening hard to reach nuts, where there isn't room for a wrench.

Fig. 1-4. A nutdriver is similar to a screwdriver, except the head is designed to hold a nut.

Although a few adjustable nutdrivers have been marketed, most hobbyists will buy a set, with each individual nutdriver designed to fit a single standard nut size.

You will need at least two *flat-bladed screwdrivers,* one with a 1/4-inch blade for most ordinary screws, and one with a 1/8-inch blade for the setscrews in most knobs and for tight work. A set of *jeweler's screwdrivers* may prove handy, but it is not an absolute necessity. You will need at least one *Phillip's head screwdriver.*

Some knobs do not have slotted-head setscrews, but have Allen head setscrews instead. There are several sizes of these, and it is best to simply buy a set of *Allen* (or *hex key*) *wrenches.*

Electric Drill and Bits

You can no more make a project without drilling holes than an omelet without eggs. It makes sense to buy an *electric drill,* preferably one with a variable-speed motor. There are some important advantages to the lower speed ranges available with modern electric drills. With slow-speed capability, you can start a hole right where you want it, without center-punching it first, and you know that the bit won't wander all over the chassis. Also, brittle materials like plastic, Bakelite, and printed circuit board drill better at low speed, with less chipping and cracking.

When you buy *bits*, you'll probably find it doesn't cost much more to buy an assortment of 10 than it does to buy one of each of the sizes you really need. However, it is worth knowing that you can get by with six basic sizes and some tiny bits for printed circuit boards. The six sizes you will absolutely need are the three sizes needed to pass the most commonly used screws, plus ¼-, ⅜-, and ½-inch sizes.

There are two ways of identifying bits: by their number sizes, which are related to their dimensions in thousandths of an inch; or by their diameters, expressed in fractions of an inch. The most common fractions-of-an-inch types are graduated in multiples of 1/64 inch and do not correspond to any of the numbered sizes. Therefore, for each screw size, there are two possible sizes of drills to buy, depending on how the drills are designated. Thus, for clearing 8-32 screws, drill a hole using either a No. 18 or 3/16-inch bit. For a 6-32 screw, use a No. 28 or a 9/64-inch bit. For a 4-40 screw, use a No. 33 or a ⅛-inch bit.

You'll need a ¼-inch bit to make holes for passing the shafts of the most common sizes of variable capacitors. The ⅜-inch bit will be needed to make holes for the bushing of potentiometers and toggle switches. It also makes the minimum size hole required to pass the jaws of the nibbling tool that will be discussed in the next section. The ½-inch bit will make holes that will pass pilot-light assemblies and some kinds of binding posts. Both ⅜- and ½-inch drills can be bought with ¼-inch shanks for use in the most common electric drills. When you use these oversize bits, however, bear in mind that you are calling on your drill to do more than it was designed to do. Expect some chatter, and brace the work firmly. Keep your fingers out of harm's way. Don't overheat your drill with heavy stock.

Most radio stores and supply houses carry special bits for printed circuit work. These bits are exactly the right size to pass the different gauges of wire you will be using. They are also designed with a different cutting angle than metal-working drills. This allows them to drill brittle printed circuit boards more easily, and reduces the chance of these delicate bits breaking.

Nibbling Tool

If ever there was a tool to gladden the heart of an electronics experimenter, the nibbler (Fig. 1-5) is it. The nibbler works on the basis of a small, tool-steel (very hard) shears and a handle with a tremendous mechanical advantage. In operation, the head of the tool is passed through a ⅜-inch hole and the jaws are positioned for the first cut. From there the tool literally

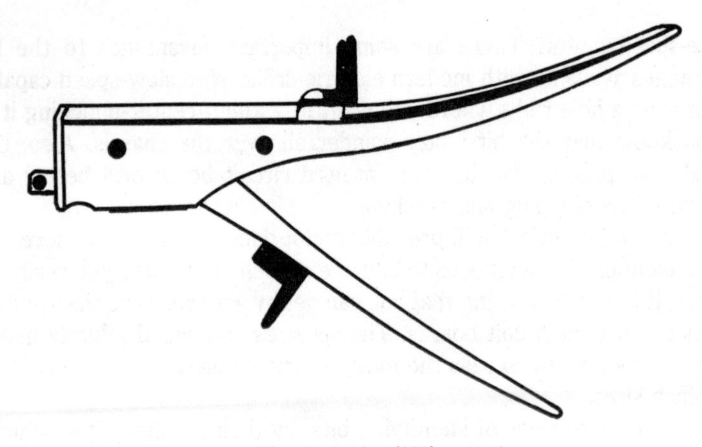

Fig. 1-5. A nibbling tool.

nibbles any desired shape in metal up to 18-gauge mild steel or 16-gauge aluminum.

Saws

Not every piece of metal or plastic can be nibbled. The shafts on variable capacitors and potentiometers, for instance, must be trimmed to size with a hacksaw. Often, plastic boxes for projects are too thick to fit in the jaws of the nibbler. This is where a keyhole saw comes in handy. One of the best and least expensive keyhole saws consists of an X-acto knife handle with a specially designed keyhole saw blade. It takes a little work to find these blades sometimes. The surest place to find them is in a store that sells to model airplane hobbyists.

Knives

It's nice to have a pocket knife that you can carry around and have handy for odd repairs, but for workshop use, the *X-acto knife* with X-acto's No. 11 blade is the handiest thing you will find. The reason for this is the extreme sharpness of the blade. This makes it easy to cut through insulation with very slight pressure, which results in fewer nicked and broken wires. One caution: the blades of these knives do wear out fairly quickly; as soon as it seems to be happening to the blade you're using, throw it out and replace it with a new one. This will save you considerable frustration, particularly when you are trimming patterns for etched circuits.

Files

Drilling holes, or sawing, or nibbling—whatever metal-cutting technique you're using—you are bound to leave burrs and rough edges. Therefore, it is absolutely necessary to file away burrs and smooth edges before painting. You will need large and small sizes of ***round, half-round,*** and ***flat files,*** as well as a ***rat tail file.*** Files do not cut well when they are clogged with metal, so you will also need a wire-bristled brush to clean your files. A ½-inch ***hand reamer*** is a handy tool to have also, for enlarging holes to odd sizes, since it will do the job faster and leave a rounder hole than a file.

A final essential for electronics is a ***soldering gun*** or ***iron.*** Proper soldering is so important to successful completion of a project that soldering tools are treated separately in the following section.

SOLDERING

Proper soldering is a very easy art to learn. However, more kits and projects are ruined by bad soldering than any other cause.

Tools

There are at least two types of soldering guns and two types of soldering irons. It is difficult to decide when to call a tool an ***iron*** and when to call it a ***gun,*** but for our purposes, an iron is any soldering implement that is on as long as it is plugged in, and a gun is any soldering implement with a trigger-type on-off switch.

The most common kind of gun has a tip that is essentially a loop of wire. This is connected to the secondary of a step-down transformer inside the gun. Another type has an enclosed heating element with a resistance that increases with temperature. This is a more efficient gun, since the current it draws varies with the head load on the tip. Its disadvantage is that replacement tips are more expensive than tips for the loop type. In fact, an emergency tip can be made for this gun out of any stray piece of heavy wire that might be in the junk box.

The advantage of the gun is that it can safely be left plugged in all the time without burning out the tip or creating a burn hazard. The gun's disadvantages are the warmup time required and its excess heat output. Most guns put out too much heat for delicate transistor work.

Soldering irons can be divided into ***low-wattage pencils*** and ***high-wattage irons.*** For anything except soldering to a heavy chassis, the pencil is the only iron suited to electronics work. A soldering pencil in the 45 W range is an almost ideal soldering tool, but for very fine work a 37½ W pencil is even safer.

Preparing Soldering Tools

To make a good solder joint, the tip of the soldering tool must be properly prepared. You cannot get a good solder joint using a tool with ⅛ inch of scale glowing cherry red on its tip. A lot of kits and projects go wrong at this point. If you are starting with a brand-new tip on your soldering tool, read on. If the tip of your soldering tool is crusty and corroded, file it down to bare copper, or replace it, and then read on.

Start with a brightly tipped soldering tool and heat it to its operating temperature. Then take your solder and liberally coat the entire surface of the working end of the gun or iron. From now on, whenever you use the soldering tool, wipe it frequently with a rag or moist sponge to keep the solder on the tip bright and shiny. When you are through with your gun or iron, wipe the tip as it is cooling to keep the tinning of solder on it shiny. This effort will repay you many times over in fast heating and good solder joints. If you have some money to spare, there is a chemically treated sponge in its own plastic holder that can be bought in most radio specialty shops and from the mail-order houses. The sponge has a number of slits in its tip which are handy for wiping the tip of your gun or iron, and the plastic holder has a nonskid base that allows easy one-hand operation.

Solder

First, never use *acid core solder.* The only kind of solder to use in electronics work is rosin core. Solder is an alloy of tin and lead. The concentrations of tin and lead in the alloy determine its characteristics. The ideal alloy is 63-37 (63 percent tin, 37 percent lead), or *eutectic solder.* This has the lowest possible melting point of any tin-lead solder, and it is nearly impossible to make a *cold solder joint* with this material. Eutectic solder isn't very common, however. The most frequently encountered alloy is 60-40. This has a satisfactory melting point and resistance to cold solder joints. Some 50-50 alloy solder is available, but its use should be avoided.

Generally, solid solder should not be used for radio work. Solder with a rosin core (or cores) carries just enough rosin with it to vapor-clean each joint as it is made. It is possible to apply rosin flux separately and use solid solder, but the technique is messy. There is one case in which solid solder and a separate paste flux is used—when it is necessary to solder to aluminum. For most low-frequency work, chassis grounds are achieved with solder lugs and screws. For high-power or high-frequency work, however, it may be necessary to solder directly to an aluminum chassis. With ordinary rosin core solder this is impossible, because the oxide coating on the aluminum resists the cleaning effects of the rosin flux. However, there is an aluminum-soldering flux that can be purchased that will clean away aluminum oxide and allow solder to flow. For this one purpose, solid (coreless) solder should be used.

Preparation for Soldering

The first thing to consider in preparing a joint for soldering is what you are soldering to. The basis for any good solder joint is some firmly fixed terminal—either a tube or transistor base pin, a solder lug, a tie point on a terminal strip, a push-in terminal on a piece of perforated board, or a pad on a printed circuit. With few exceptions, never connect two components except by soldering them to a fixed support.

If one of the components to be soldered is an insulated wire, it will be necessary to strip the end of the wire. This is best done with a stripping tool of some kind, although in a pinch a knife will do. It is very important not to nick the wire, especially with solid wires. Nicking a wire creates a stress concentration at the nick. You can demonstrate to yourself how serious this is by taking an ordinary copper wire and making a small nick in it. You will find that only four or five sharp bends will cause the wire to break at the nick.

Once you have the ends of your wires and components prepared, the next step is to fasten them firmly to the support point, as illustrated in Fig. 1-6.

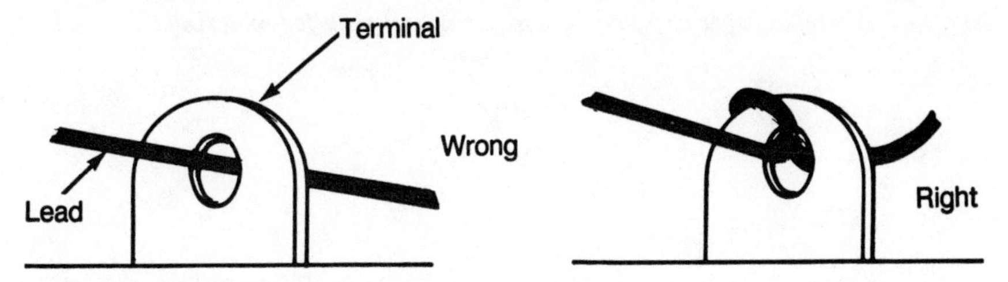

Fig. 1-6. Before soldering, you should have a good, strong mechanical connection.

Bend the wires and crimp them slightly. The mechanical strength of the joint must come from the physical attachment, not from the solder. The purpose of the solder is only to assure a complete *electrical connection,* and nothing more! On printed circuits, push the lead through the hole and bend the wire against the copper side of the board to secure it.

Next, before applying any heat, ask yourself, what will the heat do to the components? If any of the leads you are soldering come from a transistor or diode, the heat of soldering could damage the part. To protect the part, hang an alligator clip or paper clip from the wire lead, or lean a screwdriver blade or the jaws of your pliers against it to provide a heatsink between the joint to be soldered and the part itself, as shown in Fig. 1-7.

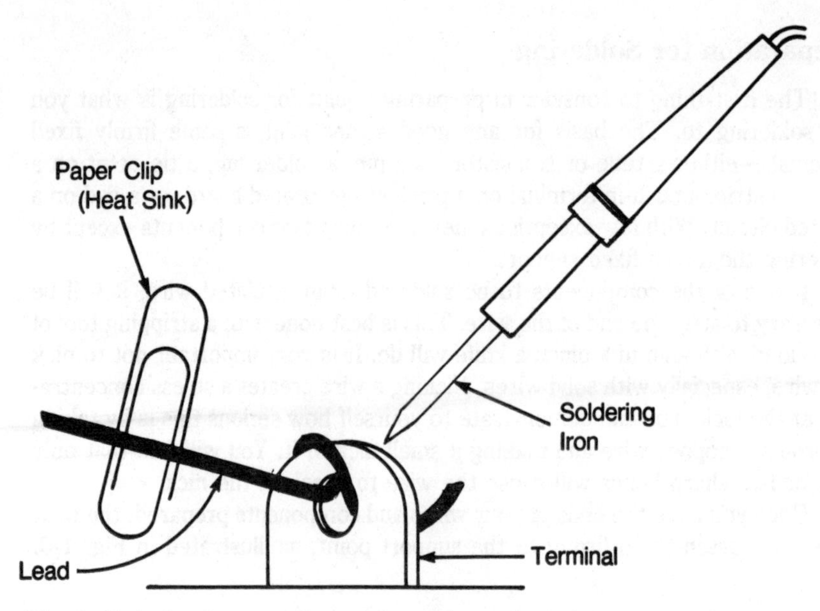

Fig. 1-7. A simple paper clip may be used as a heat sink for soldering.

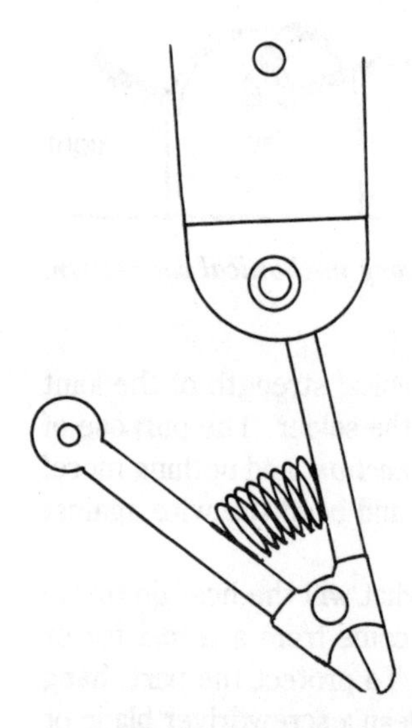

Fig. 1-8. A clip-on heat sink.

You may prefer to use a specially designed clip-on heat sink, like the one illustrated in Fig. 1-8. These spring-loaded devices are inexpensive, and very handy to use. When clipped on a lead, they remain securely in place.

Now you are ready to actually apply heat to the joint. The object is to heat the joint so that when the solder is touched to it, it will flow, as shown in Fig. 1-9. Note that you *do* not apply solder to the iron tip and let it flow down onto

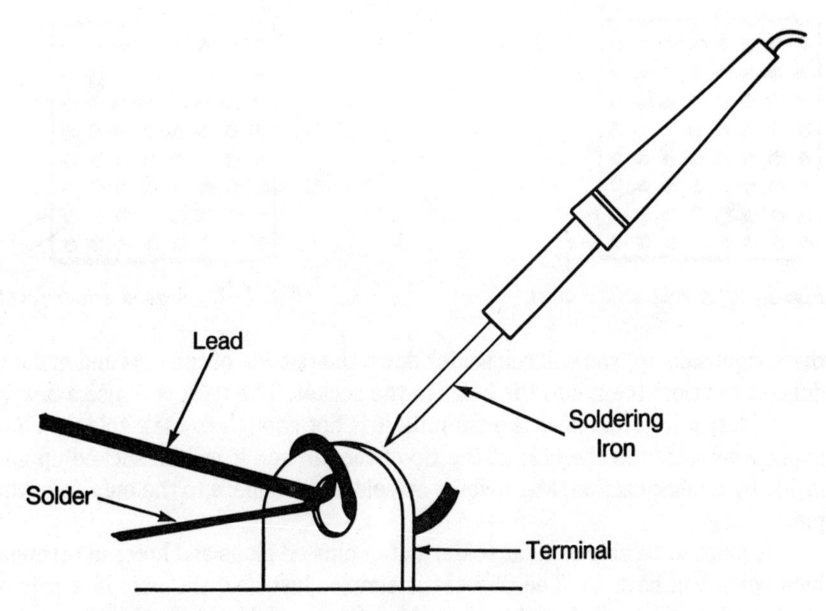

Fig. 1-9. Heat the joint so that the solder will flow when it is touched to it.

the joint. Sometimes, it is convenient to apply a surplus drop of solder to the tip of the tool to aid in heat flow, but the bulk of the solder should be applied directly to the joint.

You can tell that the solder is flowing properly by observing the way it covers the wires and support. It should wet all the surfaces it comes in contact with, exactly as if it were a drop of water. If it stands up and beads on the surface, then the surface isn't hot enough and the solder connection is incomplete. This is the most frequent cause of *rosin joint* failure. At first glance, the joint looks solid, yet all that is holding the parts together is a film of rosin flux.

Once the solder has flowed onto the joint, remove the heat; do not move any wires in the joint. If you allow any of the wires to move, you will end up with the dull gray surface on your solder that is characteristic of a cold solder joint. Cold solder joints can plague you with high resistance in your connections, or even worse, they can work as diodes to cause unwanted audio from nearby radio stations to appear in audio amplifier circuits. Figure 1-10 shows a cold solder joint.

Do not use too much solder. Use just enough to wet the whole joint. Globs of solder can interfere with pins in sockets and can cause unintended shorts. Figure 1-11 shows a proper solder joint.

Sometimes it is necessary to solder wires inside pins for different kinds of connectors. There is a technique for this that can save a lot of frustration. The technique is based on the *surface tension* of the solder and the principle of *capillarity.* The way to solder these pins is upside down. If you try to solder

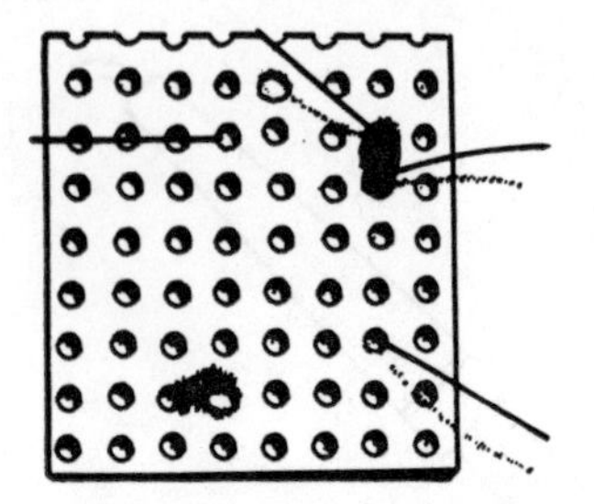

Fig. 1-10. A cold solder joint.

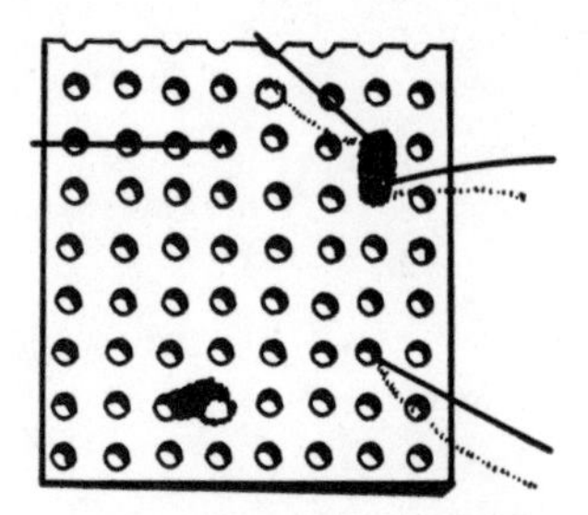

Fig. 1-11. A good solder joint.

them right side up, you will run solder down the outside of the pins and make it difficult to insert them into the holes in the socket. The trick is to use a nearly dry soldering iron and heat the pin until it is hot enough to melt solder. Then apply your solder to the hole at the tip of the pin and it will be sucked up and inside by capillary action. Meanwhile, no solder will adhere to the outside of the pin.

It helps to be able to clear solder out of pins on plugs and holes in terminal lugs when you have to. The process is simple. Just hold the part in a pair of pliers and heat it until the solder is molten. Then quickly strike the part or the base of the pliers on the table and the solder will shoot out. Be careful it doesn't splash on you, though. A variation of this technique is to blow sharply on the molten solder. This works better with solder lugs than with pins, however, since the surface tension in the pins is quite high.

With the ever-increasing miniaturization of electronic components, the potential problem of solder bridges is something every electronics experimenter must constantly be on the look out for. Solder bridges occur most frequently on printed circuit boards, particularly with integrated circuits (ICs). If you are not very careful when soldering, a little solder may ooze over and create an unintentional short between two adjacent leads, or printed circuit traces. It is good practice to examine every newly built circuit very carefully before applying power for the first time. Look at the finished circuit board carefully under a good light. If possible, use a magnifying glass. Even a very, very tiny bit of solder can create problems. A solder bridge will, at the very least, cause the circuit to function improperly. It could even damage some components.

A related problem often shows up in projects that have been working for awhile, then start behaving strangely or stop working altogether. These problems are often intermittent. That is, sometimes the circuit works correctly and sometimes it doesn't. In many cases the cause of the problem is a tiny loose glob of solder that has broken free, or a bit of a cut-off component lead. If such a stray bit happens to land where it can act as a short circuit, almost anything may happen. Such problems can be very frustrating to repair, especially if you don't know what to look for. A close visual inspection should be the first step in any repair job.

LAYOUT

It is always best to follow the same arrangement of leads and components described in your kit manual or in the article describing your project. In some cases—for instance, in many circuits in semiconductor handbooks and manufacturers' application notes—there is no information given as to the physical layout of the project. In these cases, there are a few simple rules you should keep in mind:

1. Be logical. Use the schematic diagram as a starting point. Position the components between the ground bus and V+ line. The closer your layout is in appearance to the symbols on the circuit diagram, the fewer mistakes you will make and the easier it will be to troubleshoot your project.
2. If you are using 60-Hz house current for power, keep all lines carrying the current well away from signal leads. If you must make a long run of signal lead, use shielded wire, or for shorter runs, twist the signal lead and its associated ground lead together. Power supply leads in high-power ham transmitters should be shielded.
3. Avoid running leads carrying two different frequencies (RF and IF, or RF and audio) closely parallel to each other.
4. Be careful about *lead dress* (positioning). Wires carrying radio frequencies should be as short as possible, with no sharp bends. All other wires should be bundled into cables as much as possible and routed along edges of circuit boards or chassis, or parallel to edges. For longer runs, use tiedown anchors.
5. Group controls for convenient operation. For critical settings such as main tuning controls or speed controls, larger knobs are easier to handle.
6. Do not let components hang by their leads. This is an invitation to vibration or shock failure after a few hundred hours of operation.

REMINDERS

Take the time to use heatsinks. It is surprising how little heat in the wrong place is required to destroy a semiconductor device.

In addition to using heatsinks, be careful with hot soldering irons. They may radiate enough heat to destroy a nearby transistor. Also be careful not to melt the insulation on wires near where you are soldering. At the very least, it makes the final product look unsightly. At worst, you may have a very hard-to-locate short circuit.

Avoid dropping any components. Hard as it may be to believe, a drop of a few feet onto a hard surface can subject a device to a shockpulse of thousands of times the force of gravity. This can have fatal effects on transistors and diodes.

Be especially careful when handling metal oxide field effect transistors (MOSFETs) that do not have integral gate protection. Static charges that build up on your body from contact with clothes can punch through the oxide insulating the gates of these devices from their channels. Always use transistor sockets with unprotected MOSFETs. Just before picking up an unprotected MOSFET, discharge yourself to a large metal object.

If all the preceding *dos* and *don'ts* haven't intimidated you, you are ready to take on the different methods of construction—perforated board and terminal, printed circuit, and metal chassis, with point-to-point wiring.

2

Construction Methods

There are six basic electronic construction methods which we will discuss in this chapter. One of these methods, breadboarding, is suitable only for temporary construction of circuits, permitting fast and easy changes in the design of the circuit. Two of the permanent construction methods, metal chassis and perforated board with push-in terminals, use point-to-point wiring. The other three methods use circuits on the surface of special insulating boards. There are advantages of special insulating boards. There are advantages and disadvantages to each of the methods. Your selection of a particular method for a specific project will depend on the nature of the project and the resources you have available. Of course, simple personal preference may also play a part in the selection process.

The different methods of construction are demonstrated in the projects in this chapter. It should be mentioned that there are additional construction methods not mentioned here. The methods described in this chapter are the most popular and most suitable for hobbyist (as opposed to industrial) circuit construction.

BREADBOARDING

In the very early days of radio and electronics, circuits were assembled on top of wooden bases. Brackets were used to hold controls and tube sockets above the surface of the wooden base, and wires and components were simply run from terminal to terminal. The similarity of these wooden bases to kitchen breadboards led to the name for this method of construction—breadboarding. No doubt, many actual breadboards were used by early home experimenters.

The term *breadboard* is still used to indicate a construction technique in which a prototype circuit can be quickly assembled and easily modified. Modern breadboarding is usually done with a special solderless socket, like the one shown in Fig. 2-1. This socket features many holes that component leads or wires can fit into snugly. The holes are internally interconnected in a specific pattern. A typical interconnection pattern is illustrated in Fig. 2-2.

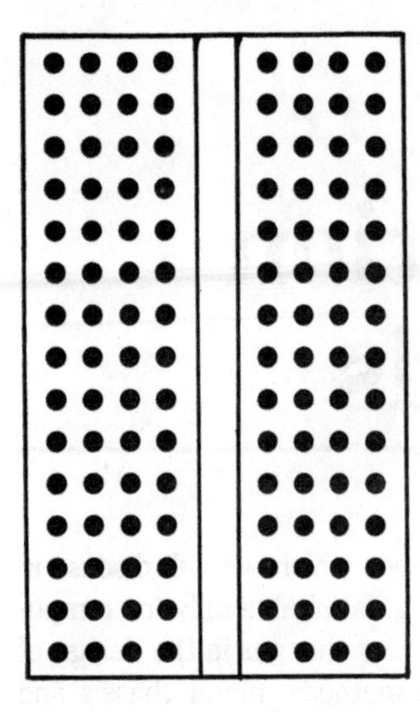

Fig. 2-1. Modern breadboarding is usually done with a special solderless socket.

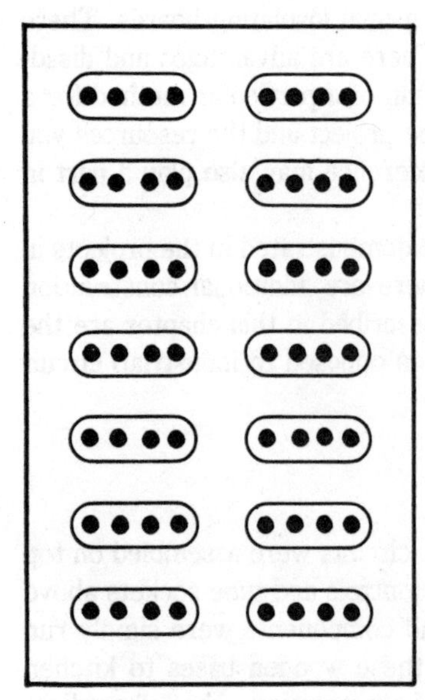

Fig. 2-2. A typical interconnection pattern for a solderless socket.

Many experimenters use a full breadboarding system. Such a system can be bought commercially, often in kit form, or it can easily be built from scratch.

It features a solderless socket, a few potentiometers and/or switches, and a common, convenient base. Usually commonly used subcircuits, such as power supplies and oscillators, are also included in the same package so that they don't have to be custom breadboarded for each individual project. A good breadboarding system can greatly speed up circuit design and modification time. A typical commercially available breadboarding system is shown in Fig. 2-3.

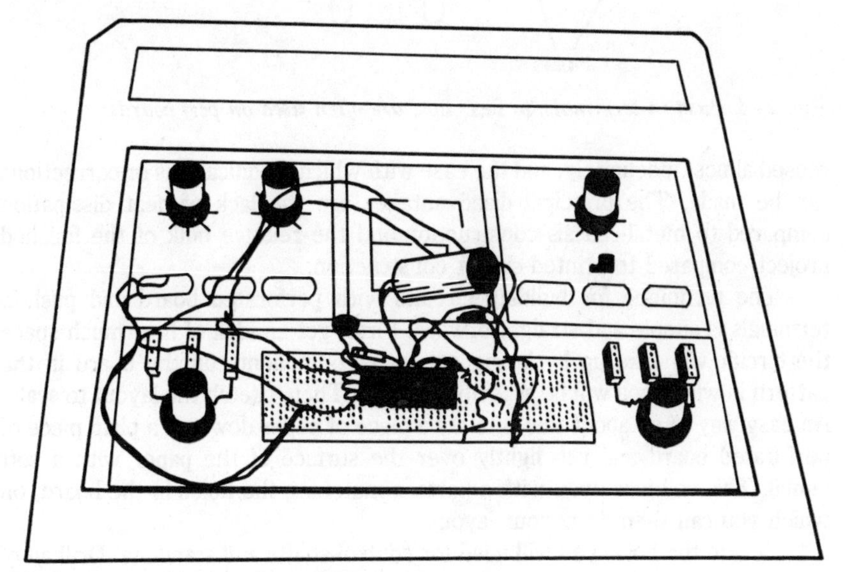

Fig. 2-3. A typical breadboarding system.

It should be remembered that any breadboarded circuit is temporary. The connections are not stable enough for long-term use.

PERFORATED BOARDS WITH PUSH-IN TERMINALS

In many ways, the modern perforated board, or perf board, is a direct descendent of the early experimenters' wooden breadboards. In some cases, component leads are soldered directly to one another. But usually, it is advisable to use push-in terminals, or *flea clips,* as shown in Fig. 2-4.

Perforated boards are available with two hole sizes, 0.093 and 0.062 inch, and several different patterns of hole spacing. There are also different kinds of push-in terminals that are made to be inserted into these holes. Most of the time, you will probably be using the kind that are designed to grip a wire while you solder to it. You may also, however, find use for the quick-disconnect type for patching together prototype circuits before you decide on a final design.

The advantages of this kind of construction are the short amount of time required to go from idea to finished circuit, the ability of the boards to be

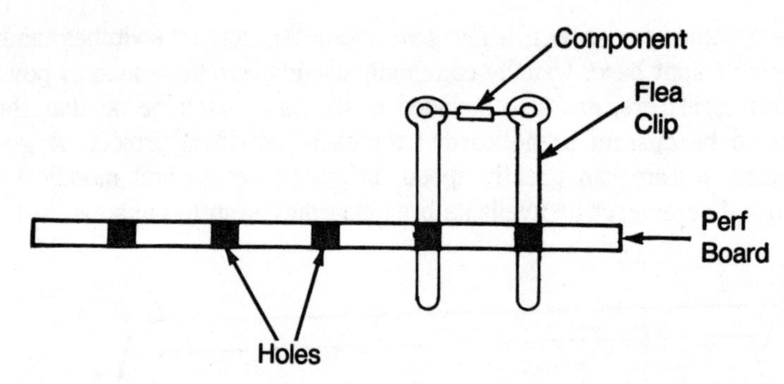

Fig. 2-4. Push-in terminals, or flea clips, *are often used on perf boards.*

reused almost indefinitely, and the ease with which modifications or corrections can be made. The principal disadvantages are the lack of heat dissipation compared to metal-chassis construction and the relative bulk of the finished project compared to printed-circuit construction.

The technique for building circuits with perforated board and push-in terminals is simple and straightforward. First, get an idea of how much space the circuit will take up by laying out your components on the board in the pattern in which you will be assembling them. Then, sketch the layout to scale. An easy way to go about this is to lay a piece of paper down on a plain piece of perforated board and rub lightly over the surface of the paper with a soft pencil. This will leave you with a pattern matching the holes in the board, on which you can then draw your layout.

Locate the holes you will need for control shafts and standoffs. Drill all of these holes before you insert any terminals.

Using your layout sketch, insert the push-in terminals into holes as required. Do not try to connect too many wires to a single terminal. Use two terminals in adjacent holes connected by a short wire.

Now mount your controls and standoffs and all the other hardware, such as transformers and sockets, that you will need. Be sure to use lockwashers under all nuts and on all control shaft bushings.

You are now ready to install your components and solder their terminals. Solder as you go. As soon as you have all of the wires for a particular terminal in place, solder them. This will help to avoid the frustration of wires springing loose.

If you are doing some original experimenting, there is a technique that can be used with perforated board that is very handy. Use a relatively large piece of board and run buses down its length for each important node in your circuit. You will have one bus at the top of the board for V+; another along the bottom for ground; another, if appropriate, for an age line; and so forth. Label each bus so you will be able to identify it easily. With this kind of arrangement, it becomes a simple matter to swap components and configurations to determine the best circuit.

PERFORATED BOARDS WITH ADHESIVE-BACKED CIRCUITS

One very handy construction technique uses conductors with pressure sensitive adhesive backing, marketed under the trade name *Circuit Stik*, among others. There are a variety of pads with proper spacing to allow immediate insertion of transistors in standard cases, ICs in dual in-line pin (DIP) housings or cans, and discrete components. There are several sizes of copper ribbon with adhesive backing that are used to connect the pads in the finished circuit. The final product is a board that looks very much like a true etched or printed circuit board (discussed later in this chapter).

The advantages of this construction method are neatness, speed of layout, and low lead inductance. You can also make corrections and modifications to your circuits. Laying out a few projects with these adhesive pads and strips will give you some good practice in laying out printed circuits—without all the fuss and bother associated with acid etching. A final advantage of this method is that you can easily transfer a successful layout to a ***printed circuit master*** for group projects.

The main disadvantage of these stick-on circuits is that they are relatively expensive. By the time you buy the board, copper strip, universal pads, and transistor pads for even a simple project, you will find that you already have a bill for $10 to $15. Another disadvantage is that sometimes pads and strips do not stick as well as they should and come loose from the board during soldering. But if you are careful to burnish the stick-on elements in place well and not to get oil from your skin on the surface of the board, peeling will not be a serious problem.

The first step in building projects using this technique is to sketch the layout you plan to follow. If you find that you cannot make the circuit without crossing wires, you can run the crossing conductors on the back side of the circuit board. Make your sketch roughly to scale.

Then begin to lay out your pads and connecting strips on the perforated board. Use the real parts for templates to determine the proper spacing between pads. For instance, you should take a resistor of the size you are using and bend its leads to the proper shape and use this to find the exact holes in the perforated board on which to place your pads. Remember that each lead from each resistor, capacitor, etc. requires its own hole.

There are two ways to mount components like resistors and capacitors. You can install them with their bodies flat against the board or, to really save space, you can use a form of what engineers call *cordwood packaging*. In cordwood packaging, one of the two axial leads of the resistor or capacitor comes straight out from the body of the component and passes through one hole in the board. The other lead is bent back parallel to the body and passes through another hole very close to the first. This practice permits components to be crowded together in a very dense manner (Fig. 2-5). The image of many

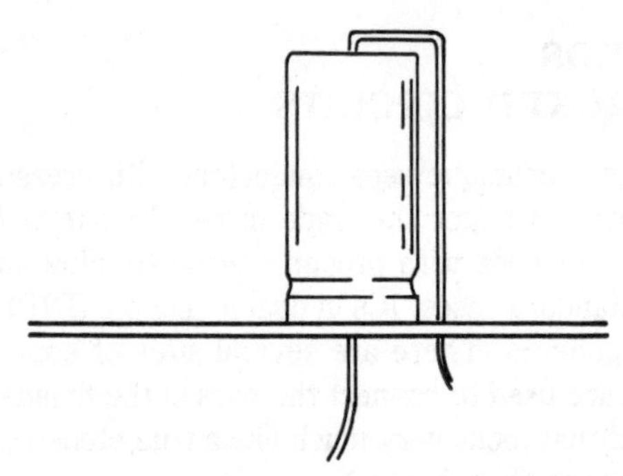

Fig. 2-5. Cordwood *packaging permits very compact circuits.*

components stacked side by side is what originally gave the mounting method its name. Another variation of cordwood packaging uses two boards, one above the other, with circuits on each board and components stacked between them. Designing these circuits is very time-consuming, but computer manufacturers, in particular, have found it rewarding.

After you have laid out all of your pads and interconnecting strips, cover the entire board with a piece of paper and burnish all the circuit elements heavily with a burnishing tool or the plastic cap from a ballpoint pen. This burnishing is very important to the sticking of the circuit elements, so make sure you bear down heavily on each one.

Now, since you have a piece of paper on top of the board anyway, take a soft-lead pencil and rub very lightly over the entire surface of the board. This will cause an image of the circuit to appear on the paper. Remove the paper from the board and transfer your initial sketch to the image of the circuit on the paper. This will tell you whether you have, in fact, provided for all of the leads from all of your components. If you have not, it is a simple matter at this point to go back and add what you need, taking care to burnish each new addition.

Following your new sketch, install your components. Push the leads of each component through the board from the top side until the body of the component is flush with the surface of the board. Then bend the leads of the component on the circuit side over until they are flush with that surface of the board, and cut all but ⅛ inch, or so off with your dikes.

Be careful not to apply any more heat than necessary, and solder each bent-over wire to its pad. Then go back and solder each discontinuity between pad and strip, pad and pad, or strip and strip. This is necessary to make sure that there are no high-resistance joints caused by the adhesive.

These adhesive pads and strips have a number of advantages. Setting up a circuit using this method is relatively fast and easy, with a minimum of mess and bother. However, there are disadvantages too. Probably the most important is that occasionally the adhesive won't hold as firmly as it should. A connecting strip can lift off the board, and break off. To prevent this, some hobbyists add a coat of varnish to the bottom of the board after all of the soldering is completed. This adds considerably to the durability of the circuit, but it severely limits serviceability. If anything goes wrong, you must somehow remove the varnish before you can do any desoldering. If you use this method, make absolutely sure that the entire circuit works perfectly for all of its functions before applying the varnish. Especially be on the lookout for poor mechanical connections at solder joints and for cold solder joints.

You could compromise by applying varnish with a small brush, avoiding the solder pads. The results will not be quite as durable as with a complete coat of varnish, but it is better than nothing, and later you will have access to the solder joints.

TRUE PRINTED CIRCUITS

The term ***printed circuit*** is actually inaccurate. So-called printed circuits are not made by any process resembling the one used to make this book. Instead, the circuits are ***etched*** in an acid bath. The starting point for an etched circuit is a piece of insulating board on which a thin coating of copper has been deposited. To make the etched circuit, portions of this copper are covered with an acid-resistant material called ***resist.*** When the board with the resist circuit on it is immersed in an acid bath, all of the copper that is not covered by the resist is eaten away, and the circuit is all that remains.

The principal advantage of etched, or printed, circuits is that the pattern of conductors can be repeated on as many circuit boards as required. This makes this construction technique ideal for large production runs. For the home hobbyist, this repeatability can also be an advantage, since groups of people can all take on the same project and be assured that they will all achieve similar results.

Other advantages of etched circuits are that they are neat, very reliable, and have low lead inductance. At ultrahigh frequency (UHF) and microwave frequencies, they can be used in stripline assemblies.

The disadvantages of etched circuits are that they are relatively time-consuming to prepare, the chemicals required are expensive and slightly hazardous, and there are no convenient ways of correcting mistakes on etched circuit boards. If you goof, you have to start over.

There are three common methods used by home experimenters for preparing etched circuit boards; all three have many steps in common. The three ways of making etched circuits are

- Direct application of resist,
- Sensitized board and photo negative, and
- Sensitized board and mechanical negative.

In the first method, resist is simply painted or drawn on the surface of an untreated copper-clad board. In the other two methods, the copper surface of the board is treated with a photosensitive chemical. When these boards are exposed to light through a high-contrast negative and developed, the parts of the board exposed to light are acid resistant, while the unexposed portions allow the acid to etch the copper.

First Steps

For all three processes, the first thing you must do is lay out your circuit on a piece of paper. Many magazine projects provide a full-size pattern for experimenters to trace or copy. If you are starting from scratch, you will have to lay out your circuit to scale, taking careful note of the size of all your components and the positions of their leads.

Many builders limit the width of their conductors and use circular pads for soldering component leads. Here are some guidelines for this technique. Five amps is the most that a single 1/16-inch conductor should be made to carry. No conductor should be narrower than 1/32 inch. Minimum spacing between conductors should be 1/32 inch. Round pads typically have a 3/32-inch radius. If you are making negative artwork allow at least 1/4 inch on each edge to allow for clamps during processing. Avoid sharp bends; use ***curves*** instead.

An alternative to narrow conductors and pads is to leave large areas of conducting copper and etch away only enough copper to form the circuit. The way to prepare your artwork, if you are using this technique, is to first lay out the circuit roughly with lines indicating conductors and then build up a series of rectangles around these lines to form large areas of conductor.

The choice of which system to use comes down to a tradeoff in complexity. Simpler circuits are more easily prepared with the block method. As the circuits get more complicated, it becomes harder to deal with the blocks and not make mistakes. As you become more experienced, you will be better able to judge which kinds of projects to subject to which layout technique.

Direct Application of Resist

After you have made your full-size layout, check it for errors, and when you are sure it is correct, tape it and a sheet of ordinary carbon paper to a clean, untreated copper-clad board. Trace over the entire design to transfer it to the board. When you remove the pattern and the carbon paper, you should find the pattern transferred to the copper surface of the board.

The next step is to apply resist to the board. You can use liquid resist and a fine-tipped artist's brush, or you can purchase a special felt-tip pen made with resist ink. With this pen, you can simply and easily draw your conductors and pads. Besides the commercially available resists, you can use exterior enamel paint or even India ink.

The resist must be applied solidly, with no small gaps. Any gaps in the ink coverage, will allow the copper foil to be etched away. The resistance of the trace could be increased significantly. In severe cases, you could end up with an open circuit.

Once you have the pattern on the board, you can follow the steps in the section titled "Etching" later in this chapter.

Using a resist pen to draw the pattern directly onto the copper-clad board is not the only method. Some experimenters prefer to use the photographic method discussed below.

Sensitized Board and Photo Negative

As mentioned earlier, this process uses a copper-clad board on which a light-sensitive chemical has been deposited. In order to make the chemical acid-resistant in the areas you want to be conductors, you must expose the board to light in these areas. To do this, you will need a negative of your circuit artwork, a piece of film in which the conductors are transparent and the background is black.

Start with your original layout. If the circuit is simple, you may want to stay with a life-size master. If the circuit is more complex, you may want to redraw the art two or even three times life size. Whichever size you choose, make your drawing very neat. If you draw, use drafting paper and a drafting pen. Do not use a pencil. If you are not used to preparing art for reproduction, do not try to use ballpoint or felt-tip pens. Actually, the simplest approach to preparing your master is to buy adhesive-backed pads and lines from a mail-order house or radio hobby store. These will produce the most professional results. Remember not to make sharp bends in your conductors.

When your positive master is complete, inspect it carefully. Check for dirt and smudges. Make sure all lines and edges are sharp. Then put your master in an envelop—don't fold it—and take it to a photo process laboratory. Ask a local printer to refer you to someone in the area who can do ***process*** or ***litho photography.*** Ask the laboratory to make a line negative from your master. This shouldn't cost more than a couple of dollars. It's a service they do regularly for their customers, and you aren't asking anything unusual. Normally, you can get same-day or at least 24-hour service. What you will get back from the process laboratory will be a very high-contrast negative of your original artwork. Take it and follow the exposing, developing, and etching steps given below.

Sensitized Board and Mechanical Negative

There is an alternative to taking your finished master to a process laboratory and obtaining a photographic negative. You can, if you wish, prepare what is called a ***mechanical negative*** without using a photographic process.

You will need to go to an art supply store and ask for an appropriately sized sheet of a material called ***Amberlith.*** Another material, called ***Rubylith,*** will also do the job, but it is darker and it's harder to see the pattern through it. Either of these materials will appear to be simply a sheet of rather thick, transparent, tinted plastic. Actually, the color comes from a very thin film on one surface of the plastic.

Tape your master artwork down on a hard surface and tape the Amberlith on top of it, ***dull side up***. Use a sharp tool to outline each conductor and pad. You can use an X-actio knife or a pin, but the best tool is a pair of draftsman's dividers, or a bow compass with a point in each leg. With either of these tools, you can adjust the spacing of the points on the instrument to the width of the conductors you will be drawing and cut both sides of the conductor at once. You can use the instruments as you would an ordinary compass to scribe circles for pads.

After you outline each conductor or pad, use a sharp implement to tease up a corner of your conductor and peel the film away from the transparent base material. Be sure that you are removing the red film from the conductor area and not from the background. When you have finished, you will have a negative image of your circuit, with conductors clear and background red or amber.

The reason you can use this transparent red material in your negative instead of an opaque-back material, as in a photographic negative, is that the photoresist chemical is not especially sensitive to light at the red end of the spectrum. Thus, to it, the red Rubylith appears as black as if it were opaque.

When you have completed your mechanical negative, you are ready for exposing, developing, and etching (see sections below).

Exposing

Before you take the light-sensitive copper-clad board out of its protective wrapper, you will have to darken the room somewhat. The photoresist isn't very sensitive, so you can have some light, but be sure there are no fluorescent lamps burning. You can use a 15- to 25-W bulb, shaded so that no direct light falls on the board, 7 ft or more away. You can also use a red darkroom *safelight*.

When you have the lights adjusted, sandwich the negative between the copper-clad board and a sheet of glass, and clamp the edges together. The copper side of the board should be against the dull side of the photographic negative or the shiny side of the mechanical negative. In other words, you should be able to see the circuit exactly as you wish it to appear in the final product. Of course, the negative and glass plate should be totally clean and free from dust specks, as each dust speck will show up as an imperfection in the finished circuit.

To make your exposure you can use either a standard 150-W reflector lamp or a No. 2 photoflood bulb in a reflector assembly. If you use the 150-W reflector lamp, make a 3½-minute bulb, make a 6-minute exposure at a distance of 10 inches. Whichever bulb you use, plan your exposure so that your fingers and other things sensitive to extreme heat are not in the direct light of the bulb. After you make the exposure, keep the room darkened until you have finished the developing step.

Developing

Use an aluminum or glass tray for the developing process. A plastic tray may be dissolved by the developing chemical. Place the exposed circuit board face up in the bottom of the tray and cover it to a depth of approximately ¼ inch with developer. You will notice that the developer is very volatile. It is best to use a sheet of glass to cover the tray to keep in the fumes. This holds down evaporation and is a desirable health precaution.

Gently agitate the developer for 2 minutes. At the end of this time, turn on the lights and tilt the tray so that all of the developer collects at the far end, and carefully remove the circuit board with fingertips or tweezer. Be very careful not to handle the surface of the board.

Allow the board to dry for 30 seconds to 1 minute. Do not wipe or shake the board, or even blow on it to hurry it up. The developer that remains in the tray can be reused once or possibly twice more before it potency is gone.

Etching

If you used an aluminum tray for developing the board, you cannot use it for etching. The etchant would attack it the same way it attacks the bare portions of the circuit board. For etching, you must either use a glass or a plastic tray.

Place the board in the tray and add the etchant until the board is covered. The etching will take anywhere from 1 to 2 hours, depending on the temperature of the solution. During this time, you must agitate the tank every 5 minutes or so.

As you watch the etching process, you will observe that the portions of the copper-clad board that are not protected with resist will take on a brownish color. If you pull the board out of the etchant at this point, you will notice that as the brown liquid gradually drains off the surface, the bare areas of the board show a color somewhat like the color of a very new penny. This is the natural color of copper, when it is not affected by a surface oxide. The etching process is not complete until there is no more of this rosy pink elemental copper on the board. Do not be fooled into taking the board out of the etch bath too early. Inspect the board periodically to find out how much of the copper remains. Be especially careful where there are narrow gaps between conductors. The etching is finished when the only thing you can see is resist-covered conductor and dull brown circuit board.

After the etching has been completed, remove the board and rinse it under running water. You cannot reuse the etchant, so dispose of it, preferably down the commode, or, if in the sink drain, with a thorough water flush. Avoid handling the board with your fingers, since the etchant stains may cause irritation.

If you have an older home, check to make sure you do not have copper pipes. If you pour etchant down a copper pipe, it will eat through the pipe, which is obviously a big disadvantage.

Finally, remove the resist from the conductors. This is essential to good soldering. Use plain steel wool, or steel wool and a solvent intended for removing resist.

Finishing the Board

Now that you have a board with a circuit, you must prepare it for the components you will be mounting on it. Each pad must be drilled in the center for the component lead that will go through it. There are special printed circuit board drills that you can buy for this purpose. They have a different angle to the cutting head than ordinary general-purpose drills. This helps them drill easily through the brittle circuit board without destructive stresses. For larger holes, such as for mounting brackets and stud-mounted components, you will have to use your regular drill bits. Use moderate pressure and a slow speed. If the drill binds, do not try to get it restarted by running the speed control up and down. You will damage the drill motor, if you do. Instead, work the board off the drill and get the drill going at medium speed without a load. Then apply the rotating drill bit to the hole, and it will cut right through.

Install the larger components first; then the smaller resistors, capacitors, diodes, and transistors. Push resistor and capacitor leads through their appropriate holes until the body of the component is flush against the board. Then bend the lead against the board and clip off all except 1/8 inch or so. Treat transistors in the same way, except leave about 1/2–3/4 inch of lead on the top side of the board. This will permit you to attach heatsinks to the leads before soldering. Use a soldering iron with low heat.

In some manufacturing operations, components are soldered to printed circuit boards with a device called a *wavesoldering machine.* In this machine, there is actually a fountain of molten solder that is kept flowing at all times. Circuit boards go through the machine much like cars through a car wash. They first pass over a flux bath to clean all the leads and conductors, and then they pass over the solder fountain to solder all of the connections in one pass.

The usual home experimenter technique is to mount the finished circuit in some sort of case, using metal brackets and screws. Some more sophisticated projects use a technique borrowed from industry, called *edge-card mounting.* In this technique, the circuit on the board includes a series of rectangular pads along one edge. These pads connect to the various inputs and outputs of the circuit. The board is intended to mate with a socket, called an *edge-card connector.* This is a long connector with fingers that make contact with the pads on the circuit board. The board itself is the plug, and there is no need for a separate male connector.

The advantage of this is that a series of edge-card connectors can be mounted adjacent to each other on a panel or another printed circuit board, called a ***mother board,*** and each stage of a device can be built separately and plugged into its own edge-card connector. This aids in both design and troubleshooting. It aids in design because each stage can be modified independently on its own edge-card connector, and comparisons can be made by just removing one design and substituting another. It aids troubleshooting because a faulty stage can be isolated quickly by simple substitution. As the technique becomes more common, you can expect to see more kits and magazine projects using edge cards.

UNIVERSAL PC BOARDS

A lot of experimenters don't care for the fuss and bother of making their own printed circuit boards. Many projects presented in magazines include an address for ordering a premade printed circuit board. But if a prepared board is not offered, or if you are working on an original or adapted project, this option is not available.

One solution is to use a universal printed circuit board. This is a board with a generalized pattern of traces, like the one shown in Fig. 2-6. It is similar to the interconnection pattern of a solderless socket. The trace pattern is general enough to be used in a great many different circuits. Provisions are made for power supply and ground connections.

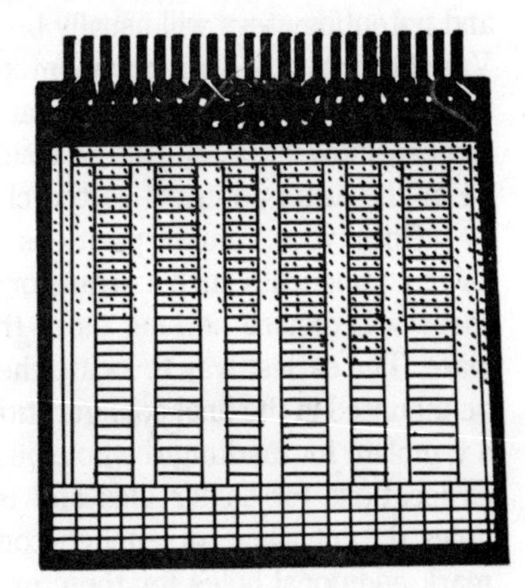

Fig. 2-6. Many different projects may be built on a universal pc board.

Two common types of universal printed circuit boards are labelled ***op amp*** and ***digital.*** These names are just indicators. You could build a digital circuit on an op amp board. The only difference between the two types is the number of busses. A digital board has two—for the power supply and the ground. The op amp board has an additional bus line for the negative supply voltage.

METAL CHASSIS

Before the days of etched circuits and perforated boards, metal chassis construction was used for virtually all electronic assemblies. Today, its use is limited to equipment that requires it. For instance, most vacuum-tube circuits require periodic replacement of tubes. This places mechanical stresses on the surfaces used to mount tube sockets. Metal chassis are better equipped to handle resultant strains than are brittle circuit boards. Vacuum tubes are also great producers of heat, which is more efficiently conducted away and radiated by a metal chassis.

Vacuum tubes are not the only components that generate heat. Diodes and power transistors carrying large currents use the metal chassis on which they are mounted as heatsinks.

Heavy metal chassis are required to support the weight of transformers and chokes in high-voltage power supplies.

Construction Techniques

Before you do anything else to your metal chassis, cover it completely with some sort of material that you can write on—masking tape is ideal. This will give you an easy way to mark holes for cutting and will protect the finish of the metal from scratches caused by chips and burrs.

Next, determine the location of all your controls. Connectors, switches, and potentiometers will usually be mounted on the sides of the chassis or box. Variable capacitors can be mounted on either the side or top surface. All shafts and some mounting brackets of variable capacitors are electrically connected to the capacitor rotor, so plan to insulate the shaft and bracket if they are not to be at the same potential as the chassis.

Once your control positions are chosen, locate the positions for other large components such as transformers, chokes, filter capacitors, etc. Most of these will require nothing more than 9/32-inch holes for their mounting hardware. The easiest way to locate the holes is to position the component as it will be mounted in the final configuration, and then use its own mounting bracket as a template for marking the hole positions. Some of these holes can also be used as locations for solder lugs and terminal strips. If the component-mounting holes do not allow for enough conveniently located lugs and terminal strips, mark additional holes for them on the chassis.

Mark positions for tube socket holes on the top surface. Some sockets are fastened in place by means of spring clips; others have brackets with screw holes. If you are using the latter type, don't mark the holes for the bracket-mounting screws just yet. Wait until you have the socket holes punched or cut, and then, with the socket in the hole, use the bracket as a template.

Once you have the positions of all the major components laid out, you are ready to start drilling. If you do not have a variable-speed drill, use a center punch to locate the drill tip for each hole. If you have a variable-speed drill, you can skip the center punch and start at a low speed. For larger holes, 1/4 inch and up, start with a smaller hole and work your way up.

The best technique for large round holes is to use a ***socket punch*** of the proper size. Drill a 3/8-inch hole first, to pass the center bolt of the punch, and then assemble the halves of the punch on either side of the chassis and tighten the bolt.

Chassis punches are best for making larger holes, but they are expensive. If you are careful, you can make an acceptable hole with a nibbler and round file. Nibble the hole just a little undersize, and then carefully enlarge it until its edges are smooth and it just fits the socket. Mark the locations of the socket-mounting holes and drill these as you have drilled the others.

When you have satisfied yourself that there are no more holes to be drilled, remove the protective paper and install the components, solder lugs, and terminal strips. Always use lockwashers between chassis and nut. These serve two purposes: they help to tighten the screw without need for a wrench on the nut, and they help to prevent the nut from loosening during vibration.

You may now wire the chassis. Work from the schematic and proceed logically, a stage at a time. Any radio frequency (RF) wiring should go directly from point to point. All other wiring should be routed so that it will be convenient to bundle the wires into cables. Cut leads on resistors and capacitors so that they can be installed with their edges parallel to edges of the chassis. It is best to wait until all wires that go to a particular point are installed and firmly bent over before soldering. If a wire is to pass through a hole in the chassis from top to bottom, use a rubber grommet or a feedthrough insulator.

WIREWRAPPING

A relatively new construction method is wirewrapping. This technique is generally used with circuits using a large number of ICs and relatively few discrete components.

Special wirewrap sockets, like the one shown in Fig. 2-7, are used. These sockets have fairly long leads with squared off edges. A narrow-gauge wire is wrapped tightly around the square lead. The sharp edges bite into the wrapped wire, making a good, solid connection without the need for any solder. A connection can easily be unwrapped to make changes in the circuit.

A special wirewrapping tool is used to construct a circuit in this method. Fig. 2-7 shows a simple inexpensive manual model. Fancy, automatic wire-

Fig. 2-7. A special tool is used for wirewrapping.

wrapping tools are also available. Most wirewrapping tools feature a built-in wire stripper of some sort.

If discrete components are used, their leads must be inserted into a wirewrapping socket. Ordinary leads are round, and simply wrapping wire around them will not make a good, reliable connection. Typically 40-gauge wire is used for wirewrapping. Complex circuits using many ICs can be constructed quite quickly and easily using the wirewrapping method.

GENERALLY APPLICABLE TECHNIQUES

There are some practices that are common to all kinds of electronic construction. They deal with controls, shafts, wire sizes, and colors.

Controls and Shafts

Some potentiometers come with shafts already installed. In other cases, pots and shafts are sold separately, so that you can choose a round, half-round, slotted, or knurled shaft to suit the kind of knob you intend to use. These shaft styles are interchangeable, but once you insert the shaft into the pot and push it until it locks, you are committed. You cannot remove one of these shafts without destroying the pot.

On-off switches for the backs of volume control potentiometers are also sold separately. If you remove the gummed paper from the back of the pot, you will expose a small finger mounted on the pot rotor that is designed to mate with the switch actuator. Match the finger with the actuator and click the assembly together.

Obviously, you will not need the whole length of control shaft as provided by the manufacturer. Use your knobs themselves to find out exactly how much shaft you will need, and cut off the rest with a hacksaw.

Many of the pots readily available in radio parts stores are intended for printed circuit boards. You can tell these pots by the absence of holes for wires in their rather narrow solder terminals. Since these pots are intended to be mounted on brackets, only one nut is provided for their bushings. If you want to install one of these pots on a chassis or metal panel, you will find that the bushing sticks out far in front of the panel and prevents the knob from seating flush. Potentiometers intended specifically for panel mounting have two bushing nuts, one for either side of the panel. The rear nut can be adjusted for panel thickness, so that the front of the bushing does not protrude past the end of the front nut.

As already noted, variable-capacitor shafts are electrically connected to the rotor plates of the capacitor. Most designs take this into account and place one side of the capacitor, the rotor (always the curved element in schematics), at ground, or chassis potential. In some circuits, however, this is not possible. In these cases, it is necessary to insulate the shaft. Sometimes, all that is necessary is to provide a large hole for the shaft to pass through without touching any sides and a plastic or Bakelite knob. In other cases, for example, when you want to avoid side loads on the shaft, you will have to use a panel bearing coupled to the shaft through a flexible coupling. These are fairly uncommon parts, even in the best stocked retail stores, and you probably will only be able to get them through the mail-order houses. These same sources can also provide flexible shaft extensions and right-angle drives, but you may go through an entire lifetime without ever needing these items.

Wire Sizes and Colors

The most common wire used in electronic projects is No. 22 stranded copper wire with plastic insulation. For most uses, stranded wire is preferred to solid wire, because solid wire breaks too easily when it is nicked.

For carrying heavy currents, No. 22 wire is too small. High-current applications require No. 14 or even heavier wire. Some applications—for instance, coil winding—require very fine wire. Solid copper wire with an insulating coating of enamel is sold for this purpose. Number 28 wire is convenient, since it is fine (0.1014 inch) yet reasonably strong. You will have to remove the enamel coating from the ends of this wire before you will be able to solder to it. Industrial users of enameled copper wire employ a paint-removing chemical to do this. You can achieve the same results, however, by scraping the ends of the wire with a sharp knife. Be sure to scrape off the enamel all the way around the circumference of the wire.

Of course, the color of the insulating jacket on the wire in your project will have no effect on the performance of the circuit. However, you can make the project look better and assist yourself in troubleshooting if you can afford to invest in several colors of wire. The colors in Table 2-1 represent standard practice for wiring different parts of a circuit.

Routing Wires and Cables

No matter what type of construction you use, you will have to run a certain number of wires or cables within almost any project. Many printed circuit boards, for example, may require jumper wires because the copper traces cannot cross over one another. Almost all projects use at least a few components which must be mounted off-board. Front panel controls and read-out devices, such as potentiometers, switches, meters, and lamps, are usually mounted directly onto the case with wires connecting them to the main circuit

Table 2–1. A Standard Color-coding Chart for Wires.

Color	Use
Black	All grounds
Brown	Heaters or filaments off ground
Red	Positive DC supply voltage
Orange	Screen grids, Base 2 of transistors, Gate 2 of FETs
Yellow	Cathodes, emitters, sources
Green	Control grids, bases, FET gates
Blue	Plates, collectors, drains
Violet	Negative DC supply voltage
Gray	AC supply leads
White	Bias and agc lines

board. In a number of complex projects, two or more circuit boards may be used, with interconnecting wires between them. As you can see, it is virtually impossible to avoid running off-board wires in the majority of your projects.

This is an area which receives very little attention in most texts on electronics. After all, running a wire is such a mundane sort of thing. Isn't it perfectly obvious how it should be done?

Well, yes and no. You're not likely to run across anything surprising as you read this section. But there may well be some points you haven't thought of. The sad truth is, a great many problems with home-brew projects are due to wiring problems. The subject bears some discussion here.

Of course, everything in your project should be neat in appearance. For one thing, you don't want to be ashamed to show anyone the results of your work. But the need for neatness goes beyond mere cosmetics. A haphazardly wired project resembles a rat's nest, and is just about as pleasant. If wires are running every which way, it will be that much harder not to make a mistake and solder one or more wires to the wrong connection points. Troubleshooting a ***rat's nest circuit*** is next to impossible, because of the difficulty in tracing the wires.

Rat's nest wiring is also subject to premature failure, because of excess and unnecessary strain on some wires. They may eventually break. Moving the circuit board, or even opening the case could also add to the strain on poorly run wires.

As a rule of thumb, try to run wires parallel with chassis sides or circuit board edges. Group wires together, and hold them together with a tie. Commercial ties are available, or you can use pieces of string or spare wire. Tie a neat, secure knot. Avoid clumsy ***granny knots.*** The tie must be secure, but not pinch the wires.

Carefully plan the wire routing before you begin actual construction. If you don't preplan you could run into trouble with wires getting caught under screws, components, access panels, and the like. Wires and cables should never get in the way of any screws, or removable parts. If it becomes necessary to remove the screw or part, the wire will have to be pushed out of the way, adding to its strain.

Except when extra length is required to open the case, keep wire lengths as short as possible. Nice, taut wires will stay neatly in place without getting tangled or snagged. But when a wire runs to a panel mounted component on an access panel, or a removable part of the case, you must leave sufficient slack in the wire, so that you can open and service the project without making additional problems for yourself.

If there is a possibility of interference from any high-frequency (RF) signals in the area, or if the wire is carrying such signals, you should use shielded cable. Interference pick-up problems tend to increase with wire length.

In some instances, you will need to run a wire or group of wires over some distance. It is a good idea to mechanically support such wires at midpoints along their length.

When a wire's path must change direction, use a smooth curve, rather than a sharp bend. Once again, excess strain or stress on any wires is to be avoided.

Sockets

There has long been a debate over the use of sockets. Some hobbyists and technicians swear by them, while others swear at them.

Sockets should be used for any component that is likely to be changed frequently. For example, a crystal may be changed to permit the circuit to operate at a new frequency. In a computer, an erasable programmable read-only memory (EPROM) chip should usually be socketed to permit the programming to be updated.

But what about other applications, when the component presumably will never be changed unless it goes bad? Discrete passive components, such as resistors and capacitors, are almost always soldered directly into the circuit. Transistor sockets are available, but tend to be expensive and do not always make a reliable connection with the transistor's leads. For the most part, the socket debate centers around IC sockets. When using a socket on an IC there may be some unreliability problems if the chip is operated on a very high frequency, but in most applications there should be no noticeable difference in performance.

It is easy to install an IC in a socket backwards, which could very well damage or destroy the internal circuitry when power is applied. On the other hand, it is just as easy to solder an IC into the circuit backwards. Whether you use sockets or not you must be careful about chip orientation and always double-check before applying power.

The chief advantage of sockets is that you are not soldering directly to the

IC's leads. Like all semiconductors. ICs are heat sensitive, and can be damaged or destroyed by careless soldering.

The strongest argument against the use of sockets is one of economics. Sockets aren't terribly expensive, but they do add to the cost of a project. In many cases, a socket may cost more than some common integrated circuit devices, like 741 op amps, or simple digital gates. Many people in the field argue that it is silly to protect a 50¢ IC with a 75¢ socket. This is a valid argument, but the other side of the coin is convenience. Sooner or later you're bound to get a defective chip, or you may make a mistake in assembling a circuit, or you may have to repair a circuit at a later date. Desoldering and resoldering 8 to 40 closely spaced pins is a tedious job at best, and can be very, very frustrating.

Using sockets can also add to the physical bulk of a circuit. However, for the vast majority of hobbyist applications, this is not likely to be a significant consideration.

Personally, I recommend the use of IC sockets whenever practical. If you are working on a high-speed circuit, or if minimum cost is a major consideration, it makes sense to avoid sockets. But generally, I feel they are good insurance against a lot of potential frustration and time-consuming extra work. In that light, they really aren't very expensive at all.

MULTI-STROBE PROJECT

This a fun project that will add excitement to parties and provide some interesting effects for amateur photographers. It's a bright, flashing strobe lamp with a variable flash rate. It operates on house current.

At parties, with a moderate flash rate, the Multi-Strobe can add a very unusual *old-time movie effect* to fast dances. Seen in the light of the Multi-Strobe, dancers appear to move jerkily between exaggerated positions.

In photography, it is possible to use the Multi-Strobe to show a moving object in successive positions on a single frame of film.

The flash of the Multi-Strobe is very bright but of short duration. This can be used to allow you to see some very commonplace things in interesting new ways. For instance, when you look at a drop of water falling in ordinary light, the image is blurred and it is difficult to tell the shape of the drop. The flash of the Multi-Strobe is bright enough to let you see the drop, but short enough so that the image is not blurred. By releasing drops from an ordinary eyedropper in front of the Multi-Strobe, you will be able to observe their true shape, which is spherical—not at all like a teardrop. Once you are used to using the Multi-Strobe, there are many other kinds of everyday phenomena you will find interesting to study.

Principles of Operation

The most important element of the Multi-Strobe is the flash tube. This U-shaped glass tube is filled with the inert gas ***xenon***. There are three electrodes: an anode, or positive terminal; a cathode, or negative terminal; and a trigger band. In the photograph, a length of enameled copper wire has been soldered to the terminal for the trigger band and wound around the tube. This is common practice with xenon flash tubes: it helps insure triggering.

Chemically, xenon is related to neon, which is very familiar in glow-discharge display tubes, and to argon, which is used in some ultraviolet light sources. The xenon flash tube differs from common neon and argon glow lamps in that the two electrodes are separated by a much greater distance. Thus, when a high voltage is applied across the xenon flash tube, it does not immediately ionize and glow.

A high-voltage spike—in the case of the Multi-Strobe, a 6 kV impulse—is used to trigger the tube into firing. This momentary high-voltage, low-current spike is obtained from a specially designed trigger transformer, much like a miniature version of the high-voltage flyback transformer in a television set.

Power Supply

Because the overall drain of the triggering circuit and flash tube is small, it is possible to use a voltage doubler power supply (see Fig. 2-8). This produces a relatively high voltage, around 300 V, from an inexpensive transformer. This transformer is not really necessary for the functioning of the circuit. However, by isolating the components from the power lines, it provides an important safety feature.

Voltage Doubler

The basic voltage doubler circuit, isolated in Fig. 2-9, is a handy thing for the experimenter to remember, since it is a relatively cheap method of obtaining a high DC voltage. Its drawbacks are that it cannot supply very much power and cannot be used with a bleeder resistor.

Both of these drawbacks stem from the way the voltage doubler works. Each of the capacitors in the circuit provides half of the output voltage. Consider that the input wire going to the junction of the two capacitors is the neutral lead. Then we may consider that the other lead has an alternating voltage of plus and minus a certain voltage. During the positive half of the cycle, diode D2 is cut off and diode D1 conducts, charging capacitor C1. During the negative half of the cycle, D1 is cut off and D2 conducts, charging capacitor C2 in the same polarity as C1. If there is very little drain on the circuit, the voltage across the combination of capacitors is twice the peak voltage of the input signal. However, if there is a significant load on the circuit, the capacitors will be simultaneously charging and discharging, or, saying it another way, the

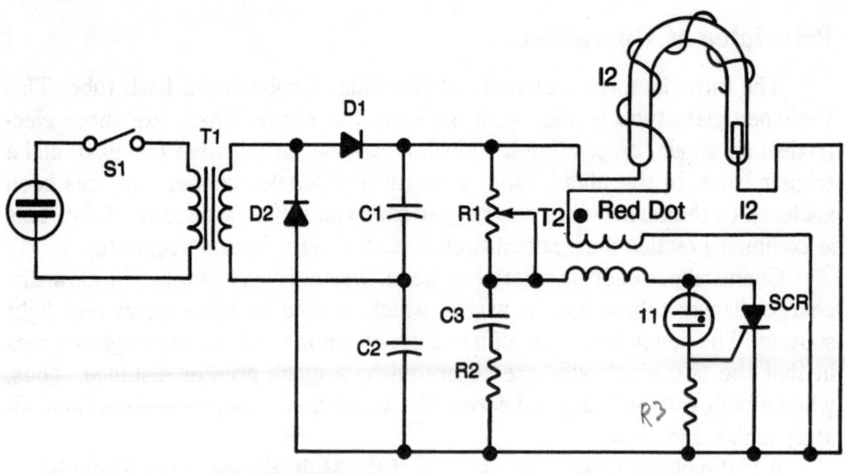

C1; C2 .20 μF, 450 V DC electrolytic
C3 4μF, 450 V DC electrolytic
D1, D2 Motorola HEP-158, IA 600 PIV silicon diode (or equivalent)
I1 NE-2 neon bulb
I2 MFT-106 xenon flash tube
R1 1M linear-taper potentiometer
R2 150K, ½W
R3 150 ohm, ½W
S1 rocker switch, spst
SCR Motorola HEP-300 thyristor (or equivalent)
T1 power transformer, 125V secondary, 15 mA
T2 TR-6 kV trigger transformer

Available from:
Great Western Aviation Co.
Electronics Products Div.
Box 20396
Denver, Colorado 80220

Fig. 2-8. Schematic and parts list for the Multi-Strobe project.

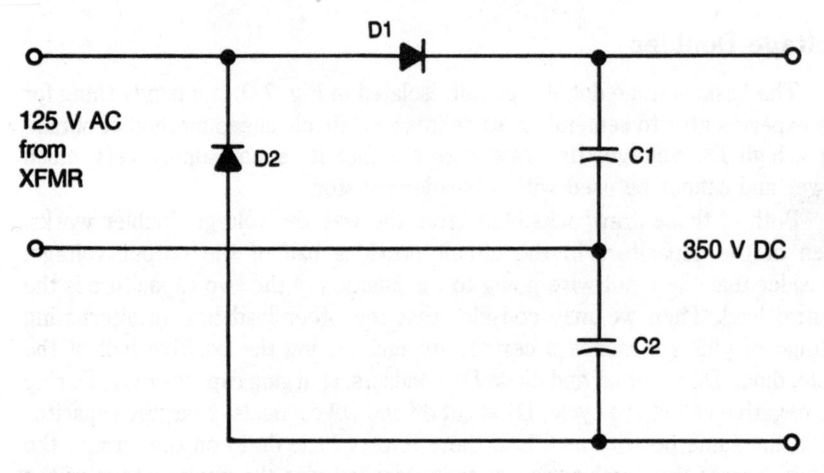

Fig. 2-9. Basic voltage-doubler circuit.

diodes will be delivering power to the external load and the voltage will be reduced.

In the case of the power supply in the Multi-Strobe, there is little load on the capacitors. The nominal output voltage of transformer T1 is 125 V rms. This means that the peak voltage is 125 times 1.4, or 175 V. Thus, the voltage doubler provides a 350 V DC voltage at no load.

The use of a voltage doubler requires a note of caution. Because there is no bleeder resistor to discharge the capacitors when the switch is turned off, there is the possibility of a lingering charge on the capacitors that can produce an unpleasant (if not particularly dangerous) electrical shock. If it is necessary to open the case of the Multi-Strobe within a half-hour or so after it has been used, it is a good idea to use a length of wire to discharge the capacitors before handling any of the components.

Timing and Firing Circuit

The voltage from the power supply is applied to the RC network consisting of R1, R2, and C3. The rate at which C3 charges is determined by the setting of R1. Neon bulb I1 *sees* the whole voltage across C3, since until it fires, it does not allow any current to flow through R3. Neon bulb I1 does not conduct until the voltage across C3 reaches its firing voltage, which is approximately 90 V. As long as I1 does not conduct, the gate of silicon controlled rectifier SCR1 is maintained at ground potential. As soon as I1 begins to conduct, current flows through R3, and the gate potential of SCR1 rises above ground. This forward-biases the SCR, allowing it to conduct. When an SCR switches to the on state, it turns on very quickly. When the SCR turns on, it discharges capacitor C3 all at once through the primary winding of trigger transformer T2. Actually, there is some high-frequency ringing at the resonant frequency of C3 and the primary coil of T2, but this does not affect our circuit.

The secondary of the trigger transformer is connected to the trigger coil on the xenon flash tube. Note that only one side of the secondary is connected to the trigger coil. The other side is connected to circuit ground. The sudden discharge of capacitor C3 through SCR1 produces a 6 kV spike across the secondary of T2, which fires the flash tube.

As soon as the voltage across the neon bulb drops below the extinguishing voltage of the bulb, it turns out and returns the gate of the SCR to ground potential. The SCR, however, does not cease conducting until almost all of the charge has been drained from C3. Then it too turns off. This allows the charging to start all over again, and the cycle is repeated as long as switch S1 is on.

Multi-Strobe Construction

Some of the components are mounted directly to a 6-x-3½-x-2-inch black plastic box, and some are mounted on an etched circuit board. The removable front panel of the box contains the flash rate potentiometer R1, on-off switch S1, and the reflector and socket for xenon tube I2. Transformer T1 is mounted to the bottom of the box.

Most of the things that could have been used as a reflector were fairly expensive. An ideal reflector, however, was found in the brightly plated cup of a 79¢ soup ladle, bought in a supermarket. The ladle's optical properties are entirely adequate for the relatively large flash area of the xenon tube.

The xenon tube has no plug as such, just three fairly stiff leads. A three-conductor socket intended for a flat-pronged plug was selected to receive these leads. This required that a rectangular hole be nibbled in the bottom of the reflector and a corresponding hole be sawed in the front panel of the meter box. The plug mounts by means of two 8-32 screws.

The on-off switch also requires a rectangular hole. The switch is mounted by inserting it into the hole until the side tabs lock in place. Its black plastic finish matches the box and makes a very attractive final product.

Figure 2-10 is a full-size pattern for the etched circuit board. The board is

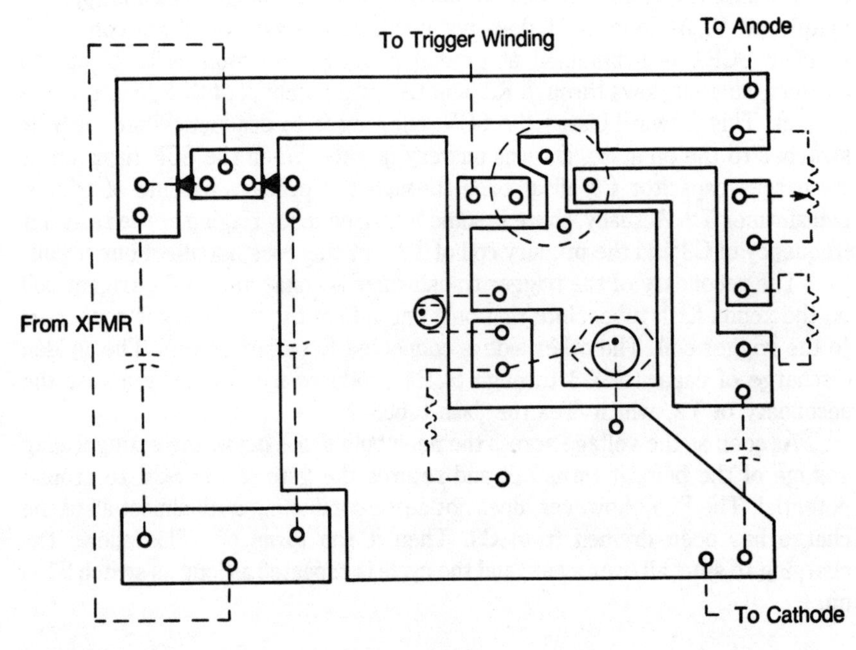

Fig. 2-10. This printed circuit pattern for the Multi-Strobe project uses conductor blocks rather than pads and ribbon conductors.

prepared as described in the chapter on printed circuits. There is some degree of nonuniformity from unit to unit in the positioning of the solder tabs on the trigger transformer. One technique for locating the position of the holes exactly is to place a sheet of carbon paper on a blank piece of paper and then to press down on the carbon paper with the transformer. This should leave four marks on the paper in the pattern of the solder terminals on the transformer. This pattern can then be transferred to the printed circuit.

The stud of the SCR is its anode, so it is important that good connection be made between the mounting hardware and the copper on the board. Use a couple of washers to limit the penetration of the stud, or you may crack the board when you mount it to the bottom of the meter box.

The circuit board mounts to the bottom of the box. Metal or nonconducting standoff insulators can be used to mount the circuit board within the box; however, in this case, a simpler method was used. Holes were drilled through both circuit board and box, and 8-32 screws were passed through the holes with ¼-inch grommets used as spacers between the two surfaces. This is simple and compact, and at the same time provides vibration and mechanical shock protection.

For an orderly appearance, be sure to bundle leads into cables and lace them neatly as shown in the photos.

Multi-Strobe Operation

The operation is simple and straightforward. Plug the Multi-Strobe into a house receptacle, turn it on, and it should begin to flash immediately. The potentiometer selected for flash rate control should vary the flash rate from about 1 flash/sec up to around 10 flashes/sec. If you have made any mistakes and the Multi-Strobe does not work, be sure to discharge the power supply capacitors before poking around with your fingers.

12 V DC TO 110 V AC INVERTER PROJECT

There are times when you would like to have 110-V, 60-Hz house current and there is none available. You might, for instance, wish to use a shortwave receiver or some home appliance on a camping trip, or you might need power during an emergency blackout. The transistorized inserter is capable of supplying a limited amount of 110-V AC to certain equipment from a 12-V DC source such as an automobile battery.

The inverter is limited in the amount of power that it can supply by the size of its transformer. The most it can supply is 25 W, given the transformer in the parts list. The transistors are good for 50 W each at the voltage levels used, however; so if you wanted to use a transformer with a higher current rating and install it in a larger box, you could expect to supply up to 100 W of output power. A second limitation is the magnitude of the load power factor that the inverter can accommodate. This means that electric motors will not work when connected to the inverter. The inverter is well suited for use with electronic equipment, however, provided power requirements are kept in mind.

The inverter uses a transistorized multivibrator to chop the 12-V input signal into a square wave (see Fig. 2-11). The transformer is then able to step up this alternating voltage to 110 V. This is the same thing that all transistorized inverters do. This particular inverter circuit differs from most in some important particulars, though. Most conventional transistorized inverters employ a saturable-core transformer with additional windings to make the multi-

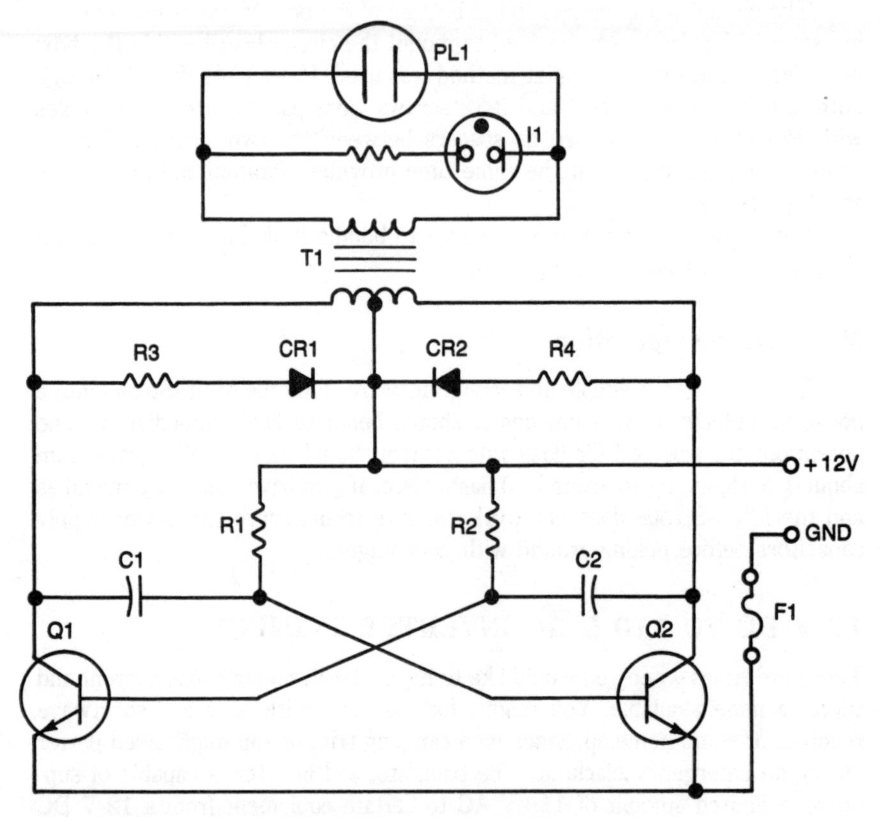

C1, C2 sprague 686X0015R2, 68 μF, 15V dc tantalum electrolytic
CR1, CR2 Motorola HEP 154, 1A, 50 PIV silicon rectifier (or equivalent)
PL1 2-prong ac plug
Q1, Q2 RCA 2N4347
R1, R2 180 ohms, 1W
R3, R4 10 ohms, 1/4W
I1 neon lamp with integral current-limiting resistor
F1 2A fuse in fuse holder

Fig. 2-11. Schematic and parts list for the transistorized inverter project.

vibrator function. These transformers are available from several manufacturers at fairly stiff prices. Our inverter uses an ordinary 24-V filament transformer, which lowers the price somewhat. The key to this simplified design is the use of ***tantalum*** electrolytic capacitors in the feedback circuit in the multivibrator.

To obtain an output frequency of approximately 60 Hz, some large values of capacitance are necessary. These values of capacitance are only available in electrolytic capacitors. Conventional electrolytic capacitors, however, are not well suited to this kind of application. They are not purely capacitive. They have, in effect, a resistance across their capacitance that diminishes their charge-holding ability. In power supply and audio coupling applications, this is not a drawback, but in timing applications, it is. Electrolytic capacitors made with the metal tantalum are a rather recent development. They are substantially better than older types of electrolytic capacitors in terms of dissipation, and they can be used successfully in timing circuits such as this inverter.

Inverter Circuit

Like the astable multivibrators used as clocks and tone generators in many low-power circuits, the inverter uses RC coupling to feed a signal back and forth between two common-emitter amplifiers. In Fig. 2-12, the inverter circuit has been redrawn a little more simply, with the transformer secondary omitted so that the use of the transformer primary can be better shown. As you can see, half of the primary—from left end to the centertap—is used as the load for Q1, and the other half—from centertap to right end—is used as the load for Q2.

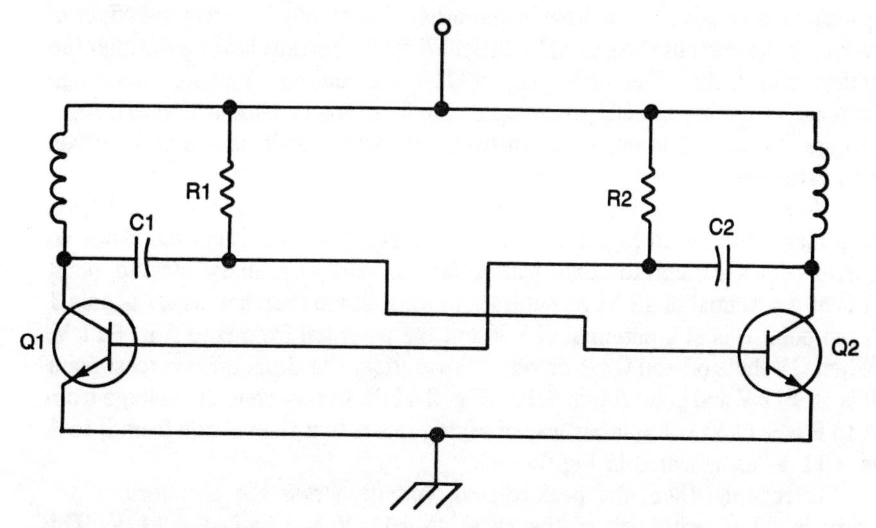

Fig. 2-12. Simplified inverter schematic showing how load transformer winding is distributed between Q1 and Q2.

Let's take a look at what happens when we apply power to the multivibrator. Initially, current flows through both transistors. It is unlikely, however, that the two transistors will be so perfectly matched that the current through each will be identical. Let us say, for argument, that more current flows through Q1 to begin with. As more current flows through the transistor, the voltage drop across its load increases and the collector voltage drops. Since the collector of Q1 is connected to the base of Q2 through capacitor C1, this dropping collector voltage causes a corresponding drop in the base voltage, and consequently, bias current of Q2. With less bias, Q2 conducts less and its collector voltage increases as current through its load decreases. This increasing collector voltage on Q2 is coupled back to the base of Q1 through capacitor C2. That naturally increases the bias current through Q1, which increases the current from collector to emitter even more. The collector voltage of Q2 continues to decrease, which causes even less current to flow through Q2; so its collector voltage increases, which makes even more current flow through Q1. All of this happens almost instantaneously, and the end result is that Q1 is fully saturated and Q2 is fully cut off.

This condition does not exist permanently, because of the effect of R1, and C1. With the collector end of C1 at essentially ground potential—thanks to the saturated state of Q1—and one end of resistor R1 connected to the +12-V supply, C1 begins to charge up at a rate determined by its capacitance and the resistance of R1.

When the voltage across C1 reaches a point that allows Q2 to be forward-biased and begin conducting, the current through Q2 causes a voltage drop across its load that, in turn, causes a reduction in the collector voltage of Q2. As in the case of the initial conditions, this drop in collector voltage on one transistor is coupled to the base of the other. Eventually, this has the effect of reducing the current through Q1 a little. With less current flowing through the collector load, the collector voltage of Q1 rises, causing an additional voltage increase at the base of Q2. Once again, the feedback conditions create a rapid change of state; Q1 is almost instantly cut off and Q2 is almost instantly driven into saturation.

What effect does this oscillation have on the transformer? Figure 2-13 will help make it clear. In Fig. 2-13A, Q1 is saturated and Q2 is cut off. Since no current flows through the right half of the transformer primary winding, point A is at a potential of 12 V. At saturation, the voltage drop across Q1 is only 1 V, so point B is at a potential of 1 V and the potential from B to A is +11 V. When Q1 shuts off and Q2 is driven to saturation, the situation reverses: point B is at +12 V and point A is at +1 V (Fig. 2-13B). In this case, the voltage from A to B is +11 V, but another way of saying this is that the voltage from B to A is −11 V, as indicated in Fig. 2-13C.

Effectively, then, the peak-to-peak voltage across the transformer primary is 22 V, with half of the cycle at +11 V and half at −11 V. The transformer, which was intended to transform 120-V AC to 24-V AC, is a

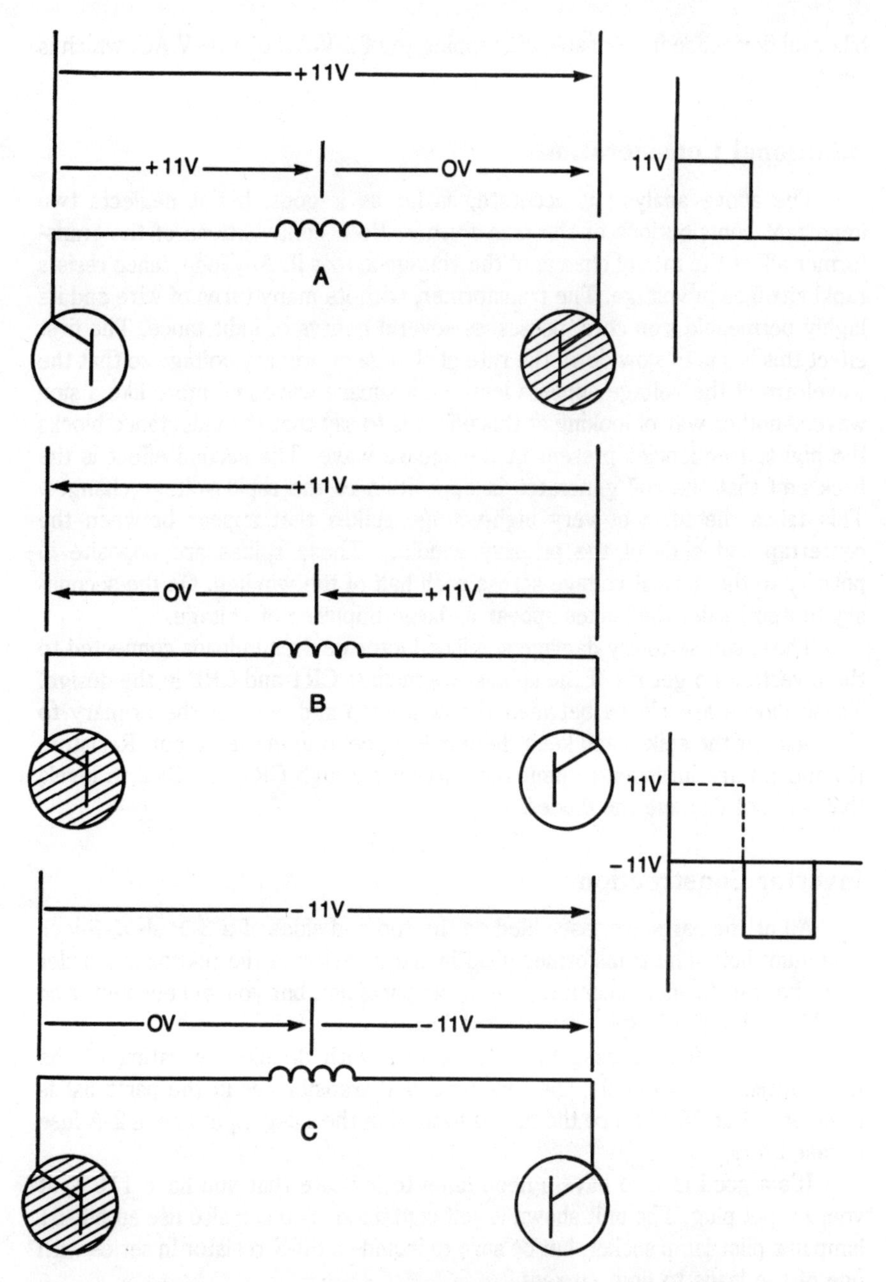

Fig. 2-13. How the chopping effect of Q1 and Q2 produces an alternating voltage with a 22-V p-p value from a 12-V DC source. The shaded transistor is cut off.

bilateral device, so it is capable of stepping the 22-V AC to 110-V AC, which is what it does.

Additional Considerations

The above analysis is accurate, as far as it goes, but it neglects two important contributions of the transformer. Both contributions of the transformer affect the rate of change of the voltage across it. Any inductance resists rapid changes in voltage. The transformer, with its many turns of wire and its highly permeable iron core, possesses several henrys of inductance. The first effect this has is to slow down the rate of change of primary voltage so that the waveform of the voltage appears less like a square wave and more like a sine wave. Another way of looking at this effect is to say that the inductance blocks the higher frequencies present in the square wave. The second effect is the back emf that the coil generates in opposition to the rapid voltage changes. This takes the form of very high-voltage spikes that appear between the centertap and ends of the primary winding. These spikes are opposite in polarity to the normal voltage across each half of the winding. On the secondary (output) side, the spikes appear as large impulses of voltage.

These can seriously damage semiconductor devices in loads connected to the inverter. To get rid of the spikes, we include CR1 and CR2 in the design. These diodes are placed between the centertap and ends of the primary to short-circuit the spikes and keep them from appearing in the output. Resistors R3 and R4 are intended to limit the current through CR1 and CR2 to levels that will not damage the diodes.

Inverter Construction

All of the parts are assembled on the top and sides of a 3-×-4-×-5-inch aluminum box. The transformer used in the inverter in the photos is smaller than the transformer recommended in the parts list, but you will encounter no trouble fitting the larger unit into the box.

A proper fuse is essential. Use a fuse with double the rating of the transformer. For example, the recommended transformer in the parts list is rated at 1 A at 24 V. Since the fuse is located in the 12-V input line, a 2-A fuse is called for.

It's a good idea to have a neon lamp to indicate that you have 110 V at your output plug. The unit shown is self-contained. You can also use an NE-51 lamp in a pilot-lamp socket, but be sure to include a 50-K resistor in series with one of the leads to limit current.

Figure 2-14 shows how to install the power transistors. The outer shell of these transistors is the collector connection, so the body of the transistor must be insulated from the enclosure. This is done with a shouldered washer in the mounting holes and a very thin mica spacer between the transistor and the top

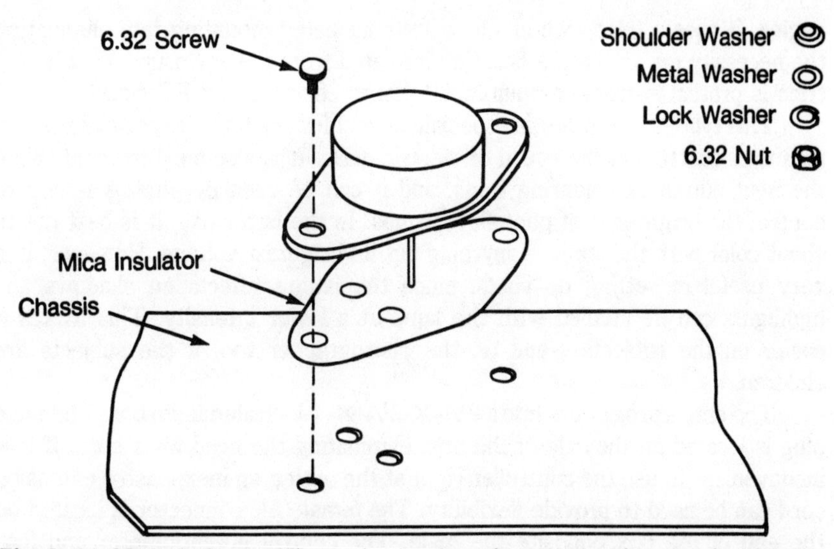

Fig. 2-14. Method of assembling switching transistors to provide electrical insulation from the chassis. A mica insulator and shoulder washer come with the transistor.

of the enclosure. There shouldn't be anything between the mica spacer and the top of the box, so you will have to take some extra care if you use contact paper to cover the box. The best way to make a neat job of it is to cover the box with the contact paper, then install the transistors without making any solder connections. Use the mica spacers as patterns and cut carefully around them with an X-acto knife. Then remove the transistors and peel up the contact paper inside the transistor mounting area. Reinstall the transistors and proceed with construction.

Be sure to use a grommet where the power cord passes through the wall of the box. This will keep the edges of the hole from abrading the cord and causing a short. Tie a knot in the cord just inside the box to provide strain relief.

You can make a very neat wiring job if you use a pair of four-lug terminal strips. One of the terminals becomes the collector lead for each transistor, and it becomes possible to make a very symmetrical pattern of connections. Remember to use heatsinks when you solder to the emitter and base leads of the transistors.

TRIAC CONTROLLER PROJECT

There are a lot of circuits for motor-speed controls and light dimmers using SCRs and triacs. This is one of the best. It uses a triac rather than an SCR, so it provides full-wave control and excellent efficiency. It employs a dual time-constant control which minimizes hysteresis and permits very small turn-on angles. It uses a triac with an electrically insulated mounting tab, eliminating the necessity for a separate heatsink insulated from the enclosure. Finally, the triac is protected from commutating voltage effects by an RC circuit.

This type of controller is especially useful for the home experimenter as a means of controlling the speed of electric drills. It is also ideal for controlling the heat output of soldering irons, and it can be used by photographers to control the brightness of photoflood lamps. In the latter use, it is best not to shoot color with the lamp at anything but its brightest voltage. However, it is very useful in setting up shots, since the lamp's effects on shadows and highlights can be studied with the lamp at a lower intensity. This makes it easier on the subjects—and on the photographer too, if the subjects are children.

The entire project fits into a 2¼-×-2¼-×-4-inch aluminum box. The male plug is located on the side of the box, eliminating the need for a cord. If it is inconvenient to use the controller right at the outlet, an inexpensive extension cord can be used to provide flexibility. The female AC connector is located on the end of the box opposite the male. The control potentiometer and fuse holder are mounted on the top of the box. A 6-A fuse was selected to provide a safety factor for protecting the 8-A triac (Fig. 2–15).

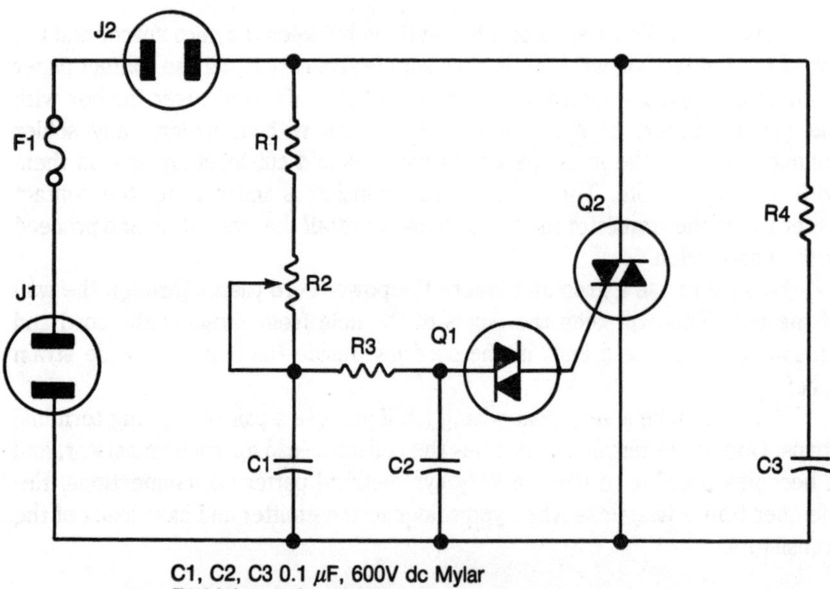

C1, C2, C3 0.1 μF, 600V dc Mylar
F1 6A fuse in fuse holder
J1 2-prong male ac bulkhead connector
J2 2-prong female ac bulkhead connector
Q1 international rectifier IRD54C diac
Q2 international rectifier IRT82C triac
R1, R3, R4 1K, 1W
R2, 100K, 1/2W potentiometer

Fig. 2–15. Triac controller schematic and parts list.

There are two terminal strips for mounting the components. The three terminals of the triac are spread out and soldered to three insulated terminals of one strip. This provides a solid base for soldering the other components. With a terminal strip on either side of the box, it is easy to mount the resistors and capacitors neatly.

Speed Controller Circuit

A triac is like a pair of SCRs connected back-to-back in parallel, with a common gate connection. An SCR is a kind of diode that does not conduct in the forward direction until the voltage across it exceeds a certain amount. Once it begins to conduct, it will continue to do so even though the voltage across it is considerably reduced. The SCR has a third electrode, called a *gate.* If a current is applied to the gate, the SCR will begin to conduct at a lower voltage. You can see that if there is a voltage across the SCR just a little less than the firing voltage, the SCR will behave like an open circuit until a current is applied to its gate, at which point the SCR will behave like a short circuit until the applied voltage falls to almost zero.

As we said, the SCR is like a diode. In the reverse direction, it always behaves like an open circuit. In AC control situations, this means that only half of the sine wave is used to power the device being controlled. How can we use both halves of the sine wave? We can use a triac. The gate of a triac controls the firing voltage in both directions.

How is its effect used to control the power going to a load? It's a question of timing. The gate is used to turn the triac on during a part of each half-cycle. If the time is very short, the power reaching the load is small. As the time increases, so does the power. Engineers find it convenient to speak of this effect in terms of ***conduction angle.*** In one complete cycle of a sine wave, there are 360 electrical degrees. Depending on the setting of the control potentiometer and the values of the other circuit elements, the triac will conduct during part of the positive half of the cycle (say, from 65 to 179 degrees) and during an equal part of the negative half of the cycle (say, from 245 to 359 degrees). Under these conditions, the conduction angle is 114 degrees during each half-cycle, and quite a bit of power is getting to the load. If we changed the setting of the control potentiometer so that the triac conducted only from 150 to 179 degrees during the positive half of the cycle and from 230 to 359 degrees during the negative half of the cycle, the conduction angle for each half-cycle would be only 29 degrees, and much less power would be reaching the load.

How does the timing circuit control the conduction angle? The timing circuit consists of a dual RC network and a device called a *diac.* The diac is like a triac without a gate. It does not conduct in either direction until the voltage across it exceeds about 30 V. At this point it begins to conduct, and it continues conducting until the voltage across it falls almost to zero.

In the circuit, the voltage across the diac is the voltage across C2. The values of C1, R1, R2, and R3 control the rate at which C2 charges. If C2 charges up to the breakover voltage of the diac very quickly, the triac will be triggered for most of the half-cycle. If C2 charges up more slowly, the conduction angle will be reduced.

Many circuits use just a single RC timing circuit, omitting R3 and C2, but by using these, we are able to control the triac down to very small conduction angles and to reduce an effect that engineers call ***hysteresis.*** In the case of lamp controllers, hysteresis refers to the difference in settings of the control potentiometer between the position where the lamp goes out and the position where it lights again.

The series resistor and capacitor across the triac is another circuit feature that does not appear in all lamp dimmer-motor controller circuits. Its purpose is to help maintain control when inductive loads such as motors are connected to the circuit. What happens without the protective circuit is that the inductive nature of the load causes the current in the circuit to lag behind the voltage. Thus, at some point after the current through the triac has passed through zero and the device is cut off, there is an applied voltage across the device. If the rate at which this ***commutating voltage*** rises is fast enough, it may trigger the triac prematurely, causing loss of control. Thus, C3 and R4 actually form a simple high-pass filter that slows down the rate of increase of the commutating voltage to a rate that the triac can easily handle.

ORGAN PROJECT

Music-making projects are always popular. In this section we will build a simple organ on an universal pc board. The board I used was one sold by Radio Shack. It was shown earlier as Fig. 2-6. Notice that this board features edge connectors which will be used in an unusual way in this project.

A universal printed circuit board is often a good choice for projects using several ICs along with a number of discrete components. As you can see from the schematic diagram of Fig. 2-16, this project certainly fits that description. The parts list is given in Table 2-2.

Top Octave Generator

This project is designed around a top octave generator. This is a circuit which accepts a high-frequency clock signal and divides it by selected values to result in multiple outputs representing the individual notes in a musical octave. Because all notes are derived directly from a single clock signal, all of the notes will always be in perfect tune with one another.

The 50240 (shown in Fig. 2-17) is a complete top octave generator in a single 16-pin IC. All you need to add is the clock input. The division factors for this chip are summarized in Table 2-3.

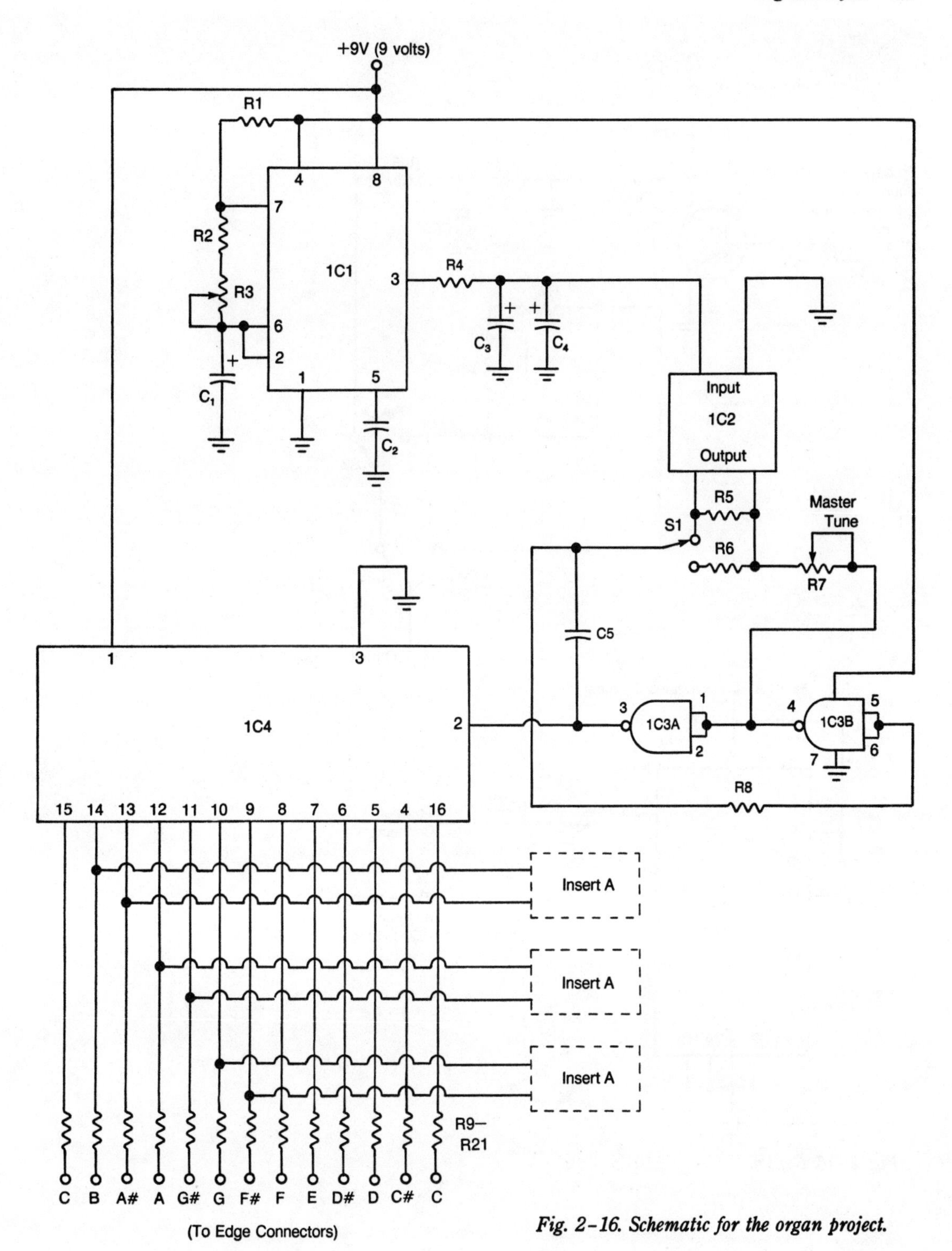

Fig. 2-16. Schematic for the organ project.

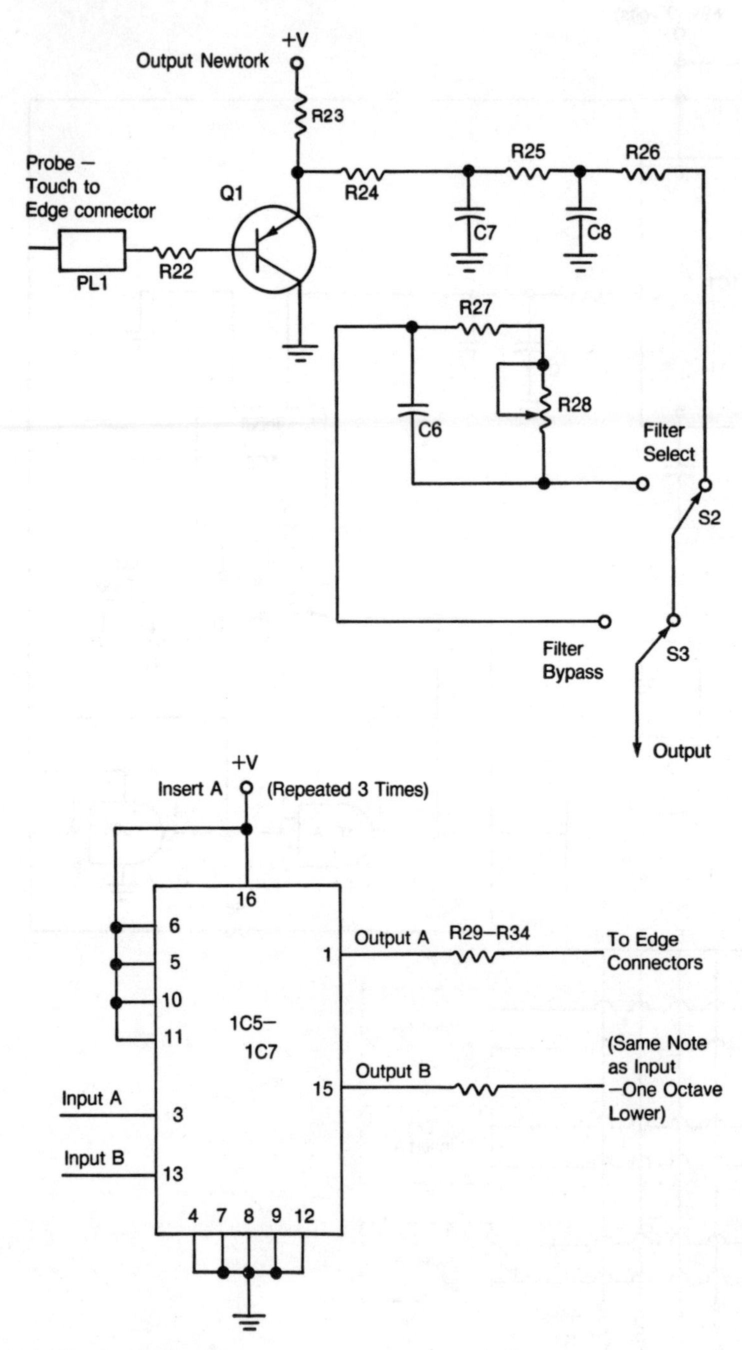

Fig. 2-16. Cont'd.

Table 2–2. Parts List for the Organ Project of Fig. 2–16.

Part	Component
IC1	555 timer
IC2	optoisolator (MOC3010, or similar)
IC3	CD4011 quad NAND gate
IC4	50240 Top Octave Generator
IC5–IC7	CD4027 dual JK flip-flop
Q1	PNP transistor (2N3906, or similar)
C1	5 μF electrolytic capacitor
C2	0.01 μF capacitor
C3, C4	500 μF electrolytic capacitor
C5	0.1 μF capacitor
C6–C8	0.005 μF capacitor
R1	470 K resistor
R2, R22	100 K resistor
R3	500 K potentiometer
R4	1 Megohm resistor
R5, R6	120 K resistor
R7	100 K potentiometer
R8	2.2 Megohm resistor
R9–R21, R29–R34	330 ohm resistor
R23	1.2 K resistor
R24–R27	1 K resistor
R28	10 K potentiometer
S1–S3	SPDT switch
	supply voltage = 9 V DC
	all resistors ¼ W 5%

The nominal clock frequency for the 50240 is 2.0024 MHz. This gives an output scale that is in tune with standard concert instruments. While the nominal value is 2.0024 MHz, you can vary this considerably, if you don't need to tune the organ to some other traditional musical instrument. The 50240 will accept a clock signal anywhere from 100 kHz to 2.5 MHz.

The 50240 is very easy to work with. A single-ended power supply is used. The supply voltage may be anything from +11.0 to +16.0 V.

The output signals are square waves with a duty cycle of 50 percent. This makes them ideally suited for digital division with standard flip-flops. This is how an expanded range including lower notes may be derived. The top octave generator produces the highest useful octave, which is where the name comes

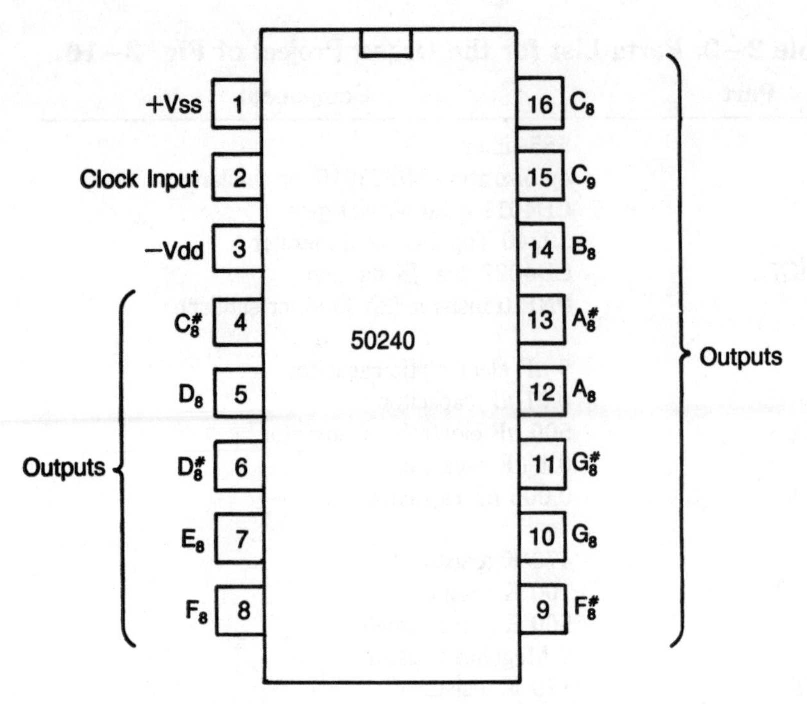

Fig. 2-17. The 50240 is a top octave generator IC.

Table 2-3. Division Factors for the 50240 Top Octave Generator.

Pin Number	Note	Division Factor
16	C8	478
4	C#8	451
5	D8	428
6	D#8	402
7	E8	379
8	F8	358
9	F#8	338
10	G8	319
11	G#8	301
12	A8	284
13	A#8	268
14	B8	253
15	C9	239

from. Another advantage of the square wave outputs is that they are rich in harmonics, so a variety of different voicings can be created by simple filtering.

Block Diagram

The circuit is broken down into its functional elements in the block diagram of Fig. 2-18. Notice that there are two oscillators. One is the main clock oscillator. It is always in the circuit. The second oscillator, which operates at a much lower frequency, may be switched in or out of the circuit. This second oscillator provides an adjustable vibrato effect. Potentiometer R3 controls the vibrato rate. The vibrato signal is coupled into the main clock oscillator circuit via optoisolator, IC2. The vibrato signal controls the intensity of the optoisolator's internal light source, which, in turn, affects the resistance across the IC's output.

The clock oscillator drives the top octave generator. At the outputs, some flip-flops are used to drop certain notes one octave, giving the finished instrument a range of about 1½ octaves. You can easily extend the range, simply by adding more flip-flops.

Finally, before being fed to the system output, a number of resistors and capacitors may be switched in or out of the circuit for various filtering effects.

Keyboard

Usually in building an organ project, the biggest problem is working up a suitable keyboard. A traditional mechanical keyboard is a fairly complex item. In this project, we will take a shortcut to create a functional, if somewhat inelegant, pseudo-keyboard.

The universal printed circuit board we are using has edge connectors. We can use the connector fingers as *keys.* The common end of the key connections (Point A in the diagrams) is carried out through a flexible wire to a small metallic plug. A submini plug (or anything similar) will do nicely. Short its two leads together.

The individual note outputs are brought out to the edge connector fingers. Fig. 2-19 illustrates how the components are mounted on the universal printed circuit board.

By touching the plug tip to the appropriate connector finger, we can play any desired note. This keyboard is monophonic. This means, that only one note can be played at a time. The 50240 is capable of polyphonic operation (two or more notes may be sounded simultaneously), so more adventurous experimenters may want to design a more versatile keyboard arrangement on their own.

If you use the edge connectors as the keyboard, you will have to specially design the case for this project. A slot must be cut into the side of the case. The

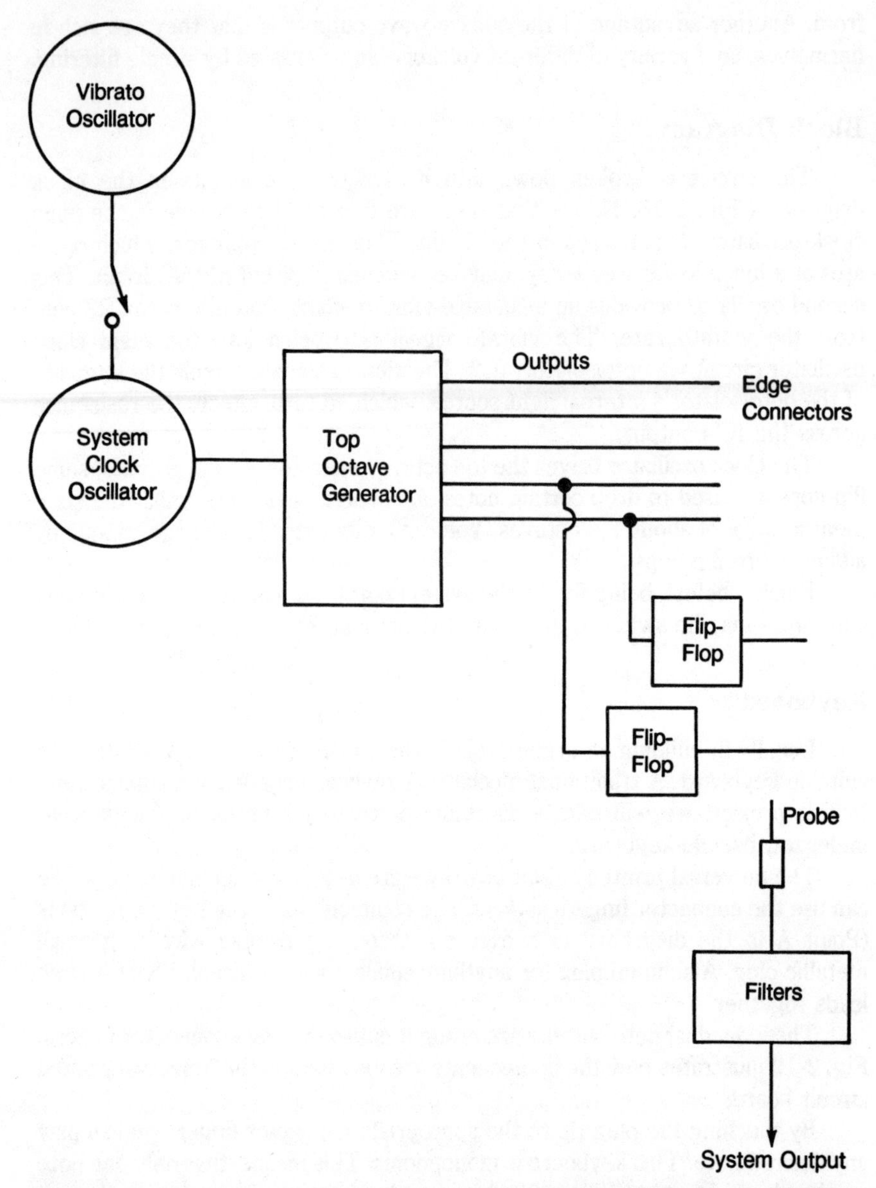

Fig. 2-18. The organ project is broken up into its functional elements in this block diagram.

circuit board is then mounted inside the case, so that the edge connectors stick out through the slot, as shown in Fig. 2-20.

The final output from this project can be fed through any standard audio amplifier. Just use a suitable plug for your amplifier at the end of the output cable.

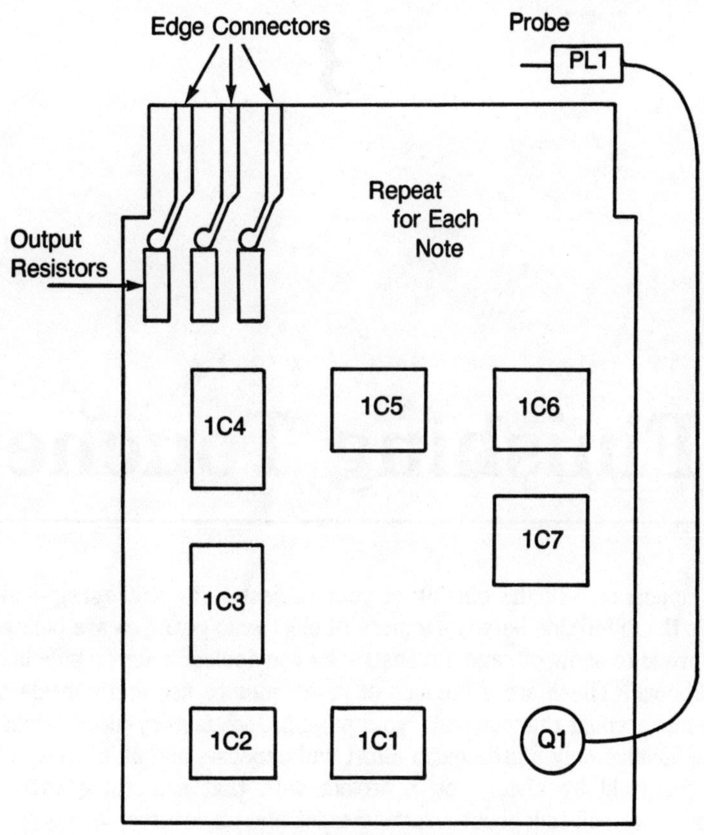

Fig. 2-19. How the components are mounted on the universal pc board.

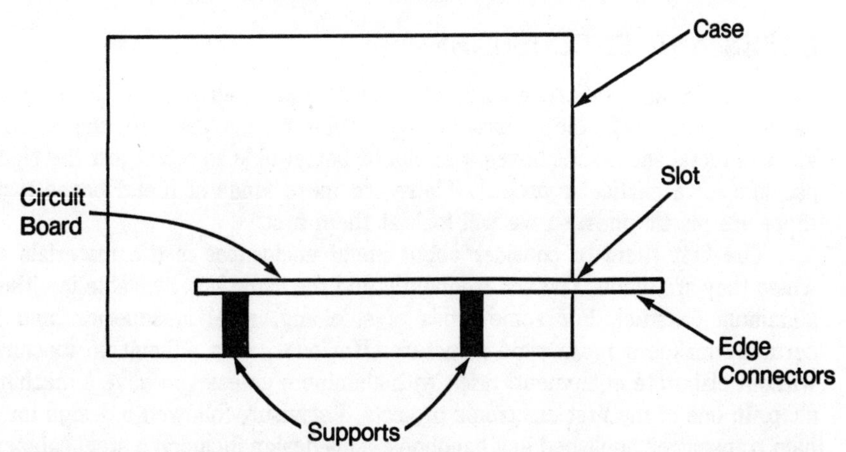

Fig. 2-20. The edge connector keyboard should stick out of a slot cut into the case.

3

Finishing Touches

The appearance of the outside of your project—its *packaging*—ultimately spells the difference between a piece of electronic gear you are pleased to use and proud to show off, and a monstrosity you apologize for or hide under your workbench. There are a number of techniques to use and considerations to take into account that can make your projects look factory-made. Most of these tricks involve only a little extra effort and expense, and all of them will repay you manyfold by giving you a project with that something extra. In this chapter, we will talk about selecting cases and enclosures; laying out a *scientific* and attractive control panel; selecting knobs, switches, and indicators; dressing up your panels with professional-appearing labeling; and making special scales for meters.

CASES AND ENCLOSURES

There are enough different kinds of cases and enclosures to fill up several pages in the average electronics catalog. If you are familiar with the various kinds of metal and plastic boxes, you will be better able to select just the right package for a particular project. There are more kinds of metal boxes than there are plastic ones, so we will look at them first.

The first thing to consider about metal enclosures is the materials of which they are made. You will frequently find the same box available in either aluminum or steel. For some kinds of shielding, steel is superior, and it certainly makes a more rigid structure. However, it is difficult to machine without elaborate equipment. Stick with aluminum unless you have a machine shop. In one of my first electronic projects, I slavishly followed a design for a ham transmitter published in a handbook. This design included a steel cabinet.

It took the better part of a month just to hack a ragged hole for a meter in the front panel. One easy way to distinguish between steel and aluminum hardware, if there is no label and you can't tell by lifting it, is the finish applied by the manufacturer. Steel cabinets are almost always coated with black wrinkle-finish paint—like something out of a 1930's horror movie. Aluminum cabinets are more often finished in gray hammertone or left ***natural.*** Some manufacturers are now offering cabinets finished in colored textured paint. These are all aluminum.

Miniboxes

The most useful style of metal enclosure for projects that do not require a lot of tuning and adjustments is the two-piece standard minibox. This is an economical design, because it is easily stamped and bent by the manufacturer and because it offers a complete six-sided enclosure with just two pieces (Fig. 3-1). One of the U-shaped pieces has side flanges that permit it to mate with the other U-shaped piece, and the flanges make the piece very rigid. Generally, all parts are mounted on the top panel and two sides of this enclosure, and the plain piece is left to serve as a cover. Although the box by itself may look a little austere, its visual impact can be improved by painting the two halves contrasting colors. Alternatively, wood-grained contact paper can make the box look like a piece of cabinetry.

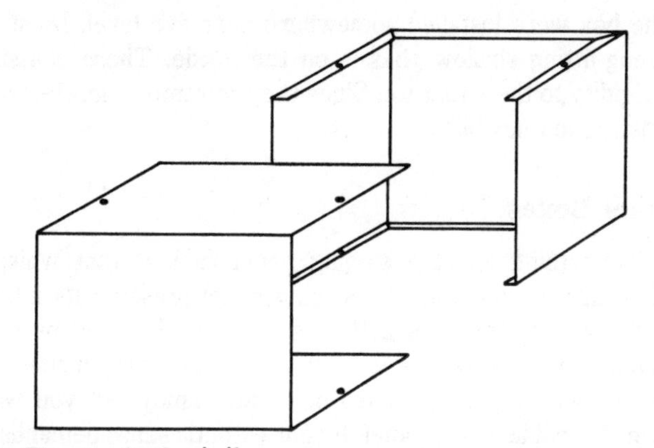

Fig. 3-1. The standard minibox.

Cowled Miniboxes

There is a new type of enclosure that is a variation of the minibox in that it consists of a pair of U-shaped metal stampings; however, its appearance is much snappier. This is due to the cowl that shades the front panel (Fig. 3-2) and the way the case sits on rubber feet. The advantages of this enclosure are all in appearance. The cowl suggests that it might shade the front panel and eliminate some glare, but it couldn't really do this unless the cowl were a lot

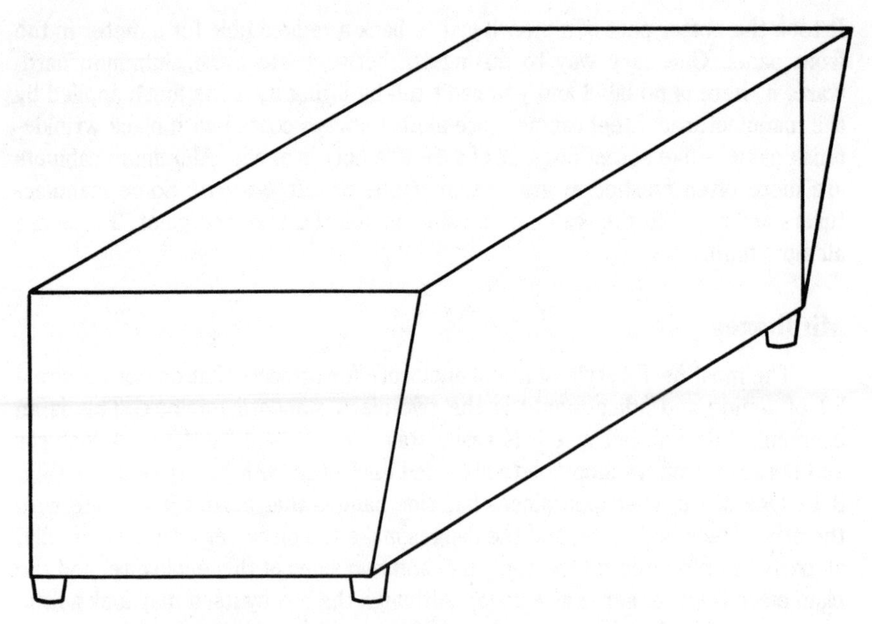

Fig. 3-2. The cowled box is more dramatic in appearance than the plain minibox, but may omit side flanges, providing less rigidity.

bigger and the box were installed somewhere near eye level. Most of these boxes have snug-fitting shallow chassis on the inside. These chassis lend a great deal of rigidity to the structure. Since they're removable, the builder has maximum construction flexibility.

Sloping-Panel Boxes

There is something about a sloping panel cabinet that whispers *test equipment.* Perhaps it's the way the sloping panel presents its information. There is no doubt that the information is there to be read precisely and accurately. Sloping-panel boxes come with and without prepunched holes for meters. The kind with the prepunched holes are handy—if you want your meter smack in the middle of the panel. If your esthetic sense demands that the meter go to one side or the other, you will have to go to work with nibbler and file. Since sloping-panel boxes are not symmetrical, there are two ways to use them. If you have a lot of controls and switches you want to arrange in a neat row, orient the box as in Fig. 3-3A. On the other hand, if you have binding posts or sockets to arrange in order, you might try swapping bottom and back, as in Fig. 3-3B.

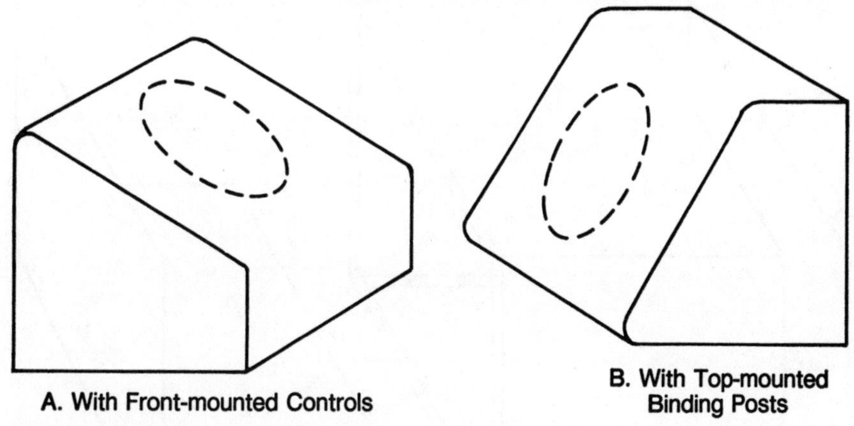

Fig. 3-3. Two ways of using a sloping panel box.

Utility Boxes

The utility box of Fig. 3-4 requires some welds, so it is more difficult for the manufacturer to make and, consequently, more expensive for you to buy. This is the kind of construction you will need, however, if you are using the conventional chassis-and-faceplate type of construction common to most vacuum-tube projects. In this kind of construction, you'll find several controls, pilot lights, switches, or connectors extending through both the panel and the front of the chassis, and these are used to fasten the two together. The faceplate is then attached to the cabinet with screws. This type of cabinet has the advantage of offering easy access to the circuits inside, since it is a simple matter to remove the backplate. Like the minibox, the utility box has a utilitarian look and benefits from contrasting paint on front and sides, or wood-grain contact paper.

There are dozens of other designs of metal cases and enclosures, and racks and panels, but you will encounter these only infrequently in your projects. You can learn more about the different kinds of boxes and their prices from the electronics parts catalogs.

Plastic Boxes

There are really only two styles of plastic boxes used for electronics projects, the black Bakelite dish with a flat plate top and the hinged folding box. Plastic boxes by themselves provide no RF shielding, but they are favorites of home builders, because they look neat and are easier to drill and cut than metal boxes.

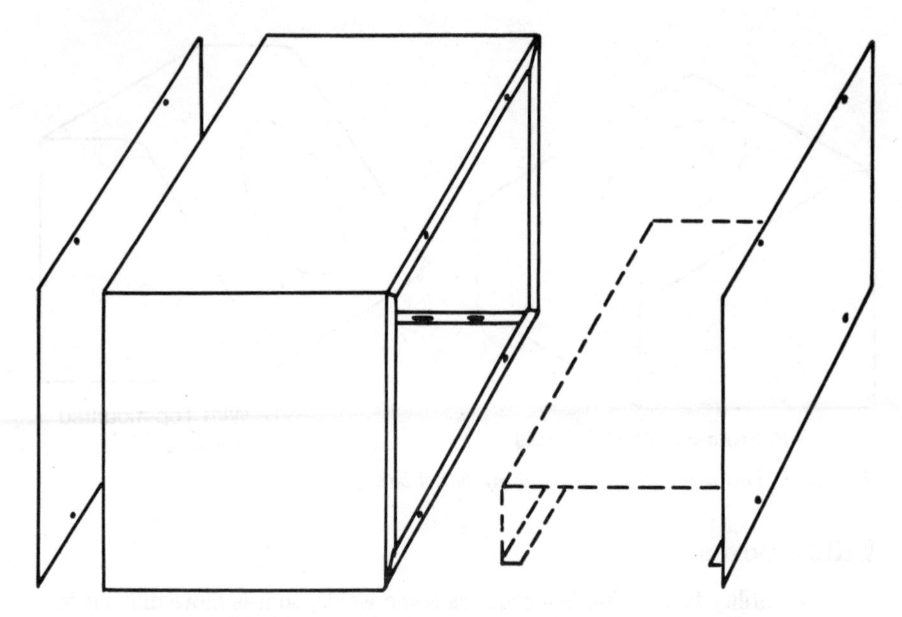

Fig. 3-4. The standard utility box provides removable panels front and back, creating a combination of rigidity and easy access. This box is often used with a separate chassis for construction of projects using vacuum tubes.

Plastic boxes should be drilled with a hand drill or an electric drill adjusted to a slow speed. High speeds can melt some of the plastics used, forming hard burrs alongside holes. Some thinner plastic panels can be nibbled, although the edges will be rough and need to be filled. It is best, however, to use an X-acto keyhole razor saw after first drilling a 9/16-inch starting hole. The saw cut should also be dressed with a file, but you will find much less filing will be required for the saw cut than for the nibbled edge.

Transparent Plastic Boxes

In the days when transistors were still a novelty and the cheapest units cost several dollars, there were many, many projects published each month for one- and two-transistor pocket AM receivers and audio oscillators. It seemed that the hallmark of all of these projects was that they were built to fit into hinged plastic boxes bought at the five and dime. It is possible that overuse of these clear plastic boxes in those days has discouraged their use today, since projects using them are rarely published. Questions of style aside, these boxes still have their uses, especially in more frivolous projects, such as the Randomizer in this book. Here, half the fun comes from being able to see the simplicity and compactness of the inner workings through the transparent sides.

Black Bakelite Boxes

The black Bakelite dish is the kind of plastic box you will find useful for serious projects. The best construction approach is to mount all of your parts on the top panel and leave the dish intact. There are different kinds of top panels available: aluminum sheet, perforated circuit board, and black plastic. The black plastic kind comes covered with a protective sheet of adhesive backed paper. Try not to remove this until you are completely finished with your drilling and cutting. The surface under the paper is very smooth—keep it protected as long as possible.

LAYING OUT A CONTROL PANEL

No matter what type of enclosure or panel you have selected, you are interested in making the best arrangement of controls and indicators on it that you can. Let yourself be guided by convenience, feedback, and appearance.

Convenience and feedback are two important ideas from the science of ***human factors.*** The meaning of ***convenience*** is clear enough: controls should be easy to manipulate and meters and other indicators should be placed where they are easy to see and well lighted. ***Feedback*** means that the way a control moves or feels tells you something about what effect it is having.

Here are some basic rules of human factors design:

1. Knobs for primary controls should be larger than knobs for secondary controls. Primary controls are those that you will be using most of the time; for example, the tuning controls on a radio receiver. Secondary controls are those you do not use as often, for example, gain controls and function switches.
2. If you are right-handed, group primary controls near the right edge of the panel; if you are left-handed, near the left edge.
3. Scale divisions on dials and meters should not be over-numbered or over-divided. On a scale from 0 to 10, there should be major divisions for each unit, and at the most, one division for each half unit. It is enough if you number every other major division. Human-factor studies have shown that too many numerals and divisions are confusing and make it harder to read a meter scale or dial quickly.
4. Try to leave enough space so that you will be able to differentiate between similar-size knobs on the basis of their positions. This is more reliable than reading the label over the knob each time you reach for it. If you can't spread the panel out sufficiently, use different-size knobs.
5. If possible, the movement of a control should correspond to the movement of the meter or indicator with which it is associated. In other words, if your tuning knob rotates clockwise, your tuning scale should rotate clockwise—not counterclockwise. Similarly, clockwise rotation

of a control should not cause a meter reading to decrease (unless you are tuning the meter for a dip).

6. If there is no outside indicator associated with a control, the control function should always increase in a clockwise direction and decrease in a counterclockwise direction.
7. Toggle switches controlling on-off functions should be wired so that up is on and down is off.

The third consideration for panel layouts is appearance. You should consider this only after you have taken the human factors into account, but there is no reason for appearance to conflict with human-factor design. The key to a neat appearance is dividing the panel symmetrically and arranging controls and switches in straight lines. Take your time, cover the entire panel with masking tape, and divide the panel into four quadrants. Then mark the position of each control so that it lines up with the other controls and is neither bunched too tightly with other controls or placed unnecessarily close to an edge. Use the entire panel.

After you have used human-factor and appearance considerations to locate all of the controls, lights, plugs, meters, and dials on the masking tape covering the panel, check to make sure that there is no interference between adjacent pieces of hardware or between controls and the edges of the enclosure. When you are sure that there is no interference, you are ready to begin drilling and cutting.

LEGIBLE LABELS

Many home builders avoid putting labels on their projects entirely. They seem to feel that since they put the equipment together, they don't need any labels to tell them what the various controls and switches do. Other builders are content to label everything with embossed plastic tape. This at least tells other people who might be interested what a particular piece of gear is doing, but embossed tape gives a project an amateurish appearance. It isn't necessary to settle for no labels or embossed tape. There is a simple, inexpensive process that will produce labels just as good as those that are silk-screened onto expensive commercial equipment.

This process employs *dry transfer letters* that are available from art supply shops and electronics retail stores and catalogs. The letters are printed with a special material on the backs of plastic sheets. When you place the sheet over a smooth surface and rub on the front side of the plastic with a blunt pencil point or a ballpoint pen, the letter transfers itself from the plastic sheet to the surface beneath. Only the letter itself is transferred, so the end result is as clean as if it has been silk-screened.

As noted, you can buy sheets of transfer letters either at art supply stores or from radio suppliers. The ones at radio suppliers are usually printed with common electronics labels spelled out right on the sheet. The ones at art supply stores just give you several repetitions of letters, numerals, and punctuation marks. It is a matter of preference, but the ones with just the alphabets seem to be more economical in the long run, and they offer you a selection of type styles.

If you are going to pick type styles, you will have three decisions to make: color, typeface, and point size.

For color, stick to white letters for dark panels, black for light panels. Other colors are available, including gold, but be careful or you'll wind up with a garish monstrosity of a panel that would be more appropriate to a Las Vegas slot machine than to your project.

Deciding on the typeface, you will face an interesting choice. Human-factor researchers have shown that type with *serifs* is easier to read than *sans serif* type. Serifs are the little strokes at the tops and bottoms of letters like the ones in this book. Common names of typefaces with serifs are Roman, Bodoni, and Caslon. While type with serifs may be theoretically easier to read, type without serifs—sans serif—looks cleaner, more modern, and less cluttered on an instrument panel. Common sans serif typefaces are Gothic, 20th Century, and Futura.

No matter which face you select, you will also have to decide on the width of the strokes in the letters. Books like this are printed in lightface type. However, you will need a bolder typeface for your panels to make the letters stand out. Select either a medium or a boldface type.

The *point size* tells you how big the type is. This book is printed in 9½-point type. Many other books use 8½-point type, which is just a bit smaller. On your projects, do not use anything smaller than 10 point or it will cause difficulty in reading the labels. You shouldn't have to use as much concentration to read a panel as you do to read a book.

The procedure for applying the letters is simple. Just place the plastic sheet in position, burnish the letter with pencil or pen point, and peel away the sheet. Here are some additional guidelines to help you achieve commercial-looking results.

1. For single words or groups of letters, select the point on the panel that corresponds to the center of the word. For instance, you may want the word *Volume* right under your volume control. The middle of the word

should be right under the middle of the control shaft. Start at this point, put the middle letter of the word, "l" in this case, right there, and work from the middle towards both ends.

2. Do not use all capitals. Combining initial capitals with lowercase letters in the body of the word gives a more pleasing appearance. For a modern, understated look, use lowercase letters for initial letters too.
3. You will save yourself a lot of headaches if you cut the sheets into sections of the alphabet. Sometimes it is even easier if you cut the sheets into separate lines. Cut-apart sheets are easier to store, also, if you put all the parts into clearly marked envelopes. You can mark the envelopes with transfers from title portions of the sheets.
4. Take as much time as you need to make your letters line up. Start your labeling in a position where you can use an edge of your panel as a guide for each letter. Continue using edges as guides, or use a label already applied to guide you.
5. Sometimes, especially on rough surfaces, letters do not transfer well. To overcome this difficulty, you can use the backing paper that comes with the transfer sheet to *prerelease* the letters. The backing sheet has a surface treatment that prevents it from picking up any transfer letters. If you put the backing sheet down on the table and the transfer sheet over that, you can rub on the letter without transferring it. This loosens the letter and helps it transfer easily to the rough surface.
6. After you transfer each letter, cover it with the backing sheet or with another piece of paper and rub over it with pencil or ballpoint to burnish it firmly in place. This may not appear necessary with every letter, but you can be sure that as soon as you think that you are doing just fine without this step, you will have a sudden accident and have to redo a whole word.
7. Sometimes, particularly on larger letters, the whole letter will not transfer the first time you rub over it. Be careful, then, when you peel the transfer sheet away from the letter that you hold it firmly in position until you are sure you have a whole letter. If you must burnish again, it is easier to reposition the sheet exactly if you have held it in place.
8. To protect your transfer letters, use a spray to cover them with a light coating of clear lacquer. A few words of caution, however: Be sure your spelling is correct before you spray, and do not spray clear plastic boxes unless you want to render them opaque.

SELECTING KNOBS, SWITCHES, AND INDICATORS

There are a great many different kinds of instrument knobs available today. Your selection will be based mainly on what appeals to your esthetic sense and what is readily available.

There are a few general observations to be made about knobs. First, do not mix several different styles on one panel. Your project will look a lot more professional if you use several sizes of the same style of knob. Second, make sure you have the right tool if you are using setscrew-type knobs. There is nothing more frustrating than finishing a project and discovering you have hex-head setscrews in your knobs and no Allen wrenches with which to drive them, or that your setscrew driver is a hair too wide to fit into the setscrew hole in the knob. Finally, if you have a rotary switch and a setscrew knob, you may find it advantageous to file a flat on the control shaft to prevent the knob from slipping.

A look through the pages of an electronics catalog will show you the kinds of knobs available. Besides ordinary knobs, you find that there are several kinds of vernier knobs that offer many turns of the knob to a single turn of the shaft. The cheapest vernier dials are planetary types with a fixed pointer and a rotating scale reading from zero to 10. Another type, which is not widely available, is a type with a moving plastic pointer and several concentric scales on which you can enter your own numbers. A third type looks, externally, like an ordinary knob. Inside, however, is a complete harmonic drive mechanism that gives a 40:1 tuning ratio with no backlash. One other type of knob is designed for use with multiturn potentiometers. This may appear to be a vernier knob, but is actually a ***turns-counting knob.*** The entire face is divided into 100 divisions over a full 360 degrees, and each time the indication goes through zero, a number in a window above the dial changes, indicating the number of turns from zero. These combinations of high-resolution multiturn potentiometers and turns-counting knobs are used primarily in analog computers.

As in the case of knobs, your choice of a switch will depend on your own personal preference and what is readily available. Generally, you will want to avoid the ***slide switch,*** even though it is the cheapest, because it requires a rectangular hole, which is hard to cut neatly. Slide switches also generally have poor reliability. ***Rotary switches*** are used when there are a lot of poles to be switched or there are to be more than two positions. There are two kinds of rotary switches, the ***shorting type,*** also called ***make-before-break,*** and the ***nonshorting type,*** or ***break-before-make.*** If you have an application in which you cannot have the input to the switch open-circuited—for example, if you are switching speakers driven by power transistors—you must use the shorting switch. If you have a single load and several sources that you want to keep isolated, use the nonshorting type.

There is an almost endless variety of pushbutton and toggle switches, with many kinds of handles and lights that come on, go out, change color, or just sit there dumbly when you actuate them. Most of them are rated for several hundred volts and at least 3 A. A few, notably the telephone type, are not designed for inductive loads and may burn away their contacts when connected to such loads. The only caution to observe with any of these switches is unless your project is very heavy, do not use pushbutton switches on vertical panels, or you may find yourself pushing a unit off the table when you only want to actuate a switch!

From a human-factors standpoint, a go/no-go *idiot light* indicator is more desirable than a meter or other indicator that has to be interpreted. In many cases, however, we want to know about trends that may be developing or whether a particular control is having an increasing or decreasing effect on some circuit parameter. In these cases, a meter is essential. The most common meters are round or rectangular, with a pointer that describes an arc along the meter scale. It has been shown, however, that these are not as easily read as edgewise meters in which the needle moves linearly along a long, narrow scale. The most effective installation for edgewise meters is vertical, so that the pointer needle moves up and down like the mercury in a thermometer.

Numerical readouts have been around for a long time in the form of neon glow tubes with stacks of numeral-shaped anodes inside. These have never been popular with home builders, however, probably because of their cost and the complexity of the logic required to drive them. New advances in light-emitting diodes (LEDs) have produced single-plane numerical displays that have minimal drive current requirements with high brightness and reliability.

At the same time, integrated circuit logic devices are becoming available that reduce the complexity of a complete decoder-driver to a single 16-pin dual in-line package (DIP). At the time of this writing, inexpensive solid-state seven-segment displays and decoder-driver packages are just becoming available to the home experimenter at reasonable prices. There is no doubt that this trend will continue, and that in a few years, use of these devices in home projects will be widespread. The Randomizer project in this book shows how these devices must be interconnected, what the various inputs are for, and what power and drive requirements are necessary.

As these components become readily available, there will be quite a rush to use them in all sorts of projects. While it is true that in terms of reading out a single number nothing beats a numerical display, it is also true that numerical displays are not as good as meters in showing whether a value is increasing or decreasing.

MAKING YOUR OWN METER SCALES

If all you are asking of a meter is to measure some common quantity like volts or amperes, you will find plenty of meters on store shelves to meet your requirements. You will even find a number of meters calibrated in volume units (VU) or decibels. However, beyond these choices, there is little selection. For example, if you want a meter that measures transistor beta and leakage, as in the bipolar junction transistor-field effect transistor (BJT-FET) transistor checker project in this book, you will not find a single meter on the market that is so calibrated.

It is neither expensive nor especially difficult to prepare custom meter scales. The main ingredients are patience, some drawing tools, and a painstaking approach. The steps are outlined below.

1. Select an appropriate meter. You will want one that spreads your scale out over the entire range and has a cover that can be separated from the works. On most old-fashioned round-faced meters, there are three or four screws on the back that can be loosened to allow the backplate and meter works to slide out of the housing. On the common, inexpensive square-faced meters, the faceplate has been snapped onto the back by means of four raised tabs on the back surfaces of the faceplate that mate with recesses on the rear shell. If you are very patient, you can shave these tabs down with a very sharp X-acto knife, allowing you to remove the faceplate. Later on, you will have to reassemble the faceplate to the rest of the meter and use a drop of cement in place of each of these tabs.
2. Take out the two tiny screws holding down the old meter scale, put them where you will be able to find them again, and remove the old scale. Be very careful at this stage not to damage the meter needle or the very fine bearings it rides on. By this point you have voided all warranties.
3. Put the meter aside and tape the scale down towards the bottom of an 8½-X-11-inch piece of paper. Very carefully line up a ruler with the lines corresponding to the maximum and minimum scale divisions and project these lines inward. The point where these two lines cross is theoretically the pivot point of the meter needle.
4. You should have already decided what scale divisions you want to use: all of them, every other one, every third one, etc. Extend a line from the pivot point on your paper through every scale division you intend to use. Extend the lines a fair distance, at least twice as far as the distance from the pivot point to the outer end of the division mark on the meter scale.
5. Use a circle template or a drafting compass to measure the radial distance from the pivot point to outermost point of the division marks on the original meter scale.
6. Double this distance and strike off an arc that intersects your projected division lines. This gives you the basis for a meter scale that is exactly twice as large as the one that was originally in the meter.
7. Remove the original meter scale and take a heavy pencil and darken each of the scale graduations, starting at the arc that you drew with the compass and carrying them inward toward the pivot point, ⅜ inch for major divisions and ¼ inch for minor divisions.
8. Place a clean piece of paper over the paper you have been using, and carefully and neatly trace the division marks you have made. Ideally, you should use a drafting pen, but you can get satisfactory results if

you use a sharp medium-soft lead pencil and are careful not to smudge. Avoid ballpoints and felt-tip pens. If you use a pencil, do not make any erasures. If you goof, start over on a new piece of paper. When you finish, spray the sheet with a workable fixative, such as Krylon, which can be purchased at an art or engineering supply store.

9. Apply transfer letters as described earlier in this chapter to identify the scale divisions and to add any other lettering you might want.
10. Carefully affix your new scale over the meter's original scale. Once again, be very careful not to damage the delicate pointer needle or its bearings. You will probably need to glue the new scale paper in place, but be very stingy with the glue. Don't use too much, or let any of it ooze out from under the paper. Don't let any glue get into the meter's works, or the meter will be ruined.
11. With the housing still open, apply a test signal to the meter to confirm that the pointer needle can move freely over its entire range without binding or catching.
12. Finally, reassemble the meter housing.

As long as you work slowly and carefully, you shouldn't run into any problems in customizing a meter using this method.

RANDOMIZER PROJECT

This project displays a random number from 0 to 9 each time it is activated. The schematic diagram for this project appears in Fig. 3-5. The parts list is in Table 3-1.

The output of this circuit is a seven-segment LED display unit, driven by IC2, a BCD to seven-segment decoder-driver.

Table 3–1. Parts List for the Randomizer Project of Figure 3–5.

IC1	HEP C3800P decade counter IC (in 14-pin socket)
IC2	SN7447N BCD to seven-segment decoder/driver
Q1	Archer (Radio Shack) 276–111 unijunction transistor
DIS1	Seven segment LED display—common cathode
C1	0.1 μF 12 V disc capacitor
C2	0.002 μF 12 V disk capacitor
R1	10 K ¼ W
R2	470 ohms ¼ W
R3	220 K ¼ W
R4–R10	330 ohm ¼ W resistor
S1	SPST momentary (Normally Open) pushbutton switch

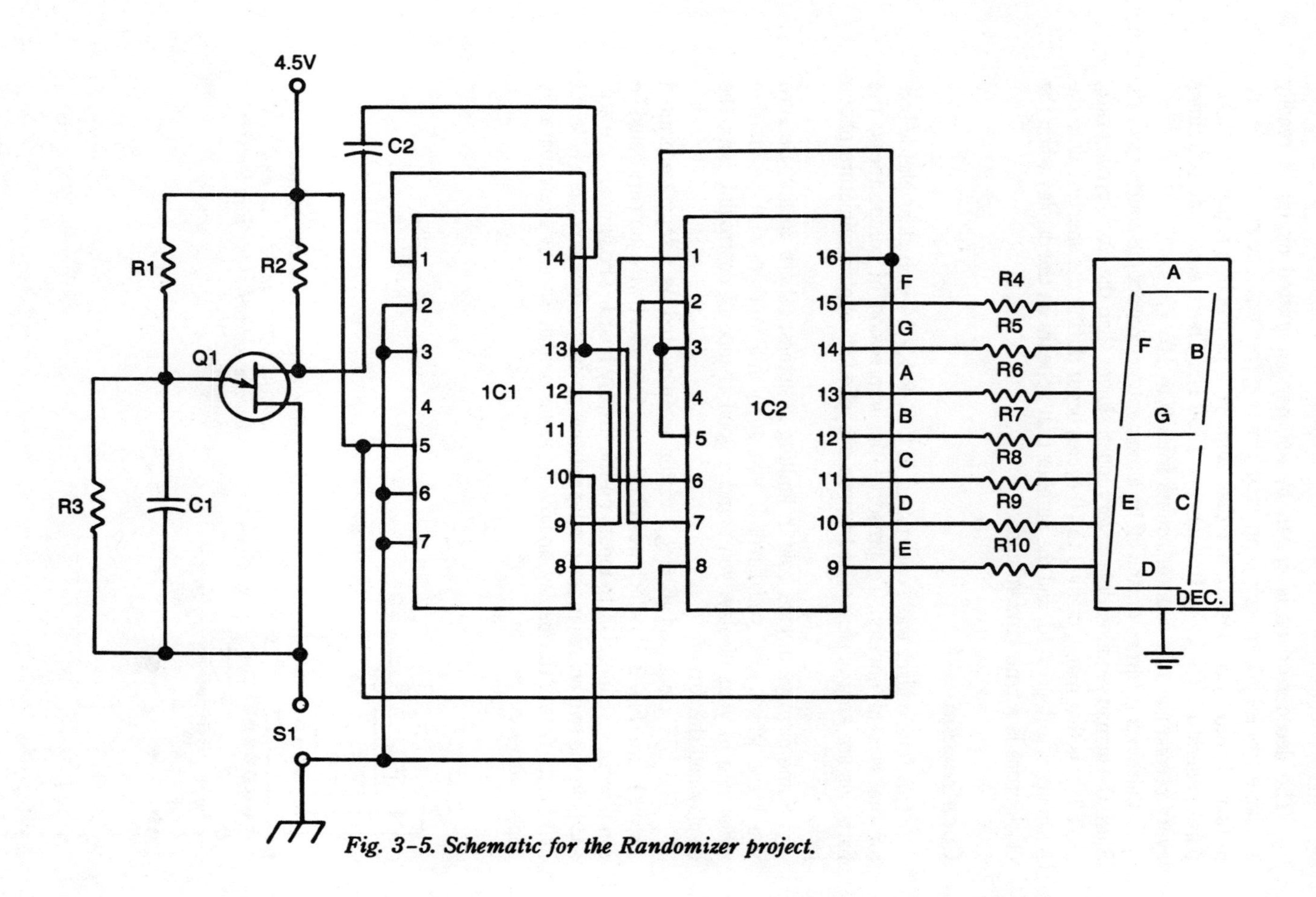

Fig. 3-5. Schematic for the Randomizer project.

The decoder-driver is driven by another integrated circuit, a readily available divide-by-10 circuit connected as a ring counter. The ring counter is, in turn, driven by a simple clock that employs a unijunction transistor (UJT) pulse generator. The clock frequency is 1000 Hz. This means that 100 times every second the ring counter counts from 0 to 10.

The clock is turned on and off by a momentary-contact pushbutton switch. Since the human reaction time is very much slower than the clock-cycling rate, the effect is that many, many 0 to 10 cycles occur during the shortest time the pushbutton is depressed, and the number that appears on the display when the clock stops is a true random digit.

Construction

The Randomizer was constructed in 2-×-2-×-1¼-inch clear plastic box for the sake of novelty. To preserve the impression of compactness, the batteries are kept separate, and power is applied through a miniature phone jack.

Some displays include built-in limiting resistors; others do not. Be sure you know which kind you have. If you apply full voltage with no limiting resistance to your display, you stand a good chance of destroying both the display and the driver.

The UJT clock and IC ring counter are mounted on a separate printed circuit board (Fig. 3-6), which was planned so that short wire jumpers could be run directly to connect it to inputs on the display board. The jumpers are stiff and hold the two boards in the proper relative positions. The box has a hole cut in it for the face of the display, and the display is held in place by a few sparingly applied drops of cement.

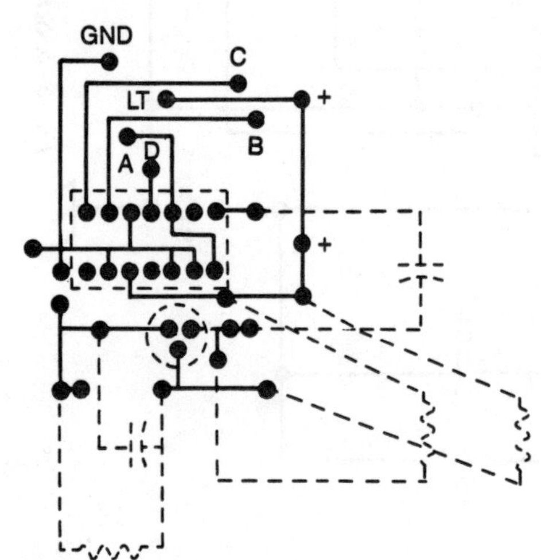

Fig. 3-6. Actual size printed circuit board for the clock and decade counter of the Randomizer project.

Randomizer Circuit

The clock is a conventional UJT relaxation oscillator. You will note from Fig. 3-7 that the clock input to the ring counter is designated $\bar{C}$. The bar over the C indicates that the clock requires negative-going pulses to trigger. To obtain negative-going pulses, the output is taken from B2 of the UJT.

The inexpensive UJTs used for this project presented something of a problem. The first couple that were tried wouldn't oscillate. Theoretically, the characteristics of UJTs do not vary much from one type to another. However, it seems that, with the low applied voltage used here, there is a difference. Check out your unijunctions beforehand, and use one that has a maximum forward resistance of 300 ohms from the emitter to each base.

Initially, the clock would run but would not trigger the counter. The addition of R3 solved this problem. Apparently, it influences the shape of the pulse. The printed circuit of Fig. 3-6 provides a place on the board for R3. In the Randomizer, R3 was ***back wired*** onto the circuit board, since there was no convenient place for it on the front.

The ring counter IC is a Motorola HEP C3800P. This is a divide-by-10 counter; its logic diagram is shown in Fig. 3-7. Each time a clock pulse is applied, the counter increases its output by one.

The output format is binary-coded decimal, or BCD. In this encoding system, there are four outputs that can be either ***high*** (*logic 1*) or *low* (*logic 0*). The outputs correspond to powers of 2. The first output is equivalent to 2^0, or 1; the second is equivalent to 2^1, or 2; the third to 2^2, or 4; and the fourth to 2^3, or 8. The number represented is the sum of the high outputs. If all of the outputs are low, the number represented is zero. If the 4 output and the 2 output are high, the number represented is 6 (i.e., $4 + 2$). You can see that numbers from 0 to 15 can be represented using this system; however, the connection of the ring counter is such that on the clock pulse after each nine count, the counter is reset to 0, so the numbers from 10 through 15 never appear.

For counting to numbers higher than 10, several ring counters can be interconnected. Figure 3-8 shows three decimal ring counters connected so as to count digitally from 0 to 999.

The key to using a seven-segment display is in the decoder-driver that takes the BCD input and decodes it to display the proper digital numeral on the LEDs. There are nine standard inputs to decoder-drivers; their functions are described below.

Ground and V_{cc}. These are self-explanatory. The standard voltage for TTL digital logic is 5 V DC ± 0.25 V. This voltage is not readily obtainable with batteries; however, TTL logic circuits do not seem to mind much if they are made to operate from 4.5 or 6 V DC. The Randomizer has been operated with both values, and the effect on the display brightness is not discernible. However, with 6 V applied, after a certain number of cycles the counter would appear to ***hang up***. This does not happen with 4.5 V, although as the batteries

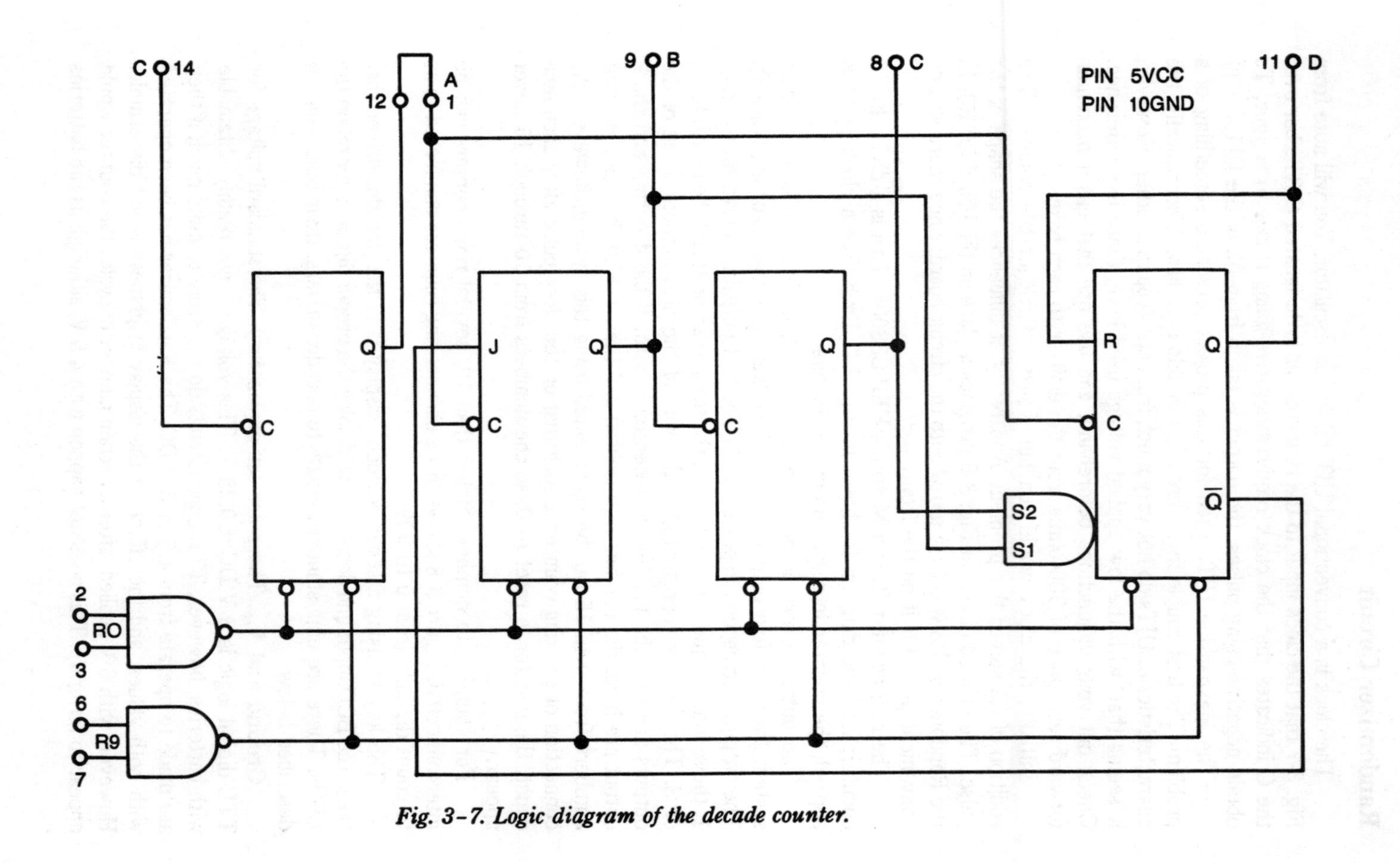

Fig. 3-7. Logic diagram of the decade counter.

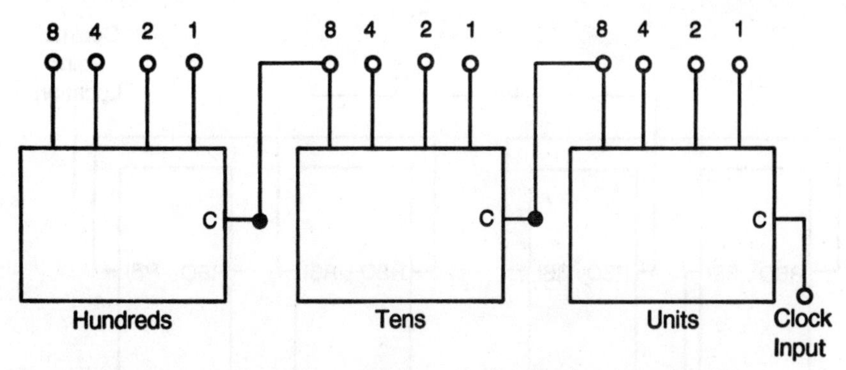

Fig. 3-8. Decade counters are cascaded to permit counting from 0 to 999.

get older and voltage drops, the oscillator may stop working until the batteries are replaced.

A, B, C, and D. These correspond, respectively, to the 1, 2, 4, and 8 outputs of the ring counter. In some cases, the ring counter designates these outputs as *Q0, Q1, Q2,* and *Q3.*

Lamp Test (LT). This input must be high for the display to show the count. If it is low, all seven segments will light up, regardless of what the other inputs are.

Ripple-Blanking Input (RBI). This input has an effect only if the BCD input is zero. In this case, if the RBI is high, the zero will be displayed. If it is low, the zero will be blanked. This is used with the ***ripple blank*** output of adjacent displays to suppress leading and trailing zeros. (See Fig. 3-9).

Blanking Input—Ripple-Blanking Output (BI-RBO). This connection can be used as either an input or output. As an input, it will blank the display whenever it is low. If it is not connected, as in the Randomizer, it will have no effect. As an output, it will be high except when the display is blanked by a low RBI, when it will be low. (Again, see Fig. 3-9 to understand how RBI or RBO work together to suppress leading and trailing zeros.)

On displays in which there is a decimal point, the decimal logic is separate from the decoder-driver. Some displays, like the one in the Randomizer, ground the decimal to turn it on. Others drive the decimal high to light it.

A final word about the clock frequency. This could have been much higher than 1000 Hz. The maximum clock rate for this kind of logic is about 18 MHz. However, it seemed easier to check the functioning of the clock if it operated in the audio range. It could then be tested for oscillation simply by connecting a high-impedance headphone across the UJT output. And the very low-speed cycling of the display has a modulating effect, so that the 8 that is displayed when the clock is running is noticeably less bright than numerals displayed when the clock is stopped. This modulation of the light output is further indication that the circuit is operating properly.

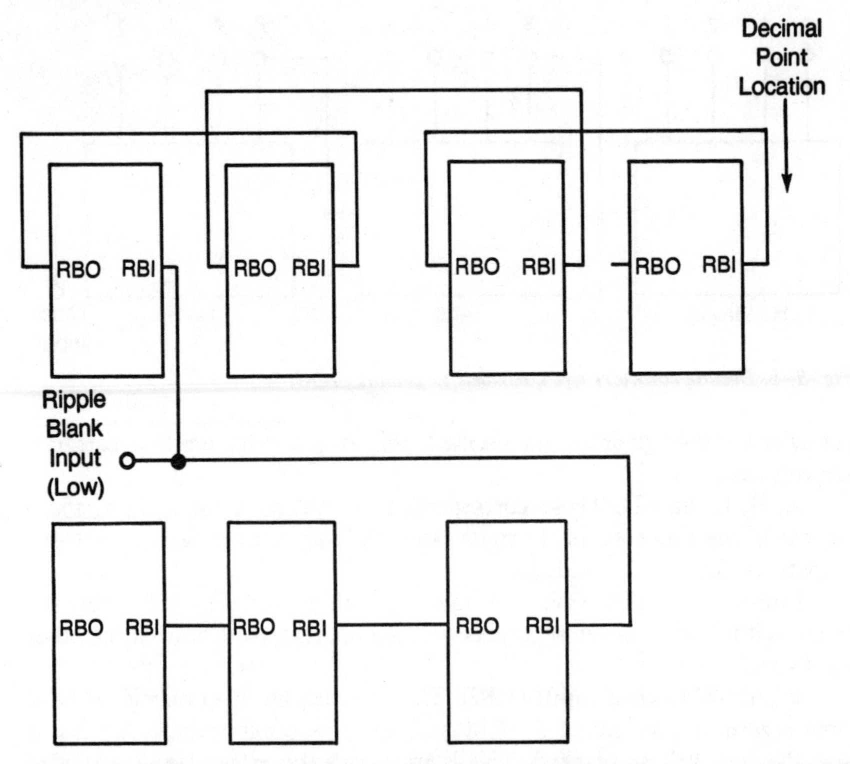

Fig. 3-9. Decoder-driver ripple blank input and output connections are used to blank leading and trailing zeros in multi-digit displays. If the input to the seven decoder-drivers were 2.03, *displays connected to the decoder-drivers would show* 2.03 *if the ripple blank input was LOW, but* 0002.030 *if RBI was HIGH.*

FET-SET SHORTWAVE RECEIVER PROJECT

With this project you can listen to distant shortwave stations in the 3.5- to 8-MHz shortwave band. The signals you will hear will include hams on the 80- and 40-meter amateur bands, domestic and overseas shortwave broadcast stations, and the 5 MHz time broadcasts of U.S. Government station WWV. The circuit takes a very old detector idea and gives it a new twist with a dual-gate MOSFET. The detector part of the circuit is actually a minor modification of a circuit that appears in the "RCA Transistor Manual." The FET-Set shortwave receiver gives you the option of listening on its own built-in speaker or on a private earphone.

Shortwave Receiver Circuit

The signal from the antenna is applied through C1 to the tuned circuit consisting of L1, C2, and C3. For the sake of stability, it is very important that C1 connect to the coil at the tap point, one-quarter of the way from the ground end of the coil. Capacitor C2 is the main tuning control, and C3 provides bandspread capability. Under the heading of "Construction," we will explain how to make the bandspread capacitor by removing plates from a larger unit. By a happy stroke of luck, the modified bandspread capacitor provides almost exactly 10:1 resolution relative to the main tuning capacitor.

The regenerative detector is an old concept in radio receivers—something quite simple, and at the same time, quite clever. Although it is the most sensitive type of detector, it has some disadvantages that limit its use to inexpensive projects like this. Its major disadvantage is that it *blocks* (loses its sensitivity) in the presence of large signals. Also, its response is less than perfectly linear (it distorts). Given these limitations, the regenerative receiver is still an excellent way to squeeze a lot of sensitivity out of a very few components.

How does the regenerative detector work? The basic concept involves an amplifier in which a little bit of the output is fed back, in phase, to the input. Think of the circuit as starting with just the gain available from the amplifier without feedback. If a little of the output is fed back into the input, this effectively makes the input signal a little bigger than it would be all by itself. In turn, this makes the output a little greater, and this boosts the input just a little more. In effect, the signal picks itself up by its own bootstraps. If the amount of signal that is fed back to the input is too great, the amplifier will go into self-oscillation and no signal amplification will take place. The amount of feedback, or *regeneration,* must be controlled so that the circuit operation is just below the point of self-oscillation.

The usual method of obtaining regeneration is by means of a separate winding, called a *tickler,* on the tuning coil. The circuit in our FET-Set uses a different approach. (Refer to Fig. 3-10). Field effect transistor Q1 is a dual-gate MOSFET; that is, it has two control elements. This kind of FET was developed for high-frequency applications, where interconnections can be made that minimize the device's apparent capacitance. For our circuit, though, we will use Gate 1 for the signal input and Gate 2 for a regeneration signal coupled back from the FET drain.

Even with all of the amplification available from the regenerative detector stage, there is insufficient audio at the drain of Q1 to the input of Q2, we should explain why the output of Q2 is coupled to the input of the audio amplifier through a matching transformer. This is necessary because Q1 and Q2 are both n-channel devices and their ground bus is negative, while the audio amplifier uses a positive ground bus.

While this project was still in the design phase, I considered designing and building an audio amplifier custom tailored for this application. It turned out, however, that there are several audio amplifiers on the shelves in the radio hobby stores that are well suited for this project and cheaper to buy than the parts for a scratch-built amplifier. The kit audio amplifier consists of a printed board with separate resistors, capacitors, and transistors, which the buyer

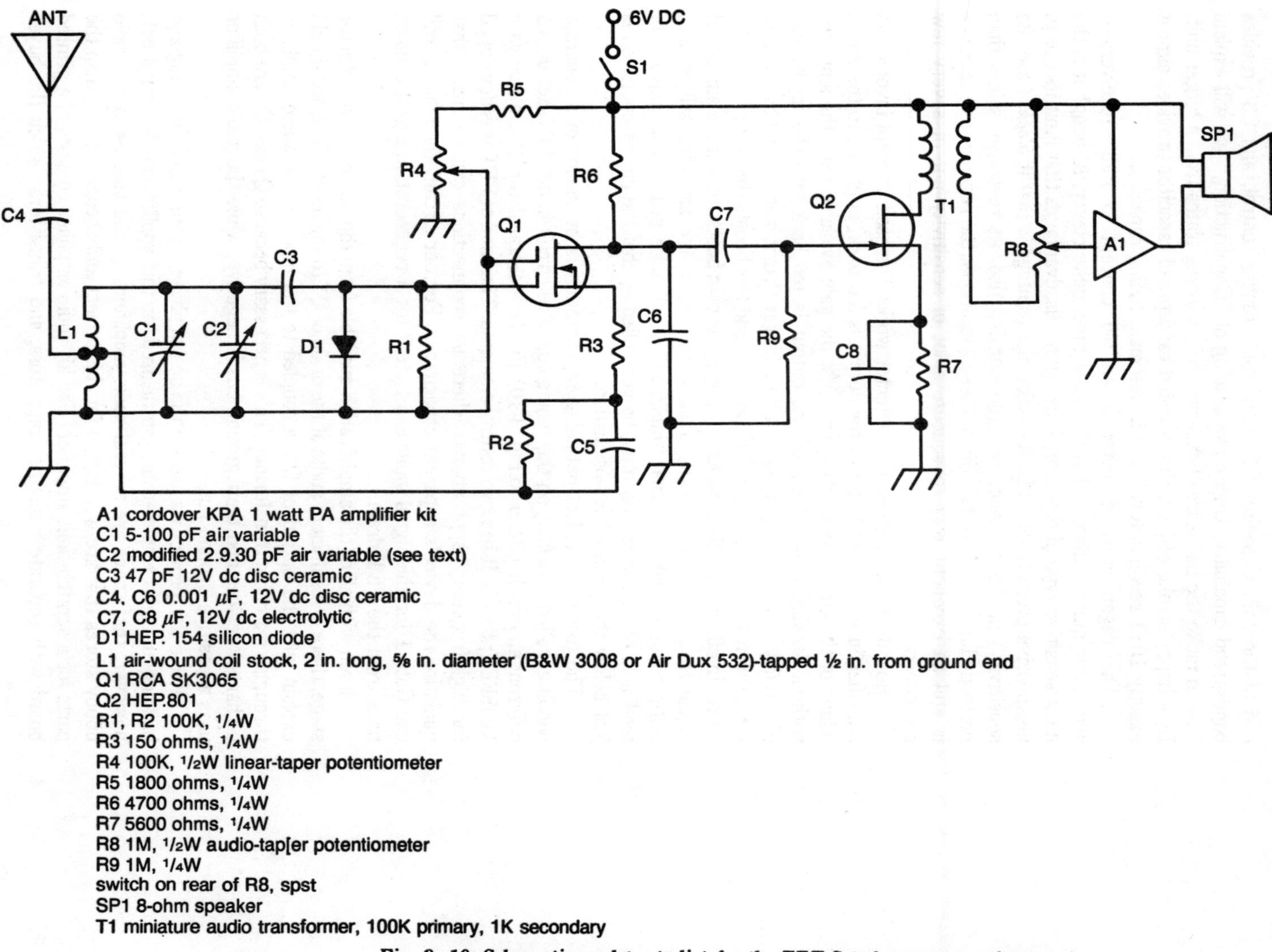

Fig. 3-10. Schematic and parts list for the FET-Set shortwave receiver project.

installs himself. Other amplifiers, equally good, may consist of a complete assembly encapsulated in black epoxy, or even a 1 W audio amplifier IC.

Construction

The exterior housing for the FET-Set is a 5-×-5-×-5-inch cowled box, but most of the components are mounted on a piece of perforated board. Without an inside chassis, these cowled enclosures are lacking in the rigidity we require for RF circuits; by mounting all of the RF components on the perforated board, we avoid the problem.

Coil L1 is made from a 2 inch length of air-wound coil stock as described in the parts list (Fig. 3-14). Two turns on either side of the tap are bent inward to provide enough space to solder the tap to the coil. This has very little effect on the inductance of the coil.

Most of the time, when you build a project using perforated board or printed circuits, you will mount the board inside the enclosure with some kind of standoff brackets. In the case of the FET-Set, though, we have the opportunity to employ a unique method of mounting the board to the enclosure. The tuning knobs for the main tuning and bandspread capacitors are planetary vernier units that mount directly to the front panel of the enclosure. On the back of each knob is a ¼-inch (inside diameter) bushing with a setscrew. The shaft of the tuning capacitor fits into the bushing. If you are very careful to make the locations of the vernier knobs on the front panel correspond to the locations of the capacitors on the perforated board, you can use the shafts of the capacitors to support the circuit board.

It isn't as hard to achieve this matching of holes as it sounds. The secret lies in the pattern of holes on the perforated board. If we select two holes on the board as the starting point for drilling the holes for the variable capacitor shaft, we can use the same two holes to mark the front panel of the enclosure with great precision. In fact, in this case, since the regeneration is mounted on the perforated board, we want to mark its location on the front panel in the same way. If we take our time and make sure that our drill bit lines up exactly with the marks on the front panel, we can be sure that the holes in the panel and circuit board will match.

The main tuning capacitor is an unmodified 100 pF air variable. The bandspread capacitor requires some modification, however. Starting with a 30 pF unit similar to the 100 pF main tuning capacitor, but with fewer plates, we proceed to remove more plates. When we are finished, we want to have only two stator plates and two rotor plates. To do this, we use long-nose pliers and our bare hands to bend each plate, in turn, away from the body of the capacitor, where we bend it back and forth until it breaks away from its mounting points. Be very careful not to damage the plates that remain on the capacitor. The last thing to do when only the treeplates are left is to check to make sure that they are all parallel and do not short together at any point in their rotation.

The battery holder for the receiver is mounted on the bottom plate of the enclosure. For batteries, C-cells were used because the audio amplifier requires appreciable current, and smaller batteries would not be able to supply this current for long periods of listening. The speaker, earphone jack, and audio amplifier all are mounted on the back panel of the enclosure.

FET-Set Operation

The key to good shortwave reception with this or any receiver is a good antenna. For these bands, you'll want a long, high wire—the longer and higher, the better. How long is long? Consider 50 ft an absolute minimum for reliable reception. You will also need a good ground for your antenna. A ***good ground*** means a solid connection to a water pipe or copper rod, or pipe driven at least 5 ft into damp soil. A barely adequate ground might be a screw attached to an electrical junction box in the room.

Having said something about good antennas and grounds, one must recognize that people have used all sorts of bad or mediocre antennas and grounds with surprising results. I have logged Radio Japan using the FET-Set with an antenna made from a short length of wire clipped to the finger stop on a dial telephone! The important thing to understand is that while bed springs and window screens may produce adequate results under some circumstances, reliable reception demands a good antenna and ground.

Once you have your antenna and ground connected to the FET-Set, turn the regeneration all the way down, the power on, and the volume up to about three-quarters of full volume. Increase the amount of regeneration until you hear a loud howl, and then back the regeneration off until the howl just stops. This is the point of highest sensitivity for the frequency to which the receiver is tuned. You should be able to tune a little ways around your starting frequency and hear a number of signals. If you tune very far off your starting frequency, you will have to readjust the regeneration. The amplifier has more gain at lower frequencies, so as you tune higher and higher, you will have to add more and more regeneration. Likewise, as you tune lower in frequency, you will have to reduce the regeneration to prevent the detector from going into oscillation. A few minutes of experimentation will make you an expert at keeping the sensitivity of the receiver at optimum without slipping into oscillation.

What Will You Hear?

At the low end of the tuning range, you can expect to hear ham signals in voice and Morse code. As you increase the frequency, you will encounter some warbling signals characteristic of Teletype, some droning signals that sound like airplane engines and are facsimile transmissions, and some commercial stations broadcasting Morse code.

At right about 5 on the vernier dial, you will encounter the regular tone and voice announcements of WWV. WWV broadcasts the same signal on 2.5, 10, 15, 20, and 25 MHz from Boulder, Colorado. The National Bureau of Standards also operates WWVH in Kauai, Hawaii. WWV uses a male voice to make a time announcement each minute, and WWVH uses a female voice a few seconds earlier. In Portland, Oregon, the FET-Set can pick up both signals. If you listen regularly to WWV, you will not only have the most accurate wrist-watch on the block, but you will also be advised of significant meteorological events during the 19th minute of each hour and of the projected quality of radio propagation across the Atlantic Ocean for the next 24 hours. This information is broadcast during the 15th minute of each hour. The broadcasts from WWV and WWVH do not make the most compelling listening on the shortwave band, but they do contain a wealth of information. You can learn more about this information from "NBS Special Publication No. 236, NBS Frequency and Time Broadcast Services." You can get it by sending 25¢ to the Superintendent of Documents, U.S. Government Printing Office, Washington, D.C. 20402.

As you tune above WWV, you will encounter several signals from the Voice of America and a number of overseas broadcast stations. Sometimes these stations broadcast in their native languages; frequently they broadcast in English. Tuning higher you will encounter the 40-meter ham band.

You will find that reception on these bands is better in winter than in summer, and better at night than during the day. The reasons for this is that in winter there are fewer thermal-type thunderstorms to create static. And at night and during the winter, conditions are more favorable in the ionosphere, which is responsible for reflecting radio waves over long distances.

4

Troubleshooting Your Projects

Like everyone else, you eventually will have the experience of building a project and then finding that it doesn't work when you turn it on. The likelihood of this happening depends on the nature of your project. If you build a kit offered by a reliable manufacturer such as Heath or Eico, chances are pretty good that your project will work the first time you turn it on. If you are building a project based on an article in a popular magazine, your chance of immediate success is good to fair, depending on how closely you follow the author's recommendations and how well the author and magazine editors check out the article. Slipups are common enough that it is sometimes wise to delay starting an expensive or complicated project 2 or 3 months to see whether any corrections appear in the magazine. If you decide to put something together on your own, based on some circuit ideas you have seen, say, in a transistor handbook, don't be disappointed if you don't get the results you hoped for the first time you apply power.

But don't let this discourage you. Engineers expect a ***debugging period*** as a part of the design of any complicated piece of apparatus. Even with established designs that are being put together on assembly lines every day, enough failures occur to keep quality control inspectors on their toes.

When you find yourself with a project that doesn't work, you will find that it helps to go through a logical series of steps to isolate the problem or problems. (Don't assume that problems only come one at a time!) This chapter will help you to proceed sensibly, one step at a time, to locate and eliminate trouble spots.

TROUBLESHOOTING EQUIPMENT

You already own some very sensitive troubleshooting equipment that hasn't cost you a thing—your senses. Use your eyes to look closely at solder joints and components. Can you smell any burnt components? Hear that transformer sizzling? Its output must be shorted. Is that resistor hot to the touch? It's carrying too much current. Some people even claim to use their tongues to check low voltages, but I have better uses for my tongue and prefer to take these reports with a grain of salt.

Once you have exhausted the possibilities of your built-in test equipment, you will find that there are some items you will have to purchase. Some are absolutely essential, and some are nice to have around but indispensable only in certain limited cases. Many of them are described below.

Meters

The most common meter for troubleshooting is the ***volt-ohm-milliammeter***, or ***VOM***. As its name indicates, this is an instrument that is capable of measuring voltage (AC or DC), resistance, and current (dc only). Prices vary, and some very good imported models sell for less than $10.

VOMs differ in the internal resistance of their voltmeter sections. Very poor VOMs have a meter resistance of only 1000 ohms/volt (ohms per volt). Better units provide 20,000 or 30,000 ohms/volt, and expensive types may provide 50,000 ohms/volt. The number of volts referred to in the ohms/volt rating is the top voltage of the meter's range. In other words, when a 20,000 ohms/volt meter is set to its 0- to 10-V range, its internal resistance is 20,000 times 100, or 2 megohms.

When we measure voltage drops across large resistances, the internal resistance of the meter is very important, since the meter can become part of the circuit and affect the voltage level. *Electronic voltmeters*, or EVMs, may be used instead of VOMs to obtain more accurate measurements. Early electronic voltmeters used vacuum tubes and were called ***vacuum-tube voltmeters***, or ***VTMs***. Some electronic voltmeters now employ field effect transistors—which, like vacuum tubes, have a high input impedance—in units similar to VTMs. These are often referred to as ***FETVMs***, for ***field-effect-transistor voltmeters***.

VOMs, FETVMs, and VTVMs are all used to measure resistance in ohms. They do this by providing an internal voltage (which is applied to the unknown resistance) and measuring the current through the resistance in question. It is important that you do not try to measure a resistance in a circuit to which power is applied. There could be enough voltage across the resistance to permanently damage the meter.

While VOMs and FETVMs are both used to measure current, most VTVMs do not have a current scale. It is very important that you understand that voltages are measured with the meter in *parallel* with the part of the circuit in question, but that current is measured with the meter in *series* with the circuit. The meter used for current measurement has some very small-value resistances, called *shunts,* inside it. Each shunt is made for a different current range and is selected by means of a switch on the front panel of the meter. The shunts are made of a few turns of very fine nichrome wire, and they are precision resistors of fractions of an ohm. The meter measures current by measuring the voltage drop caused by the current as it flows through one of these shunts. If you connect one of these shunts across any appreciable voltage, or in a circuit greater than the shunt design capacity, the shunt will overheat and burn up.

While *digital multimeters* (DMMs) were once very expensive and exotic items (not too many years ago), they are now on many hobbyist workbenches. Improved technology has dropped the prices and improved the accuracy and reliability. Many have a number of special features, which may or may not be useful, depending on just what type of electronics work you do.

Many hobbyists are drawn to DMMs because they look nifty and very modern. There are more practical advantages to using a DMM. Because the readout is in direct digital form, it is easier to read than trying to judge the exact position of a meter pointer. There is less ambiguity with the digital readout. Instead of a reading of about 5 V, you may have a digital reading of 4.97 V. The least significant digit may wobble somewhat on a DMM, but the accuracy is still higher than with a mechanical meter.

Many DMMs have an auto-ranging feature. The user does not have to worry about setting the controls to the appropriate range. The DMM does it for you. If you use a too low range on analog meter, the pointer needle may be abruptly pinned at the high end of the scale, and possibly damaged. If you set the range too high, the meter will be hard to read. The pointer may not noticeably move at all. No damage will be done in this case, however. Whenever you are making measurements with a meter without auto-ranging, always make it a practice to start out with the highest available range, and work your way down until you get the best visible reading.

A DMM does have its shortcomings, however. In many instances we will be less concerned with an exact value reading then with patterns of value changes. For example, a large capacitor can be tested with an analog ohmmeter. First the capacitor leads are shorted together to fully discharge the component. The ohmmeter is connected across the capacitor's lead. The resistance will quickly jump down to a fairly low value, then it will slowly increase in value as the capacitor is charged. This test can not be made on a DMM. You will just get a meaningless blur of digits, changing too rapidly to be read. If you can afford it, a well-stocked workbench should have both an analog multimeter and a DMM.

When making tests with any multimeter, the ohms-per-volt rating of the meter must be considered. If you are comparing your readings to reference values (often included in schematic diagrams), you must use a meter with the same rating as the meter used to determine the reference values. The resistance of the meter will affect the reading. Except when making such comparisons to a specific standard, a meter with the highest available ohms-per-volt rating will give the most accurate readings.

Signal Generator

You can do a lot of troubleshooting with just a VOM or EVM, but you can speed up your checking considerably if you have a signal generator. If you are testing audio circuits, you can get by with a single-tone audio generator, although high-fidelity testing may require a precision generator with a calibrated output so that you can observe the frequency response of an amplifier.

If the project you are troubleshooting has radio-frequency stages, you will want an RF generator. There are some simple generators that produce an audio square wave and, consequently, a series of harmonics into the radio-frequency range, but you will frequently want a tunable RF signal generator. A project in this chapter shows how you can build a simple signal generator that provides a fixed-frequency audio output and a variable-frequency RF output.

You can buy more complicated signal generators that provide a variable-frequency output. These are called *sweep generators,* because the output frequency is swept from one selected frequency to another at a set rate. These are essential for alignment of FM and television receivers and transmitters. Some even more elaborate signal generators are used by TV servicemen and technicians to generate color bar and dot patterns needed to align color TV receivers.

Signal Tracers

You may want to know if a signal is really present at some stage in an amplifier. Frequently, you can check using a pair of earphones with a high input impedance. It may help if you use a capacitor in series with one of the leads to provide DC isolation. Sometimes, with very low-level signals, you can use an audio amplifier with a high input impedance to listen for signals. If you want to check for RF signals, you can do so—if the signals are modulated in some way—with a diode in series with one lead of your high-impedance headphones or audio amplifier. At other times, this may not be satisfactory, and you will find it necessary to use a radio receiver tuned to the frequency you are interested in. Disconnect any outside antenna from the receiver and use in its place a probe connected to the receiver's antenna terminals through a piece of coaxial cable.

Bench Power Supply

For an occasional bout of troubleshooting, you can get by with some battery holders hooked up to give you 3, 6, 9, and 12 V. If you do a lot of building and experimenting, however, you will find that a regulated, variable-voltage DC power supply will come in handy. This will save you the frustration of discovering, after 3 hours of futile checking, that the reason your project doesn't work is because your batteries are dead.

Many deluxe power supplies feature current-limiting. A built-in meter displaying the output current is a popular extra. In some cases, the output current can be adjusted through a front panel control on the power supply.

Transistor Checker

You may be able to isolate a problem to a particular stage of a project, and you may direct your suspicion to a particular transistor. The only way to be certain is to remove that transistor and either insert a substitute or run some kind of a test on the suspect part. Substitution is the best way to test whether any part is working, but you may not have two of every component in your project. You can make a simple test of a common bipolar transistor with just an ohmmeter. The resistance from emitter to collector should be at least a kilohm, and more important, the effect of the transistor base-to-emitter and base-to-collector junctions should be very obvious. That is, for a ***pnp transistor,*** you should get a low resistance with the negative lead from the voltmeter connected to the base and the positive lead connected to the emitter or collector, and a high resistance with the positive lead connected to the base and the negative lead connected to either of the other two elements. For a ***npn transistor,*** the indications would be reversed.

You can tell more about a transistor's condition with a ***transistor checker.*** This instrument tells you the actual DC current gain of the device. There isn't any simple ohmmeter check that will tell you much about the condition of a field-effect transistor, but the BJT-FET transistor checker described in this chapter can tell you about the condition of FETs as well as bipolar transistors.

Integrated circuit (IC) testers are becoming increasingly available. These devices are necessarily more complex than transistor testers. Not all ICs can be tested on a single tester. The tester must be designed for the specific type of IC. Digital IC testers tend to be quite versatile, and are capable of testing a great many different types of devices.

Dip Meters and Wavemeters

Experimenters who do a lot of work at radio frequencies—particularly hams who like to try out different antennas—find use for dip meters and wavemeters. The original dip meter used vacuum tubes and was called a ***grid-dip meter.*** Modern equivalents use bipolar or field-effect transistors, which, of course, have other control elements than grids. No matter what the device inside, the idea of all dip meters is the same. Each consists of a ***variable-frequency oscillator*** (VFO) with an LC tank circuit. The inductor in this tank circuit is exposed and may be coupled to another resonant circuit, such as a tank circuit or an antenna. When the dip meter is tuned to the same frequency as the external resonant circuit, some of the energy from its coil is coupled to the external circuit, resulting in a dip of a meter needle as the dip meter is tuned through resonance. This reveals the frequency to which the external circuit is tuned.

Most dip meters are designed so that they can also be used as wavemeters. In this mode of operation, the oscillator portion of the dip meter is cut out, and a radio signal of unknown frequency is coupled to the coil. In this case, the meter shows an increase when the tuned circuit in the meter is adjusted to the unknown frequency.

Oscilloscope

An oscilloscope is really a kind of voltmeter that lets you see what happens to a voltage over a period of time. An oscilloscope displays the actual waveshape on a television-like cathode-ray tube (CRT) screen.

At one time or another, every electronic hobbyist has asked, "Can I make an oscilloscope out of an old television set?" The answer is no. In an oscilloscope, the deflection of the electron beam is controlled by the electrostatic charge on two pairs of plates in a CRT. One pair controls the up-and-down motion of the beam, and the other pair controls the side-to-side motion of the beam. Since no current flows between the plates, the drive requirements are very low. In a TV picture tube, the motion of the beam is controlled by a set of coils. There just isn't any practical way to use these coils for deflection at rates other than the ones they were designed for (15 kHz for horizontal deflection and 60 Hz for vertical deflection). Too bad. It was a nice idea. But don't be embarrassed, it has occurred to most of us.

Oscilloscopes are rated for the highest frequency signal that can clearly resolve. Signals at higher frequencies can be monitored, but the waveshape will be blurred. This may or may not be important, depending on the specific application.

Better oscilloscopes offer dual-trace displays. Two signals may be simultaneously displayed on the CRT screen. This is a very useful feature for checking input and output signals, checking for phase shifts, and other comparison tests.

In the past, the only type of hobbyists who were really interested in oscilloscopes were radio hams, who used the scope to assess the quality of modulation of their transmitters. In today's high-tech world, the oscilloscope's importance has increased. Many modern devices require precisely shaped signals for the best performance. With the oscilloscope, distortion is easy to spot.

If you do a lot of work with advanced digital circuits, you will need a scope with a very high frequency rating. Before investing in an oscilloscope, you should think carefully about the type of work you will be doing. For some hobbyists, an oscilloscope may be indispensable. For others, it may be an unnecessary luxury. Everyone who works with electronics could use an oscilloscope, at least some of the time. The questions are, how much will you need it? Can you get away without it? Is it worth the price? Oscilloscopes do tend to be fairly expensive devices.

SYSTEMATIC TROUBLESHOOTING

No matter how much test equipment we have, it doesn't do us much good unless we use it in some logical, systematic fashion. The steps summarized below aren't the only way to go about finding what's wrong with a project, but they represent an approach that will work, and one that can be modified to fit nearly any circumstance.

Was the Job Put Together Right?

Take the time to match the circuit, one stage at a time, to the schematic. Use a pencil to mark off each part of the circuit as you check it.

Are the components the proper value? Check the color code on each resistor and the value printed on each capacitor. Do they match the schematic?

Are the right transistors in the right locations? One transistor looks pretty much like another. Check again to make sure Q2 isn't where Q4 is supposed to be, for example.

Are the transistor lead connections right? Remember, transistor base diagrams read from the bottom. Have you interchanged collector and emitter accidentally?

Remember that many components are polarity sensitive. That is, one lead must be positive, and the other negative. Typical examples are diodes and electronic capacitors. Make sure that all polarized components are connected in the proper direction.

Check to see if any ICs are installed backwards. This is very easy to do if you're not careful. Generally, Pin 1 is marked on the case of the IC to permit you to orient the device correctly. Sometimes one or more of an IC's pins may get bent under the body of the device and not make actual contact with the circuit. This, of course, will prevent the circuit from operating correctly. Since an IC's pins are so close to one another, it is very easy for a tiny solder bridge to short two or more pins together. You may need a magnifying glass to check this because all it takes is a tiny dot of stray solder in the wrong place to create problems.

How Does the Project Look?

Inspect your solder joints. Do any seem to be cold solder joints? Do some joints have globs of solder on them? This could mean that there is no solder on the joint, just hardened rosin. Also, big globs of solder could cause a short—look closely.

Are there any lost nuts, lockwashers, or bits of clipped wire that could cause a short? Pick the equipment up, turn it upside down, and shake it vigorously to get rid of any junk. If you are afraid that turning the equipment upside down and shaking it will break something, you haven't built your project strong enough.

Are there any items of mounting hardware that could be causing a short? Could they short when the covers are screwed on? If you have any feedthrough insulators or binding posts that must be insulated from chassis or panel, are they shorted out?

Did your soldering iron burn through the insulation on any wires, creating a short-circuit hazard?

How do your components look? Are the exterior surfaces of your capacitors smooth, or are they bubbled and darkened from excess heat? Do the color bands on your carbon resistors show signs of discoloration? If a resistor looks bad, tap it with the blade of a screwdriver. A resistor that has failed because of too much current passing through it will often break in two when you tap it. Inspect chokes and power transformers. Excessive current causes heat that may cause potting compound to liquefy and run out, or it may cause lacquer to overheat, making the part smell burnt. The windings on RF chokes subjected to too much current may be visibly burnt. Usually, when transistors fail, there is no visible sign, but sometimes they may heat up and literally explode. Vacuum tubes that have lost their vacuum may show a whitish deposit inside. If you look at vacuum tubes that are operating and see a bluish discharge inside, the tubes may be gassy or operating under too high a voltage. (Voltage regulator tubes are another story. They're supposed to glow. Some glow orange and some glow blue.)

Is DC Power of the Right Polarity Going Where It Should?

Disconnect the DC supply from the equipment. What is its output voltage? Reconnect the DC supply. What it is voltage now? Does the equipment cause the supply voltage to fall below the required level? If it does, is this due to a short in the equipment, or to a supply with inadequate current capacity?

If the fuse blows when you apply power, you are in luck—all you have to do is to find out why it blows. This isn't a facetious statement; a gross short circuit is more easy to locate than a more subtle failure.

Never, never, ever replace a fuse with one of a higher rating, or bypass a fuse—not even for a second or two. If the specified fuse keeps blowing, there is a reason. Find the cause and correct it. If you substitute a higher rated fuse, there is a good chance that another (probably more expensive) component will blow out to protect the fuse, which is obviously not what you want.

If a new fuse blows as soon as power is applied, the problem is almost certainly a short circuit somewhere in the circuit. If the circuit works for awhile with a new fuse, but then the fuse blows something has changed value somewhere in the circuit. It is possible that the problem is heat sensitive, and the component has to heat up before it acts up.

A fuse is a safety device. It not only protects the circuit, it protects you. Never use a bridge to bypass a fuse. The result could be a dangerous or even deadly electrical shock, or a fire hazard. Don't take chances. Never defeat the purpose of a fuse.

With the on-off switch off, measure the voltage across its terminals. The voltage should equal the full supply voltage. If you can't get a reading, there is an open somewhere between the supply and the switch, or the switch and ground. Now turn the switch on and again attempt to measure the voltage. The potential drop across the switch should be zero. If you read the full supply voltage, the switch is broken.

If the project has an AC supply, measure the voltage at the output of the transformer. Is it what it is supposed to be? If it isn't, the supply voltage may be down, the transformer may be defective or mislabeled, or the current drain of the circuit may be too great. The last case indicates, among other things, that the fuse to the transformer is not working or has the wrong value. Measure the output of the filter with the circuit disconnected. You should get a value around 1 to 1.4 times the nominal value of the transformer output voltage (unless you are measuring a voltage multiplier circuit), depending on the size of the bleeder resistor. Leave the circuit unconnected and switch the meter to the AC voltage scale. Any evidence of a substantial AC signal at this point suggests an open filter capacitor. Now reconnect the circuit to the power supply and again measure the DC output voltage. It should be only slightly lower than it was in the no-load case.

Turn the power off. Refer to the schematic and check for continuity (zero resistance) between all ground connections. Make sure that the ground bus is connected to the proper (either negative or positive, depending on the circuit) terminal of the power supply.

Leave the power off and check all of the points connected to the other side of the power supply. Turn the power on and make sure that each point is at the same potential. At each transistor collector, FET drain, or vacuum-tube plate, measure the voltage at each end of the load resistor (if one exists). In general, if there is no change in voltage across the load resistor, the transistor (or other active device) is open-circuited. If the entire supply voltage drops across the load resistor (or across the load and bias resistors), the active device is shorted out.

What Is the Condition of Each Stage of the Circuit?

If your project has an audio output to a speaker or an earphone, use a signal generator to apply an audio tone to the input of the device. Can you hear anything? If you do not have a signal generator, use a battery and a pair of wires to apply a DC voltage to the speaker or earphone. You should hear a click each time you make or break contact if the device is functioning.

If your project has a lamp, LED, meter, or other optical device as its output, remove the device from the circuit and check to see whether it is working. Incandescent lamps and LEDs can be tested with an ohmmeter. Incandescents should read a dead short; LEDs should have low resistance in one direction and high in the other, just like other diodes. When the resistance is high, the negative lead of the meter is connected to the cathode of the diode. Gas discharge tubes, such as neon bulbs and xenon flash tubes, cannot be tested easily except by inserting them in a circuit you know is good. Incidentally, the firing voltage of a neon bulb will increase over a period of time if the bulb is not exposed to light.

Do not check a meter with an ohmmeter. If the questionable device is an ammeter, select a combination of voltage and resistance to give a known current in the meter's range, and use this to verify proper meter operation. If the suspected device is a voltmeter, find a known voltage in the proper range and measure it with the voltmeter. Do not apply AC to a DC instrument.

If your project uses a digital device with an integral decoder-driver, first check all of the voltage levels at the ***lamp test, blanking, ground,*** and V_{cc} terminals. The proper levels for these are covered in the Randomizer project. If all of these are as they should be, disconnect the A, B, C, and D inputs to the decoder, and input these manually, observing the effect on the display. Figure 4-1 shows the output of a seven-segment display in response to all binary inputs from 0 to 15.

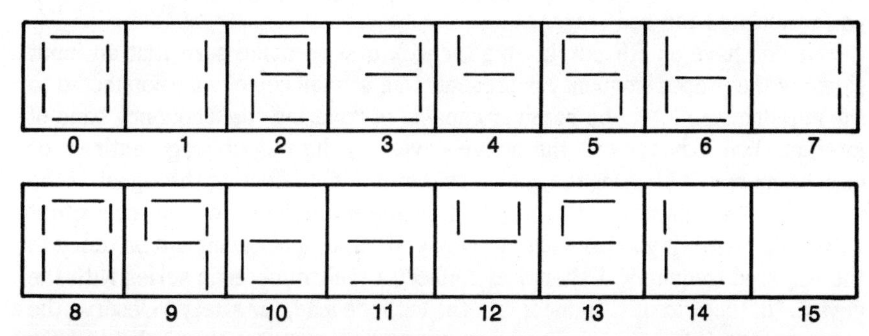

Fig. 4-1. Output of a seven-segment display in response to BCD inputs from 0 to 15. Outputs higher than 9 may cause you to mistakenly think that some segments aren't working.

If your project is a signal generator, VFO, exciter, or transmitter with an RF output, is your antenna or dummy load a satisfactory match for the output stage? Has the output been somehow short-circuited? If you are using a radio receiver to look for your signal, is it possible that the receiver is too closely coupled to the output of your project? This could result in blocking the receiver. Also, check again to make sure that the receiver is tuned to the same band as the anticipated signals from the project. If you get many different signals, do not assume that they are all in the output of your project. It is likely that a lot of these are *birdies,* spurious signals created inside the receiver by its own heterodyning action. You can get rid of most of them by moving the receiver farther away, until it barely picks up the signal from the project. Then look for the lowest, strongest signal you can find. That will be the true output from your project.

The above steps account for troubleshooting most of the different kinds of output devices you will encounter. The inverter project in this book has one of the more unusual output devices you will come across—a power transformer. In this case, you should first check the electrical continuity of the transformer primary and secondary. Then, disconnect the transistor circuitry, connect the 110-V output to the house current supply (be sure to use a fuse and wrap all of the connections to prevent a short), and measure the AC output voltage at the 24-V side. Remember to use the AC range of your VOM. If the transistor passes this test, it's okay.

If you have an audio output stage, use an audio signal generator to inject a signal at both input and output. Can you hear the difference caused by the gain of the last stage? If the output from the signal generator is too high, the difference may not be obvious. Use the lowest audio output that will produce an audible signal at the speaker when injected into the input of the final stage. If you do not have a signal generator, you may be able to use an audio amplifier or a high-impedance earphone to check for a signal at the input to the final amplifier stage. If the signal is there but is not coming out the other end, you've isolated the bad stage.

If you have an RF output stage, check first to make sure that an input signal of the proper frequency is present. Use a small coil of wire connected to the antenna terminals of a receiver capable of covering the frequency band of interest. You may remove the active device in the output stage entirely or disconnect one of its leads to make sure that it is not affecting the signal. If the input signal is present and the output stage has a tuned tank circuit, check first with a dip meter, if you have one, to verify that the tank circuit can be tuned to the required frequency. Otherwise, connect a milliammeter in series with the device. (In tube circuits, connect it in the cathode lead, for safety.) Observe the current as you tune the tank circuit through resonance; there should be a pronounced dip in current. If there is not, it is possible that the active device has internal capacitances that prevent the circuit from resonating. If this was a potential problem, the kit manufacturer or the author of the article should have some instructions for *neutralizing* the device. If this is your own design, you'll find an extensive discussion of neutralization in the "Radio Amateur's Handbook."

If you have a digital output stage with a separate decoder-driver IC, follow the procedures described above for digital output displays with integral decoder-drivers.

If your output stage contains an SCR or triac, you'll need a DC supply with an output voltage of 50 V or more, an adequate load resistor, and some external means of triggering the gate of the device. (A wire connected to the minus supply is fine, but be careful not to lay it down where it could short out the supply.) Connect the DC voltage across the device in both directions without triggering the device, and observe the voltage drop. The full supply voltage should appear across the device each time. Then trigger the SCR or triac. The voltage drop across a triac should fall to nearly zero, regardless of the polarity of the applied voltage. An SCR, however, should show a high resistance, regardless of triggering, when the positive end of the supply is connected to its cathode and the negative to its anode, and a low resistance, after triggering, when it is forward-biased by the supply.

If the output stage is functioning properly, you will have to work backwards through the preceding stages. For each type of circuit, follow the same procedure that you used for the output stage.

Is the detector, mixer, or modulator working? In the ***modulator***, the inputs to the circuit are a RF carrier and an audio (or composite video) signal containing information to be impressed on the carrier. An ordinary modulator fed with a single tone provides an output consisting of the audio signals, the original RF carrier, and two additional RF signals, or ***sidebands***—one with a frequency equal to the carrier frequency plus the audio frequency, and one with a frequency equal to the carrier frequency minus the audio frequency. The amplitude of the sidebands is proportional to the audio amplitude. Another kind of modulator, called a ***balanced modulator***, suppresses the carrier frequency so that only the sidebands and audio signal are present in the output. Usually, there is a simple filter to eliminate the audio in both kinds of modulators. Single-sideband transmitters use another filter to suppress either the upper or lower sideband.

In the ***mixer***, used most often in superheterodyne radio receivers, one input signal is the actual radio signal received at the antenna. This consists of the carrier and both sidebands, or in the base of a single-sideband transmission, just one sideband or the other. Another input to the mixer is the signal from the receiver's local oscillator, which is tuned to a frequency above or below the frequency selected by the antenna tuned circuit. The amount by which this frequency differs from the input signal frequency is called the receiver's ***intermediate frequency*** (IF). The output from the mixer consists of all the input frequencies, the intermediate frequency, and two sidebands on either side of the intermediate frequency. These sidebands bear the same relationship to the IF that the original sidebands bore to the RF carrier. Usually, mixers have sharply tuned output stages that filter out the RF and local oscillator signals.

In the ***detector***, the input is just some RF or IF carrier plus one or two sidebands. The detector output contains carrier and sidebands, and the audio signal that originally produced the sidebands. Generally, the detector output is bypassed (filtered) to suppress the RF signals and leave just the audio.

We will use the same technique for troubleshooting modulators, mixers, and detectors. Refer to the discussion of each mode above, and determine what inputs and outputs should be present. Then use a suitable device to check for them. Without a device called a *spectrum analyzer*—which not too many people ever get to see—you will not be able to observe sidebands directly, but you will be able to check for RF signals using a separate receiver tuned to the proper frequency. If you can hear the audio in the output of the second receiver, the sidebands are there.

You can also check for the presence of the IF even though you do not have a receiver that tunes that low, as long as your receiver has the same IF. Just make a coil of several turns of wire and connect this to your probes. Place the coil near the tube or transistor that serves as the first IF amplifier in the receiver, and the signal should be audible.

Audio frequencies are most easily checked by means of an audio amplifier or a high-impedance earphone connected to the appropriate terminals of the detector, mixer, or modulator.

If the appropriate signals are present at the appropriate terminals of the circuit, you can dismiss the detector, mixer, or modulator as the source of your troubles.

Check audio oscillators with an audio amplifier or a high-impedance earphone across the output. Use a probe connected to the antenna terminals of a suitable receiver for RF oscillators. If you cannot observe any signal, disconnect the oscillator from the preceding stage and check again. If you get oscillations now, something in the next stage is loading down the oscillator.

It is an unfortunate fact of life that sometimes, in some circuits, one tube or transistor of a given type will not oscillate, when most others of that type will. If you are down to the hair-tearing stage, try substituting another device of the same type. The odds for success aren't great, but you may find that this solves your problem.

Nonsinusoidal oscillators like the multivibrator may be checked in the following way. First, apply power and determine which transistor is conducting and which is cut off. The voltage drop across the conducting transistor will be very low, while the drop across the cut-off transistor will be the full supply voltage. If both transistors are conducting or both are cut off, one or both are bad. If you have determined that one is indeed in the on state and one is in the off state, try to get the multivariator to flip into the opposite state. Do this by sorting out the resistor that is connected to the base of the nonconducting transistor. If you are unsuccessful, suspect a failure of either the conducting transistor or the capacitor connecting its collector to the nonconducting transistor's base. If the multivibrator *flips*, but *flops* right back, the characteristics of the two transistors are too far apart; one of them will have to be replaced. If the multivibrator flips and stays flipped, the capacitors you are using have too much leakage.

If a nonsinusoidal oscillator using an SCR or triac triggered by a neon lamp or diac refuses to oscillate, first check the SCR or triac as outlined above. If that checks out OK, the problem is either a faulty neon lamp or diac, or too low a voltage on the high side of the lamp or diac. This could result from a bad power supply—which you should have detected before this—or from an open resistor or leaky capacitor in the timing circuit.

What Components Are Causing the Problem?

Let's assume that you have followed the steps described in the preceding paragraphs and have isolated a faulty stage. Your next task, if you are building with discrete components, is to isolate and replace the faulty component in that stage. If you are using ICs, your problem is solved, since you simply have to replace the faulty IC. Don't be too hasty about replacing that IC though, give it one last functional check when it is out of the circuit, to see whether some problem with another stage hasn't been giving you a false indication. And try to see why the IC failed. Was its supply voltage too high? Was it wired incorrectly? Is the current drain of the next stage excessive? Could the IC have been damaged by excessive heat during installation?

Let's look at some of the tests we can perform on discrete components.

Resistors. As noted earlier, you can spot some burned-out resistors by discolored code bands. If there is any doubt, though, it is best to measure the resistance with an ohmmeter. To get an accurate reading, it frequently will be necessary to disconnect one end of the resistor from the circuit.

Capacitors. Some capacitor failures you will be able to hear. In a radio receiver or audio amplifier with a 110-V AC supply, an open-circuited filter capacitor will betray itself by a loud hum audible from the speaker and independent of volume control setting. Short-circuited capacitors are revealed by a simple resistance measurement, but be certain the capacitor is disconnected from the rest of the circuit before you decide it is shorted. Be careful also, when dealing with large capacitances, that you discharge the capacitor before you touch it. The only way to check small capacitors that do not appear to be shorted is by means of an instrument like the Capaci-Bridge described later.

Inductors. The easiest test you can perform on an inductor is to measure its DC resistance. On most coils, this should be very low. If you have established that the coil is not open, you still cannot tell whether the coil has the proper inductance without measuring it, and this requires some fairly sophisticated equipment.

If the coil is made from air-wound coil stock, or if you wound it yourself in a single layer, you can calculate its inductance to verify that it is indeed the proper value. Figure 4-2 is a graph that gives the relationship between the ratio of a coil diameter to coil length (d/1) and a form factor (F). To calculate an

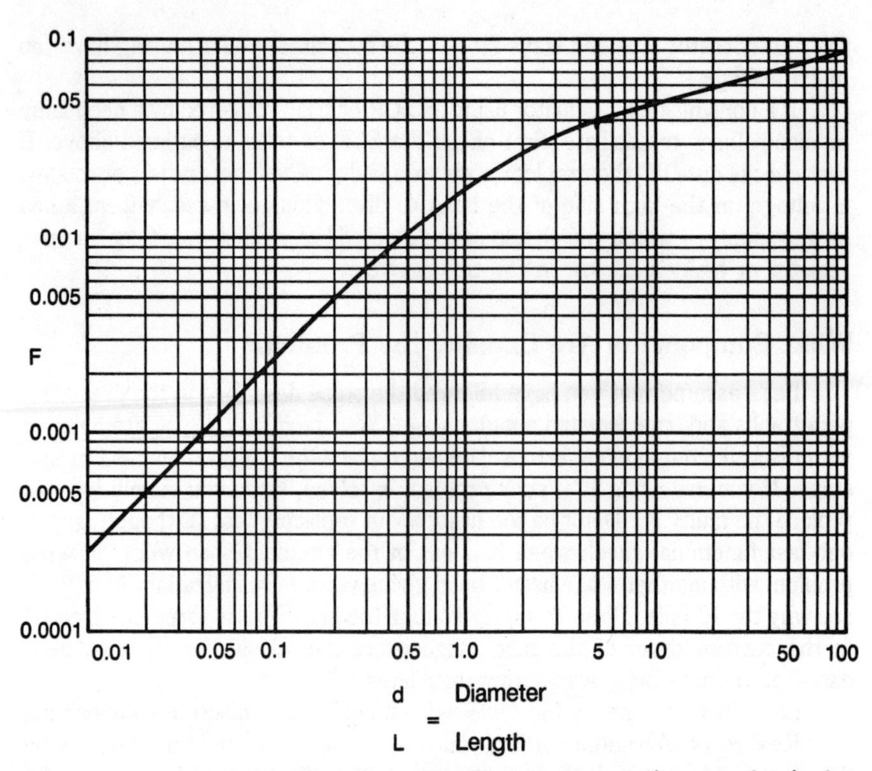

Fig. 4-2. Graph for finding form factor (F), given length and diameter of a single-layer coil. The value of F is then used to calculate inductance.

inductance, first measure the length and diameter of the coil and calculate d/1. Then refer to Fig. 4-2 to find F. The inductance in microhenrys is given by

$$L = Fn^2d$$

where n is the number of turns and d is the diameter of the coil.

Semiconductors. Diodes, of course, can be tested with an ohmmeter. Most diodes are made either with the circuit symbol printed right on them or with a band of paint around the cathode end. In the case of the circuit symbol, the arrowhead is the anode end, and the bar is the cathode end.

Bipolar transistors can be given a simple test using an ohmmeter, as described in the section of this chapter dealing with transistor testers. The BJT-FET transistor checker project provides a simple device that can check both kinds of transistors.

Vacuum Tubes. You can check whether a tube filament is burned out by measuring the resistance between the two base pins connected to the filament.

For other tests, there is no substitute for a tube tester. The supermarket variety, known as ***emission testers,*** are adequate to detect gassy tubes and most common failures. More expensive, laboratory-quality testers, or ***trans-conductance testers,*** give more insight into a tube's condition. High-power tubes such as TV horizontal output tubes, transmitting tubes, and audio power-amplifier tubes cannot be tested on any kind of tester for anything but gas or open filaments.

Integrated Circuits. ICs are generally tested as "black boxes." It's usually a waste of time to try to take measurements of their internal circuitry. Just check the inputs and outputs. If the inputs are correct, and one or more outputs are not correct, then the chip is presumedly bad. If the inputs and outputs are correct, it is reasonable to assume that the IC is OK.

AFTER YOU FIND THE TROUBLE

Once you have isolated the cause of the failure, your troubles are almost over. Before you replace the defective component, ask yourself this: was it bad when I installed it, or did it fail because of some other defect in a nearby circuit? Don't install that new part until you are certain it will not follow its predecessor down the path to destruction.

Even at this stage, you may still not have met with success. There is no guarantee that there will be only one bug in a project. If the equipment still doesn't work, stay calm—go get another cup of coffee—and continue with the troubleshooting procedures I have discussed. Look on the bright side; think of all the potential trouble areas that you have already eliminated.

CAPACI-BRIDGE PROJECT

It is relatively easy to test a capacitor of several microfarads with just an ohmmeter. If you connect the ohmmeter across the capacitor, there should be an initial jump of the needle, after which it should settle back to nearly infinite resistance. There is no such easy test for smaller capacitors. This can make locating a faulty capacitor a difficult job. The Capaci-Bridge is useful in checking capacitors and in determining the value of capacitors that are unlabeled. The circuit (Fig. 4-3) is basically a 1-kHz bridge using a high-impedance crystal earphone as a detector.

Capaci-Bridge Theory

The basic bridge circuit is much like the one drawn in Fig. 4-4. Consider the bridge to be just two parallel voltage dividers. One voltage divider is composed of R_a and R_s, and the other is composed of R_b and R_u. The source applies a signal across the pair of voltage dividers and the detector is placed between the junction of R_a and R_s and the junction of R_b and R_u.

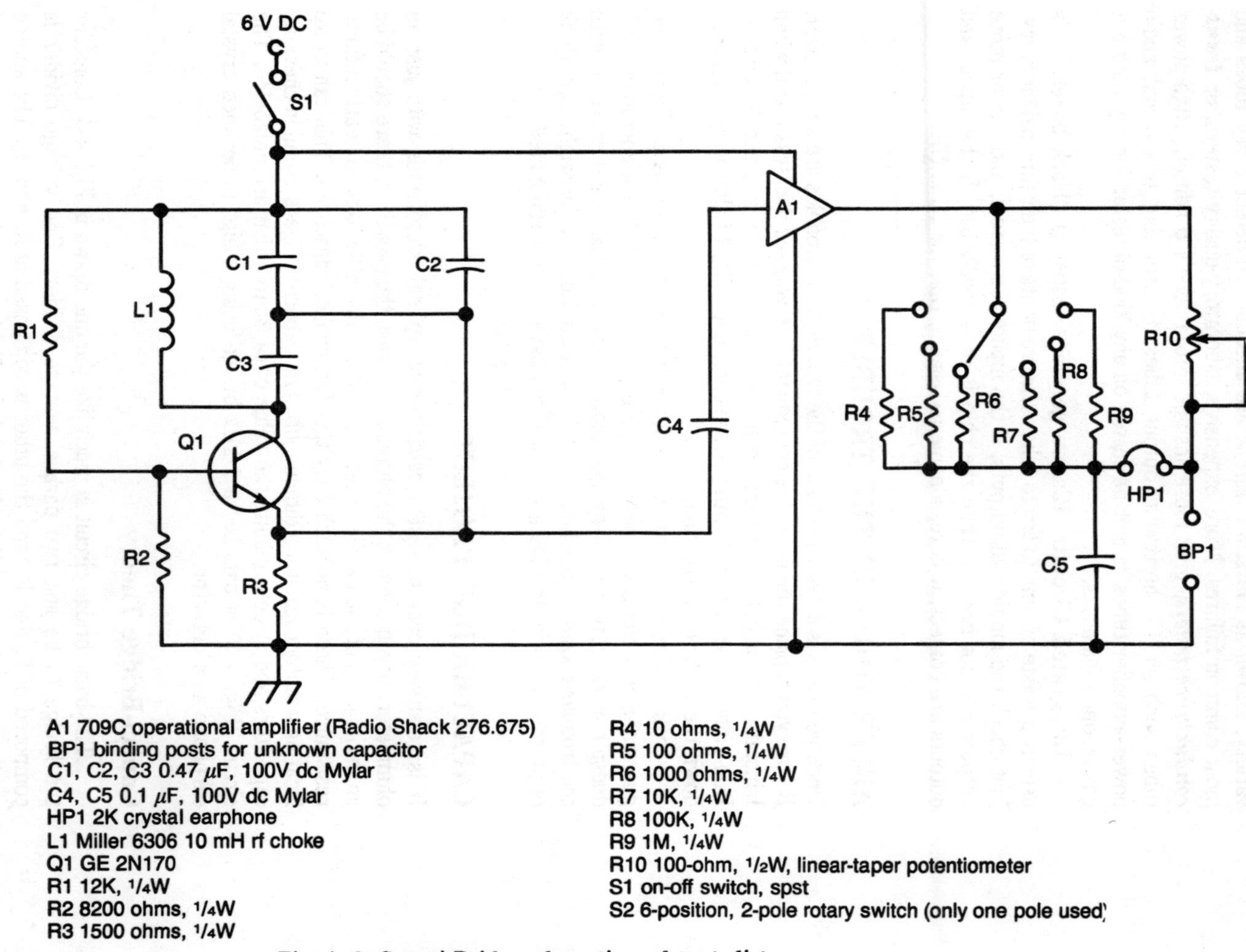
6 V DC
S1
C1
C2
C3
L1
R1
Q1
R2
R3
C4
A1
R4
R5
R6
R7
R8
R9
R10
HP1
BP1
C5
A1 709C operational amplifier (Radio Shack 276.675)
BP1 binding posts for unknown capacitor
C1, C2, C3 0.47 μF, 100V dc Mylar
C4, C5 0.1 μF, 100V dc Mylar
HP1 2K crystal earphone
L1 Miller 6306 10 mH rf choke
Q1 GE 2N170
R1 12K, 1/4W
R2 8200 ohms, 1/4W
R3 1500 ohms, 1/4W
R4 10 ohms, 1/4W
R5 100 ohms, 1/4W
R6 1000 ohms, 1/4W
R7 10K, 1/4W
R8 100K, 1/4W
R9 1M, 1/4W
R10 100-ohm, 1/2W, linear-taper potentiometer
S1 on-off switch, spst
S2 6-position, 2-pole rotary switch (only one pole used)

Fig. 4-3. Capaci-Bridge schematic and parts list.

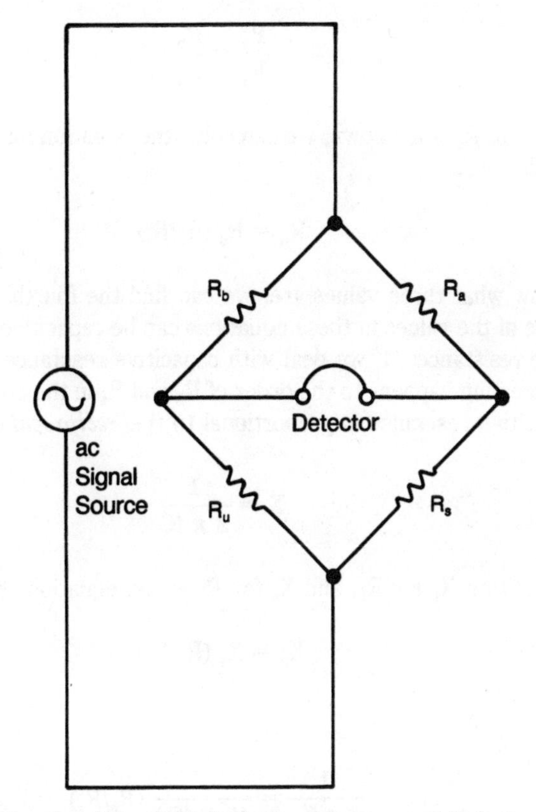

Fig. 4-4. Basic bridge circuit in which R_u is the unknown, and R_s is the standard to which it is compared. By varying R_s and R_b we can balance the bridge and discover the value of R_u.

To understand what happens in the bridge, imagine for the moment that R_s is twice as big as R_a. This means that at the junction between R_a and R_s there is a signal with respect to ground that is equal to two-thirds of the signal with respect to ground that is equal to two-thirds of the signal from the source. If R_u is zero and R_b is any value greater than zero, the signal across the detector will be essentially two-thirds of the source signal. If we gradually increase the value of R_u, the signal across the detector will gradually decrease. It will reach a minimum value (theoretically, zero) when R_u is twice as big as R_b, because then the voltage at each end of the detector will be the same. If we continue to increase R_u, a signal will again appear across the detector and will increase as R_u increases.

What we are interested in is the situation when the output of the detector is nulled. From the explanation above, we can understand that the value of R_a is to the value of R_s as the value of R_b is to the value of R_u. Stated algebraically,

$$\frac{R_a}{R_s} = \frac{R_b}{R_u}$$

If the value of R_u is unknown, we can solve the equation for R_u in terms of the other resistances.

$$R_u = R_s\ (R_b/R_a)$$

If we know what three values are, we can find the fourth.

Some of the values in these equations can be capacitive reactances rather than pure resistances. If we deal with capacitive reactances, however, something interesting happens to the order of R_b and R_a in the equation. Remember that capacitive reactance is proportional to the *reciprocal* of capacitance.

$$X = \frac{1}{2\ \pi\ fC}$$

If we substitute X_u for R_u, and X_s for R_s in the equation above, we have

$$X_u = X_s\ (R_b/R_a)$$

or

$$\frac{1}{2\ \pi\ fC_u} = \frac{1}{(2\ \pi\ fC_s)}\ (R_b/R_a)$$

We can cancel ½ π f from each side of the equation, leaving

$$\frac{1}{C_u} = \frac{1}{C_s}\ (R_b/R_a)$$

If we move the capacitances to the numerator, we get

$$C_u = C_s\ (R_a/R_b)$$

This tells us some interesting things. It tells us that the action of the bridge is not influenced by the frequency of the source and that an unknown capacitance can be compared to a known, standard capacitance times the ratio of two resistances.

You might ask whether the same theory could be applied to inductors. It could, except that in the case of inductors some practical considerations get in the way. The values of inductance we are most interested in measuring are in the microhenry range. If we use an audio-frequency generator for our bridge, the inductive reactance of these coils is very small; on the same order as the lead resistances in our bridge. Obviously, then, we cannot measure small inductances with an audio-frequency bridge. To successfully measure small inductances, we need a *RF bridge.* However, this creates much more potential for design problems. The shielding requirements, in particular, require considerable time to solve. That is the reason this project is a capacitance bridge rather than an inductance-capacitance bridge.

Capaci-Bridge Circuit

The bridge circuit is described above. Resistor R_a is a 100-ohm potentiometer, and R_b is a switch-selected fixed resistor of 10, 100, 1,000, 10,000, 100,000, or 1,000,000 ohms. Capacitor C_s is 0.1 μF. This combination of values lets us measure unknown capacitances from a few pF to 1μF.

The oscillator (see Figs. 4-3 and 4-5) is a 2.5 kHz Colpitts. To keep the bridge from loading down the oscillator, a 709 operational amplifier is used as a buffer. These operational amplifiers are compact, cheap, and have high input and output impedances. The operational amplifier is installed on a printed circuit board (Fig. 4-6). Only one input is used, and the frequency compensator terminals are left open.

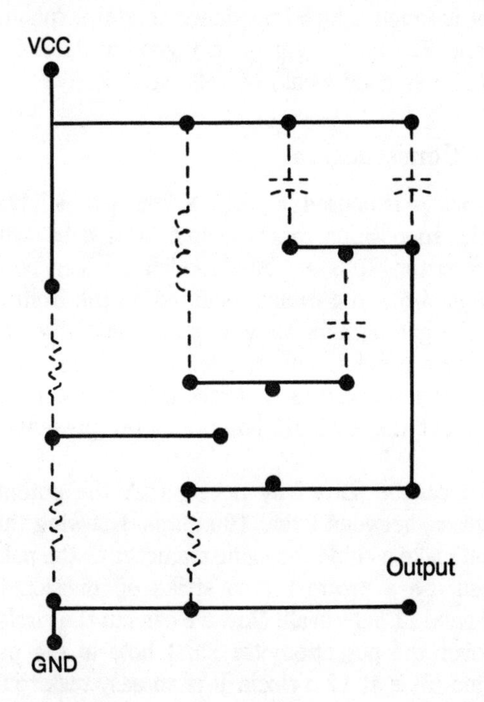

Fig. 4-5. Oscillator printed circuit (full-size).

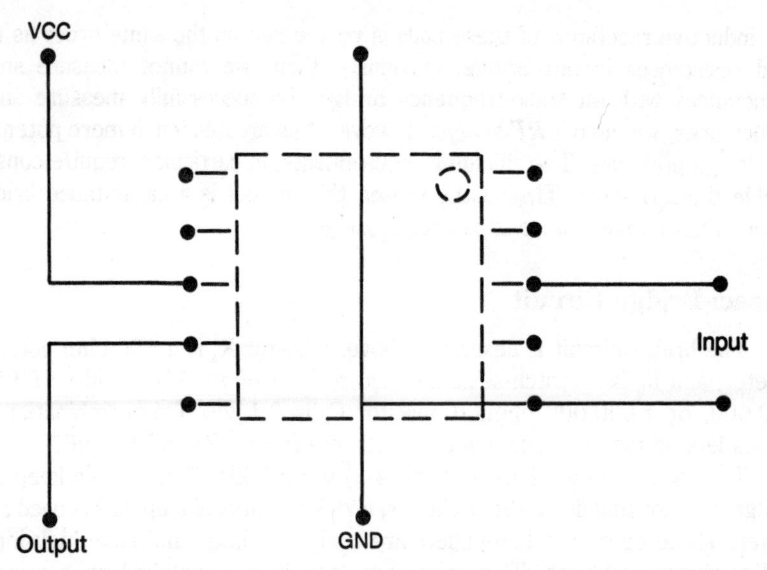

Fig. 4-6. Printed circuit pattern for a flat-pack 709 operational amplifier buffer (4 times actual size).

The detector is simply a high-impedance crystal earphone. This is simple, cheap, and efficient. The human ear is very good at detecting nulls. A meter circuit that would be as good would be relatively costly.

Capaci-Bridge Construction

The Capaci-Bridge is housed in a 4¼ × 7½ × 2 inch black Bakelite box. Half of a two-pole, six-position rotary switch is used to switch the resistors making up R_b. Resistor R_a is a 100-ohm, linear-taper potentiometer. The circuit boards (Figs. 4-5 and 4-6) are mounted on the front panel with right-angle brackets. The photographs show more detail of the construction.

The metal front panel is covered with wood-grain contact paper and labeled with transfer letters. It is no great trick to locate the labels for the multiplier switch, but the VALUE potentiometer presents a more difficult problem.

The problem can be solved by noting that the potentiometer rotates through 315 degrees, between 0 and 100 ohms. Knowing this, we can take a piece of paper and make a circle the same diameter as the pointer knob for the pot. We can then use a protractor to strike off division lines every 31.5 degrees. We will need 11 lines in all. Now we can cut the circle out of the paper and position it over the potentiometer shaft hole in the panel, so that the middle division line (5) is at 12 o'clock. It is an easy task to use the marks on the paper as guides for placing the required numbers.

Capaci-Bridge Operation

Turn on the Capaci-Bridge and insert the earphone in your ear. You should hear an audio signal. (You may not get a signal if the potentiometer is at zero. Pick some point in the middle of its range.) Connect the capacitor to be tested across the unknown terminals. In each position of the multiplier switch, rotate the VALUE potentiometer through its full range. If the capacitor is good, at some point you will hear the signal in the earphone fall through zero and rise again as you continue rotating the pot. Center the pointer in the middle of the null and read the value. Multiply this by the value shown on the multiplier switch to find the value of the unknown capacitor.

BJT-FET TRANSISTOR CHECKER PROJECT

How often have you found yourself in this situation? You've constructed a project according to some plans you've found in a magazine or a book like this, you've applied power, and . . . and nothing happens.

You begin a thorough logical analysis. Yes, the circuit matches the schematic. Yes, all of the solder joints are solid. Well, how about the components? Are the transistors functioning? How can you tell?

If you have this BJT-FET Transistor Checker, you can tell in a hurry. It's actually a fairly simple device, nowhere near as complicated as transistor testers can get, but it will test your transistors with about the same reliability as a drugstore tester checks your vacuum tubes.

As we said, this is a fairly simple project. All you need is a meter, a light, one transistor, a handful of resistors, and three switches. Yet in spite of its simplicity, it can be one of the handiest items of test equipment you will ever build.

THEORY OF TRANSISTOR CHECKER

The BJT-FET transistor checker performs two measurements on conventional BJTs and two go/no-go tests on FETs.

BJT Tests

Measuring leakage current from emitter to collector with the base open (Fig. 4-7) is a good check of a transistor's output resistance. This is the first thing to examine. For the most part, a good transistor will cause only a barely perceptible movement of the meter needle. If the leakage current indication moves out of the GOOD region, be very suspicious. Of course, a short-circuited transistor can peg the meter needle.

The leakage test will smoke out a short-circuited transistor, but it cannot tell whether a transistor has opened or not. For that, we need a *gain test.* This

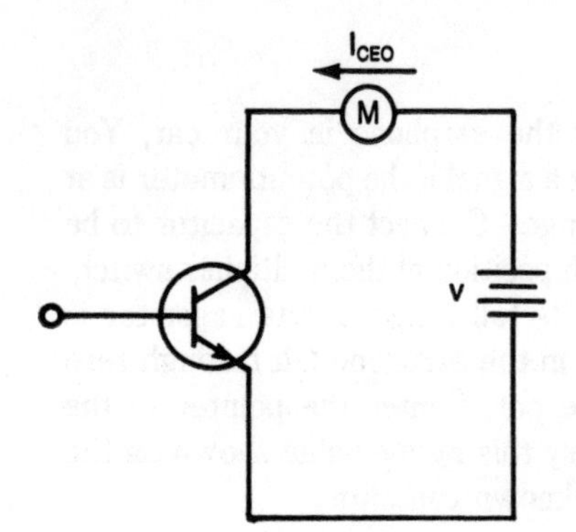

Fig. 4-7. Measuring I_{ceo} reveals whether the transistor is shorted and gives an indirect indication of transistor quality.

not only tells us whether the transistor is open, it also gives us an idea or how good an amplifier the transistor is.

The gain we are measuring is the *DC current gain,* or *beta,* of the transistor, generally referred to as H_{FE}. The current gain is equal to the collector current divided by the base current. In a good transistor, the gain can be anywhere from 20 to over 100, but it averages around 50.

To measure gain, we forward-bias the base of the transistor with a fixed current. In this case, our supply voltage is 9 V and we use a 300 K resistor in the base lead, so our base current is effectively 30 μA. With gains of from 20 to 100 or so we can, therefore, expect collector currents of from 0.6 mA (at a gain of 20) to 3 mA (at a gain of 100). Obviously, then, our meter must read from 0 to 3 mA.

While it is possible to buy such a meter, it is not as common as the 1 mA meter used in the BJT-FET transistor checker. To make the meter read full scale when the collector current is 3 mA, the meter is inserted in a current divider (Fig. 4-8) made up of a 5 K and 2.5 K resistor in parallel. Ammeters have a very low internal resistance (in contrast to voltmeters, which have a very high internal resistance), so the meter does not affect the current divider circuit. Of the total current flowing through the collector, two-thirds flows through the 2.5 K resistor and one-third through the 5 K resistor and meter.

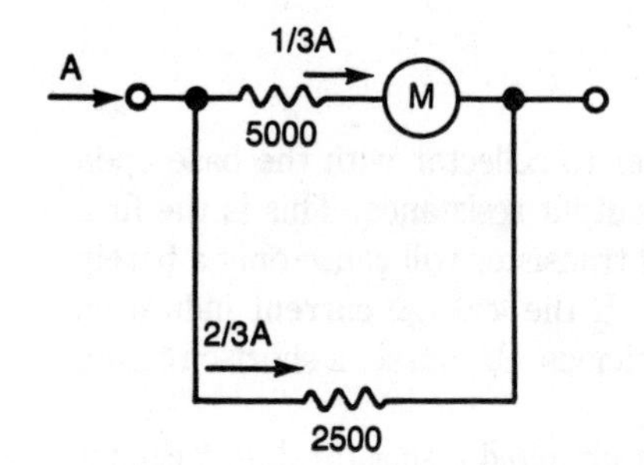

Fig. 4-8. This current divider causes a 1 mA meter to read full-scale when the total current, A, is 3 mA. The 5000-ohm resistor also protects the meter from excessive current.

This has a very helpful effect from the standpoint of calibration, since the gain can be read directly from the meter scale. How did this happen? Note what occurs when the gain is 100. The collector current in this case is 100 times 30 μA, or 3 mA. This is *full scale* on the meter, or 1. When the gain is 60, the collector current is 1.8 mA. One-third of this, or 0.6 mA flows through the meter, so the needle points to 0.6. You can see that the gain is simply 100 times the meter reading.

In the transistor checker, the meter scale has been modified as described in Chapter 3, and the multiplication by 100 has been incorporated in the scale, so that it reads gain directly.

Field-Effect Transistors

To help understand the tests that the BJT-FET transistor checker makes on FETs, let's review what goes on inside the FET.

Think of the FET as having a channel from source to drain. There are two ways electricity flows through this channel. If the FET is made with an *n-channel,* electrons carry the current flow, and the drain is customarily made positive with respect to the source. If the FET is made with a *p-channel,* so-called *holes* in the crystal lattice carry the current flow, and the drain is made negative with respect to the source.

In either case, the current flow is controlled by means of the gate electrode. For an n-channel FET, the greater the negative voltage (with respect to the source) on the gate, the less current flows in the channel. For a p-channel FET, the greater the positive voltage on the gate, the less current flows in the channel. In either case, the gate voltage can cut off all current flow in the channel, if it is large enough. Figure 4-9 illustrates this effect.

There are three kinds of FETs, each of which can be either n-channel or p-channel. Two of the types of FETs—the junction FET or *JFET,* and the depletion metal-oxide FET, or *depletion MOSFET*—function pretty much alike. In both types, the channel is normally open. For n-channel FETs like these, the gate must be more negative than the source to cut off the channel. Similarly, for p-channel FETs like these, the gate must be more positive than the source to cut off the channel.

The third kind of FET, which may also be n-channel or p-channel, is the enhancement metal-oxide silicon FET, or *enhancement MOSFET.* In this device, the channel is not open unless there is a voltage on the gate. In n-channel enhancement MOSFETs, the gate must be more positive than the source for current to flow; current is cut off when the gate is grounded to the source. Conversely, in p-channel enhancement MOSFETs, the gate must be more negative than the source for current to flow; current is cut off when the gate is grounded to the source.

If you find all of this confusing, that's not surprising. We're dealing with six different devices and several levels of gate voltage. Table 4-1 should help clarify these matters.

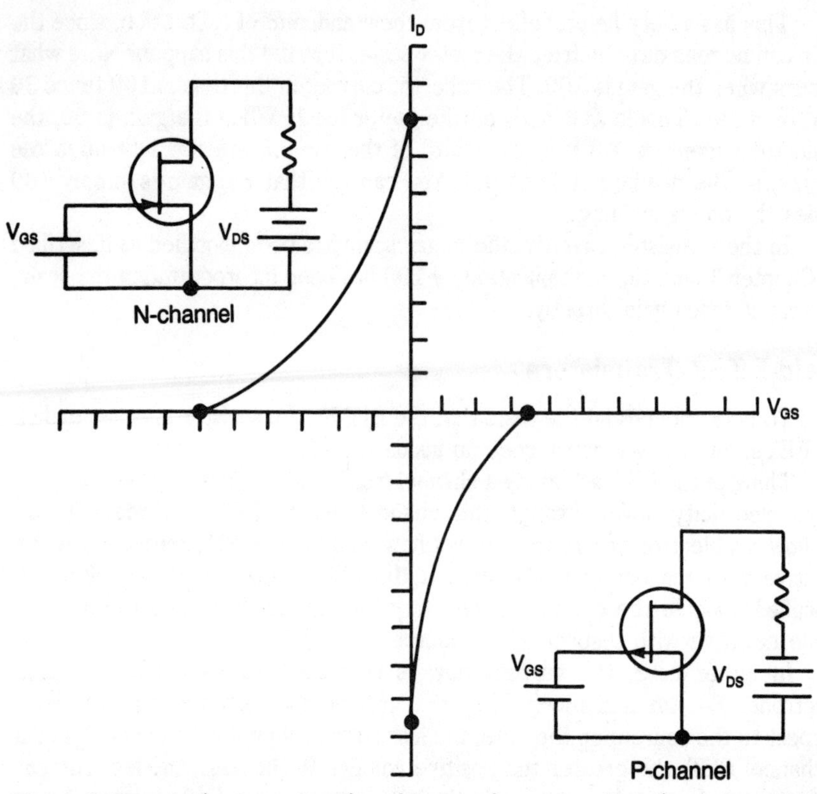

Fig. 4–9. How drain current varies with changing gate voltage in n-channel and p-channel FETs (V_{ds} is held constant).

Table 4–1. Biasing of Various FETs.

Channel	Device	Gate-to-Source Voltage		
		Positive	**Negative**	**Same Potential**
n	JFET	Conducts	Cut off	Conducts
	Depletion MOSFET	Conducts	Cut off	Conducts
	Enhancement MOSFET	Conducts	Cut off	Cut off
p	JFET	Cut off	Conducts	Conducts
	Depletion MOSFET	Cut off	Conducts	Conducts
	Enhancement MOSFET	Cut off	Conducts	Cut off

FET Tests

Table 4-1 suggests a way of checking FETs, and this is exactly the test that the BJT-FET transistor checker performs. What we have to do is apply positive and negative gate voltages to the FET under test and short its gate to its source. We can make the magnitudes of the voltages large enough to insure that we have complete cutoff or complete saturation in each case. If we can find out what the condition of the channel is for each applied voltage, we will know whether an FET is good or bad.

The current through the FET channel during saturation may not be very great. To get a positive indication, we use a junction transistor (Q1 in Fig. 4-10) as a lamp driver. When the channel is cut off, no current flows through R5 and R6, and the base and emitter of Q1 are at the same potential. When current flows through the FET, R5 and R6 form a voltage divider, and Q1 is biased on, lighting lamp I1. Resistor R7 is selected so that at the lamp's operating current of 60 mA, the voltage across the lamp is 2 V.

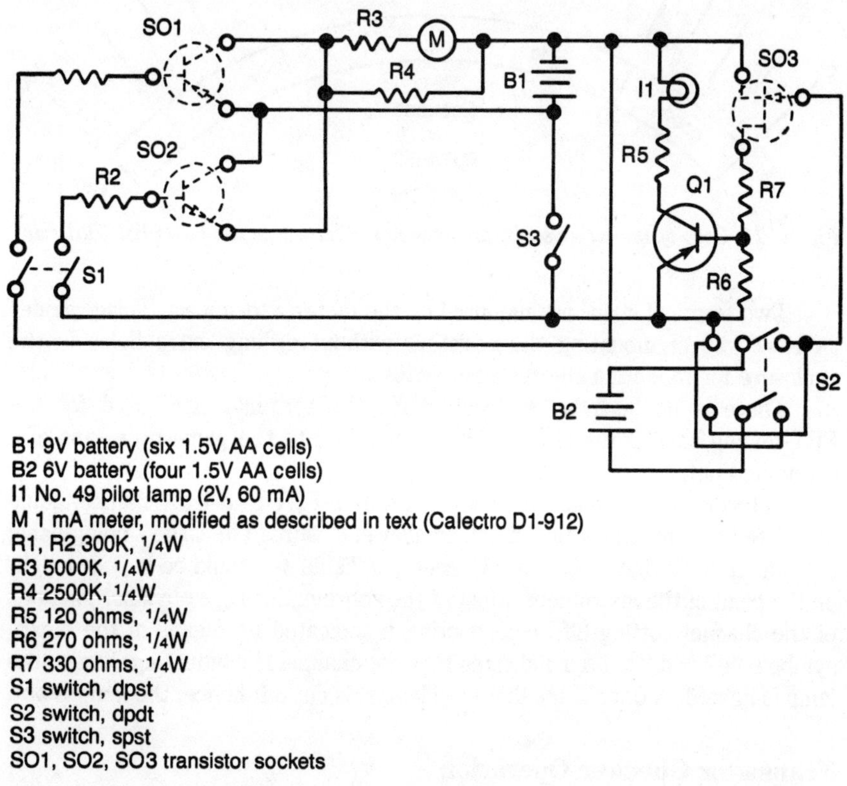

Fig. 4-10. Schematic and parts list for the BJT-FET transistor checker project.

In the transistor checker, I used an incandescent lamp. However, an even better indicator would be an LED, such as the Radio Shack No. 276-026. This is brighter, and will last many times longer than an incandescent lamp. It is also more rugged and requires a smaller mounting hole. Remember, the LED is a diode and must have its cathode connected to the ground bus.

Transistor Checker Construction

The unit in the photos was constructed in a 4⁵⁄₁₆-×-7-×-4¹⁄₁₆-inch sloping-panel box. The meter scale was modified as described in Chapter 3. Figure 4-11 is a life-size copy of the scale used on the meter. Remember, if you use a different meter, you will have to make your own scale. (This isn't very hard.)

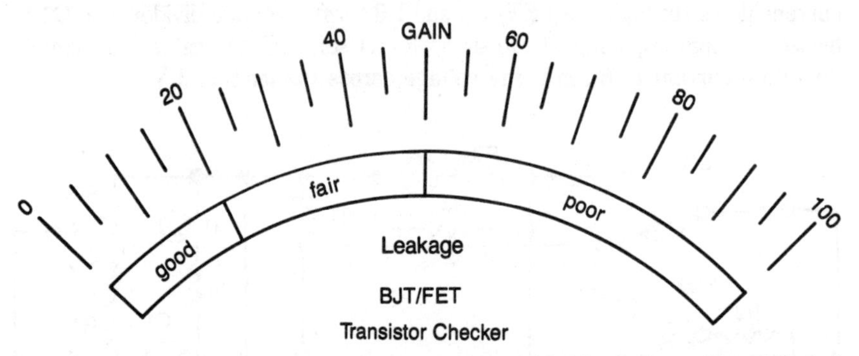

Fig. 4-11. Copy of the meter scale for the meter specified in the parts list (full size).

Two terminal strips are mounted on the meter's terminals. This provides an easy way of mounting the resistors without drilling extra holes in the enclosure for mounting the terminal strips.

There is no such handy location for the terminal strip used for the FET-testing part of the circuit. This is attached to the panel by means of a screw and nut.

In terms of electronic components, the BJT-FET transistor checker could have been assembled within a much smaller enclosure. The large sloping-panel box was used so that a condensed version of Table 4-1 could be included right on the front of the instrument. Most of the abbreviations are obvious. The idea of the channel cutting off or conducting is indicated by means of the binary symbols "0" and "1." A 1 indicates that the channel is conducting; hence, the lamp is lighted. A 0 indicates that the channel is cut off; hence, the lamp is out.

Transistor Checker Operation

If you understood the section on theory, you should have no trouble with operation of the BJT-FET transistor checker. For bipolar transistors, select the proper sockets—either NPN or PNP—and plug in the transistor. Select the leakage test first, and turn the power switch on. If the transistor is shorted, the meter needle will move up into the BAD region of the leakage scale. If the transistor is not shorted, the needle will move only slightly. In fact, with good transistors you may not even be able to see the needle move at all.

If the transistor is not shorted, move the function switch to the GAIN position and read the h_{FE} of the transistor on the GAIN scale of the meter. Be suspicious of any transistor that reads below 30, and write off any that indicates below 20. For most transistors, you can find the normal range of gain for that type in the transistor handbook issued by the transistor's manufacturer.

To check field-effect transistors, plug the unit into the FET socket. Use the same source, gate, and drain connections for both n-channel and p-channel devices. (The direction of current through the channel isn't important from the standpoint of these tests.) Turn the power on and observe the lamp with the BIAS switch set, in turn, in both positions and in the middle. Then use the table, or LAMP MATRIX, as it is labeled on the unit, to determine whether the transistor is good or bad. In general, if the lamp stays on no matter what position the switch is in, the FET is shorted. If the lamp stays out no matter what position the switch is in, the FET is open.

SIGNAL GENERATOR PROJECT

A good signal generator is a useful troubleshooting adjunct. With such a device you are able to inject an audio or radio-frequency signal into any stage of a malfunctioning receiver and observe the effect of the signal. The output of a signal generator is a known quantity; it eliminates guesswork in alignment and repair.

The signal generator circuit in this project is the most complicated discrete-component circuit in this book; yet if you carefully duplicate the printed circuit board (Fig. 4-12) you will find that it is difficult to go astray.

One headache with any signal generator project is calibration of the output. How do you correlate dial settings with specific frequencies? In a little while, we'll explain how you can completely calibrate this signal generator with just two items: a conventional broadcast band receiver and an ordinary FM receiver. You can even calibrate two of the bands without an FM receiver if you have a television set handy.

Many commercial signal generators are complicated by elaborate bandswitches. This project avoids that complexity by using plug-in coils. It has most of the other important features of its big brothers, including separate audio- and RF-level controls. The big, easy-to-read dial with the see-through plastic pointer looks professional, but you'll find you can make it yourself in a few minutes.

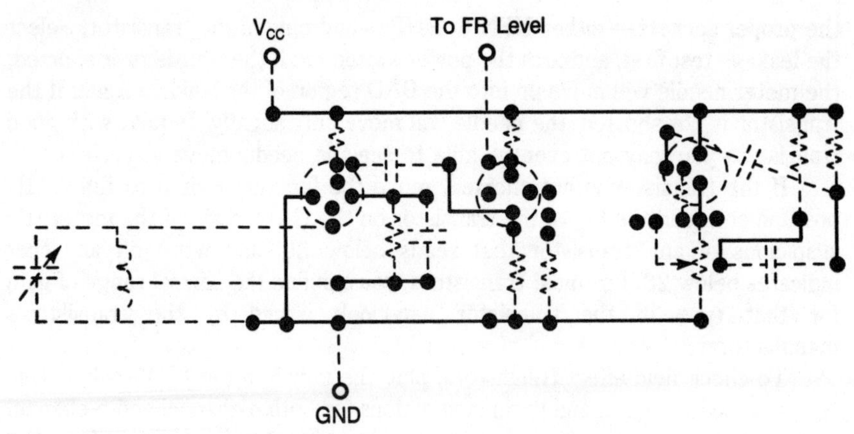

Fig. 4-12. Full-size signal generator printed circuit board.

Theory of Signal Generator

Figure 4-13 shows the circuit of the signal generator. The oscillator is a Hartley-type, with a centertapped inductor. This lets us use a standard-broadcast 365 pF variable capacitor for tuning. We used an FET for the oscillator, because it simplifies the design and has good high-frequency characteristics.

Since you will be connecting the output of the signal generator to quite a few different circuits and don't want these circuits to affect the oscillator frequency, the project uses an untuned FET amplifier stage as a buffer. You may notice that this is the same amplifier used in the FET shortwave receiver. This is a good, simple, reliable, and most versatile circuit. The output of this stage is applied to a 1 M potentiometer, and the slider on the pot is connected to the center conductor of the output jack.

That accounts for the RF part of the output. We still have to provide modulation for RF and an audio signal for audio circuits. Bipolar transistor Q3 together with its phase-shift network provides an audio signal of around 400 Hz. We use this signal to modulate the RF signal by applying the output of Q3 to the source of Q2, the output stage. The modulation level control is actually the load resistor for Q3, and the level is controlled by varying the position of the slider on the pot. We can turn Q3 completely off by means of the switch on the back of the audio level control.

Most modulation circuits use a tuned output in the modulator stage or in one of the stages that follow it. This effectively suppresses the audio signal in the output. However, since the output stage of this signal generator is untuned, the output contains not only the modulated RF signal but the audio signal as well, so that this is both a variable-frequency RF and a fixed-frequency audio signal generator.

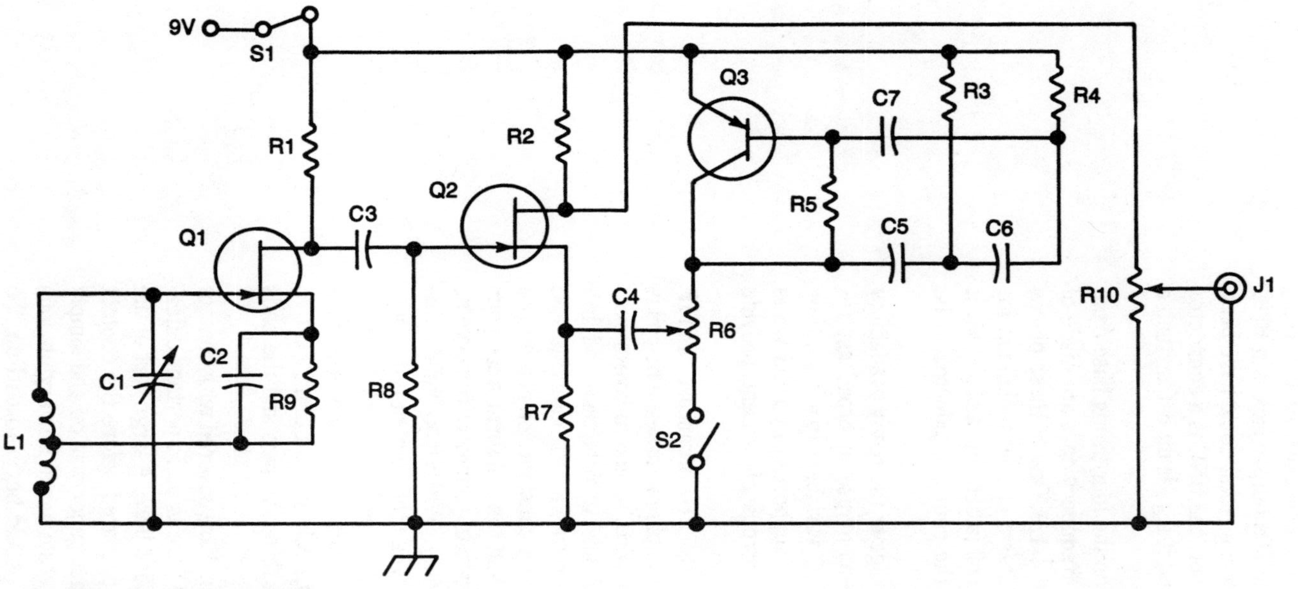

C1 365 pF (standard broadcast) air variable
C2, C3 0.01 μF, 12V dc ceramic disc
C4 0.1 μF, 12V dc ceramic disc
C5, C6, C7 0.05 μF, 12V dc ceramic disc
J1 female BNC bulkhead connector
L1 band 1: 200 turns (overlapped) No. 28 AWG enameled copper wire on 5/8 in. diameter, 1 1/4 in. long coil form (Tap at 100th turn.)
Band 2: broadcaast band antenna coil (Archer 270-1430)
Band 3: 32-per-inch pitch, 5/8 in. diameter, 1 1/8 in. length of air-wound coil stock, centertapped (B&W 3008 or Air Dux 532)
Bank 4: 8-per-inch pitch, 1/2 in. diameter, 2 in. length of air-wound coil stock, centertapped (B&W 3002, Air Dux 408)
(Each coil is mounted on plug-in base from 5/8 in. diameter by 1 1/4 in. coil form.)
Q1, Q2 Motorola HEP 801
Q3 Motorola HEP-251
R1, R2, R3, R4 4700 ohms, 1/4W
R5 470 kilohms, 1/4W
R6 3300-ohm, 2W audio-taper potentiometer (AF LEVEL)
R7 5600 ohms, 1/4W
R8 1M, 1/4W
R9 1000 ohms, 1/4W
R10 1M, 2W linear-taper potentiometer (RF LEVEL)
S1, S2 switch, spst (ON-OFF)
S2 switch, spst on back of R6—MODULATION OFF

Fig. 4-13. Signal generator schematic and parts list.

Signal Generator Construction

All of the small components are mounted on the printed circuit board (Fig. 4-12). The board—along with the tuning capacitor, coil socket, switch, pots, and output jack—are mounted on the front panel. The output jack is a BNC female bulkhead connector. A less expensive type of connector could have been used; for example, an RCA-type phono jack. However, the BNC is a connector that stays mated until you remove it on purpose, and it stands up better than a cheaper connector in use.

There are four plug-in coils. Each is mounted on the four-prong base from a 5/8 inch diameter plug-in coil form. The two high-frequency coils are made of air-wound coil stock, as noted in the parts list (Fig. 4-13). One of these pieces of coil stock has a very fine pitch, which necessitates a special technique for making the centertap connection. One loop of the coil is bent inward so that it is separate from the adjacent turns. This allows the center conductor to be easily soldered.

For the broadcast frequencies, we took advantage of the ready availability of prewound loopstick antennas made for replacement use in broadcast receivers. Since these are designed for use with 365 pF variable capacitors, they are ideal for our signal generator. The loopstick is supported by solid wires which make connection to the base plug. A few drops of cement provide additional support.

The fourth coil is necessary to obtain the low 455 kHz broadcast band IF signal. This is the only coil that uses the form with the plug-in bases. The coil is composed of 200 turns of No. 28 enameled copper wire, wound in layers over the 1/4 inch length of the coil form. To wind the coil, unscrew the base and drill a small hole in the side of the form. Pass one end of your wire through this hole and start winding. The best way to wind is to keep a constant tension on the wire and wind it up as if you were reeling in a fish and the coil form were your reel. When you reach the 100th turn, twist a big loop in your wire to serve as the lead from the centertap, and continue winding. When you finish, use cement or nail polish to hold the windings in place. When the cement is dry, you can screw the base plug back onto the coil form and make all of your connections to the pins on the base. Be sure to scrape the enamel coating off the ends of the wires before you attempt to solder them. Check with an ohmmeter to make sure you have good connections between each of the pins and the ends and center of the coil.

The outside of the utility box and the front panel are covered with contact paper, as in several other projects in this book. There is an important difference, however, in the way the contact paper is applied here. A large part of the front panel is given to the dial, and this must be prepared before the contact paper is applied. If we waited until after the contact paper was on and simply glued the dial scale in place, the edges of the dial scale would be very obvious right from the beginning, and after a while, they would begin to curl up. To avoid this, the dial paper is applied first, and the contact paper is placed over it. This guarantees that the edges of the dial will remain neatly in place.

To make the dial, first glue a sheet of white paper over the entire front of the panel. Then locate the center of the hole for the tuning-capacitor shaft and make a mark at this point. Use a felt-tip pen—or better yet, a drafting pen—and a large circle template to draw a series of semicircular arcs at 2, 3, 4, and 4½ inch diameters. The arcs should terminate on a line that passes through the center of the tuning-capacitor shaft hole and runs parallel to the top and bottom edges of the front panel. At each point where an arc ends, extend the arc downward in a straight line about ½ inch. You'll need this extra ½ inch to identify the scales as kilohertz or megahertz. Finally, draw a line across the bottom of the dial scale.

Now you are ready to cover the entire front panel with contact paper. Lay it down neatly, but don't rub it in place just yet. If the lines on the dial face are good and dark, you should be able to see them through the contact paper. Use an X-acto knife and your circle template to score a line around the outside of the outermost semicircle on your dial scale. Finish up along the bottom with X-acto knife and ruler, and carefully peel away the contact paper covering the dial. The contact paper should peel off, without taking any of the dial scale paper with it.

The dial-pointer knob is easily made from an ordinary round knob and a piece of clear plastic from a dime-store plastic box. You can saw the plastic to shape with a keyhole or coping saw, but you will find it just as easy if you score the plastic several times along the line to cut with an X-acto or other sharp knife and then just break the plastic with your fingers on one side of the score line and a pair of pliers on the other side. You'll find that this way the edges are actually straighter and neater than if you had sawed them. Of course, you'll still need to give them a final dressing on the edge of a file. Use the file to round off all of the corners also.

The index line down the center of the pointer is made by scoring the line with a sharp knife and then applying India ink to the line. Use a tissue to wipe away the ink that beads up and you'll be left with a thin black line down the middle of the pointer.

Glue the pointer to the back of an ordinary round knob and, when the glue is dry, drill out a hole for the capacitor shaft. If the shaft of the capacitor is flatted, be sure that you locate the index line of the pointer diametrically opposite the setscrew on the knob.

The cable for the output probe is RG-58A/U coaxial cable. The center conductor of the cable can be extracted through the side of the braid to provide a probe lead and a ground.

Calibration

We'll start our calibration with the broadcast band, since that offers a direct correspondence between frequencies and points on the dial scale of the signal generator. Begin by turning the generator on, with the broadcast band coil plugged in and the tuning capacitor turned fully counterclockwise. Put the probe of the signal generator near the antenna of your broadcast receiver, turn the receiver on, and tune until you find the lowest signal on the broadcast band. Leave the generator capacitor fully counterclockwise, and adjust the slug in the loopstick until the frequency of the generator corresponds to the lowest dial marking on the broadcast receiver. Now match each dial marking on the broadcast receiver with a setting of the generator tuning capacitor. Use transfer numerals to identify each point.

To calibrate the next higher band, you should use the broadcast band calibration marks you have just made and the intermediate frequency of an FM receiver. All FM receivers in the U.S. use an IF of 10.7 MHz. If you adjust the signal generator so that it is putting out a signal of 5.35 MHZ, the second harmonic of that signal will fall right on the FM receiver's IF. From our design calculations, 5.35 MHz is somewhere within the frequency range of the second coil. To find that frequency, bring the generator probe near the FM receiver's antenna, turn the receiver on, and tune for signals. The signal generator is rich in harmonics, so there will be a lot of signals, but there will be only one that will be independent of the FM receiver's dial setting. When you find that signal, the generator is tuned to 5.35 MHz.

Well, that's just one frequency, you say. How do you calibrate the entire dial? There's a trick. If you grind through the algebra, you will discover that no matter what the inductance, the frequencies corresponding to two settings of the tuning capacitor will always fall into the same ratio.

Let's see what that means. Take one setting of the tuning capacitor corresponding to 1400 kHz with the broadcast coil plugged in. Take another setting corresponding to 700 kHz, or one-half the original setting. Now it doesn't matter what the actual values of capacitance are—whatever coil you plug in, the frequency at the 700 kHz setting of the capacitor will always be exactly one-half the frequency at the 1400 kHz setting.

Here's how to use this mathematical relationship to calibrate the generator. Once you have found the dial setting corresponding to 5.35 MHz, look down and see what broadcast frequency corresponds to this setting. As an example, let's say it corresponds to 1070 kHz. To find out the broadcast band setting for any other frequency, all you have to do is multiply that frequency by 1070/5.35. For example, the broadcast frequency setting that corresponds to 5 MHz is 5 − (1070/5.35), or 1000 kHz. Similarly, the broadcast frequency setting that corresponds to 4 MHz is 4 − (1070/5.35), or 800 kHz. Naturally, you cannot rely on 5.35 corresponding to exactly 1070 on your generator, so you'll have to check it yourself and make your own calculations.

The high-frequency band is even easier to calibrate, because 10.7 MHz itself appears in this band. Once you find it, proceed as you did with 5.35 MHz, except this time the multiplier is 10.7/broadcast band setting.

You can calibrate this band, if there is no FM receiver available, with a television set. The idea is the same as it was with the FM receiver. All TV receivers made in the last 15 years use an IF of 45 MHz. (Earlier ones used 21 MHz, but there was too much interference from the 15-meter ham band.) Half of 45 MHz is 22.5 MHz, a frequency that will fall within the range of the high-frequency coil of the signal generator. As before, connect the generator probe to the receiver antenna and search for a signal that is independent of channel selector setting. It's interesting to watch the effect of the generator. As before, connect the generator probe of the generator's output very easily, because the audio modulation produces a bar pattern. This is often more obvious than the audio signal, since the video detector responds to AM, and the TV-audio detector responds to FM.

If you experiment, you will note that harmonics from the midband (2.7 to 7.8 MHz) coil of the generator will come through the TV IF. Unfortunately, however, these are such high-order harmonics that the fundamental frequencies are very close together and cannot be differentiated well enough to allow us to calibrate the midband coil with the television receiver.

There is one more calibration to make, the AM broadcast IF of 455 kHz. It would have been nice if this could have been included in the calibration of the broadcast-band frequencies, but it lies too far below the broadcast band to permit us to do so. This is the reason for the fourth coil, the one we had to wind by hand. The procedure for finding 455 kHz is the same for the other intermediate frequencies. Look for a signal that is unaffected by the dial setting of the broadcast receiver. You can calibrate all of the other frequencies on this band.

Signal Generator Operation

The functions of the front-panel controls are self-explanatory. Here's how to use the generator to align an AM broadcast receiver.

Turn the receiver on and set the dial to some position where there is no signal. Adjust the output of the signal generator to 455 kHz and turn the audio and RF-level controls up full. Clip the ground lead to the receiver chassis or ground bus of the printed circuit and position the probe near the receiver's last IF stage until you can hear the audio signal in the receiver speaker. Reduce the generator RF level until the signal is barely audible. Adjust the slugs in the second IF transformer to produce the maximum signal at the speaker. (It will be easier if you can measure the voltage at the speaker with the AC range of an electronic voltmeter.) Now move to the first IF amplifier and repeat, peaking the slugs in its transformer. Always operate with the minimum signal generator level you can hear.

After adjusting the receiver's IF, return the generator to 540 kHz and set the receiver dial to the same frequency. Now adjust the antenna and local oscillator for maximum signal. On older sets, the antenna-tuning adjustment is a trimmer capacitor on the side of the tuning capacitor, adjacent to the large section, and the oscillator adjustment is another trimmer on the tuning capacitor, adjacent to the small section. On more recent transistor sets, the antenna coil and oscillator coil are adjusted by slugs located within the coils. In either case, the antenna adjustment is usually quite broad and the oscillator adjustment is quite sharp.

After peaking the receiver at the low end, reset the generator and receiver to 1600 kHz and repeat the procedure. If you find that you have to change the antenna and oscillator settings very much, go back to 540 Hz and recheck your settings there. You may have to "split the difference" to achieve optimum performance and dial accuracy across the entire band.

More sophisticated receivers require more sophisticated alignment procedures. Refer to repair manuals written specifically for these receivers for alignment instructions.

DELUXE LOGIC PROBE PROJECT

In many ways, standard analog test equipment, such as multimeters and the like, isn't very practical for troubleshooting digital circuitry. Exact voltage levels usually aren't particularly important in most digital circuits. Our main concern is the state of the digital input and output signals.

A digital signal may take on just two possible states—high or low. A low signal is typically close to ground level. A high signal is generally just below the supply voltage. For a steady-state (unchanging) digital signal, an analog multimeter may be used, but it is almost overkill for the application. Many digital signals are in the form of very brief, or rapidly changing pulses. At low frequencies, an analog voltmeter is awkward to use. At high frequencies it will be absolutely useless.

A logic probe is a useful piece of test equipment specifically for digital work. It is a device that indicates the logic state of a digital signal. A basic logic probe is nothing more than an LED in series with a current-dropping resistor, as shown in Fig. 4-14. The ground clip is connected to a ground point in the circuit under test. The probe tip is touched to the point in the circuit to be tested. This will usually be an IC pin (either an input or an output). If the signal at the test point is high, the LED will light up, otherwise, the LED will remain dark.

This is adequate for certain tests, but its value is severely limited by its extreme simplicity. A lit LED is a fairly clear indication of a signal at the test point. A dark LED is more than a little ambiguous. It could mean a logic low signal, or it could mean that there is no signal at all getting through to the test point. Even a lit LED doesn't really tell you too much. Is it indicating a steady-state high logic signal? Or is it indicating a string of high-speed pulses, causing the LED to blink on and off at a rate too fast for the eye to distinguish the individual pulses? A lit LED could even indicate a short circuit to the positive supply voltage.

A simple logic probe will be helpful for certain simple test situations, but

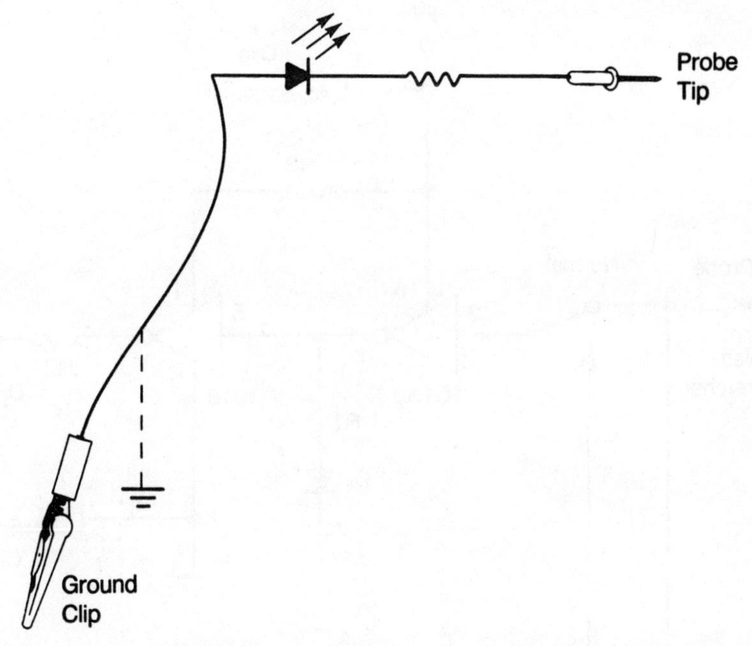

Fig. 4-14. A simple logic probe can be made from an LED and a current dropping resistor.

for serious work with digital circuits, a more versatile piece of test equipment is required.

Fig. 4-15 shows the schematic diagram and parts list for a deluxe logic probe you can easily build. Notice that the circuit is still pretty simple. You can use almost any construction method you prefer for this project.

This deluxe logic probe has two LEDs to indicate the monitored logic state. This set-up unambiguously indicates four possible conditions:

- Both LEDs off—no signal
- A on/B off—high
- A off/B on—low
- Both LEDs on—pulses

The deluxe logic probe also features a pulse stretcher (IC2, and its associated components). In many cases, we will need to monitor a steady-state signal for an isolated, brief single pulse. You might easily miss the brief blink of the LEDs. It might even be too fast for the eye to catch.

When the pulse stretcher is switched into the circuit, the input signal is used to trigger a 555 monostable timer. When triggered, the output LED will remain lit for a period of time determined by the values of capacitor C and resistor R.

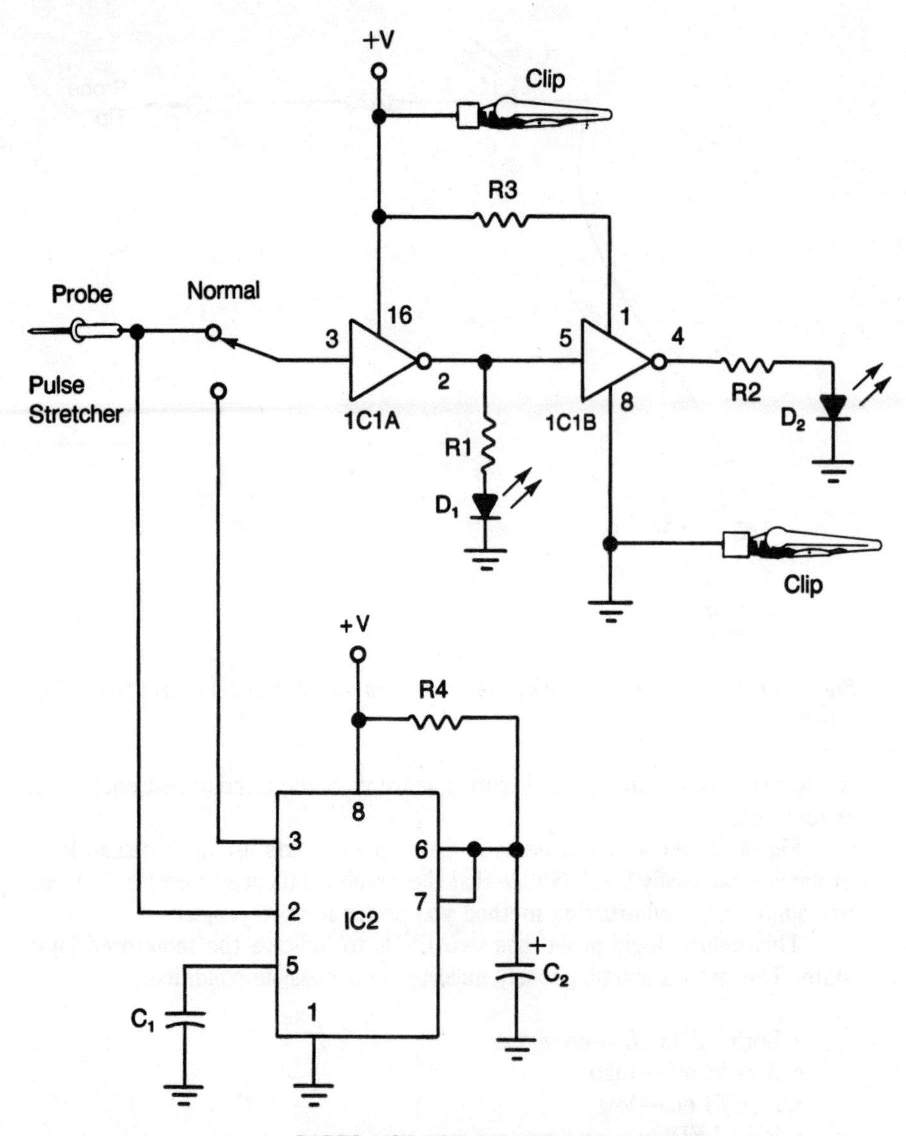

PARTS LIST FOR DELUXE LOGIC PROBE

1C1	CD4009 hex inverter
1C2	555 timer
D1, D2	LED
C1	0.01 μF capacitor
C2	10 μF electrolyte capacitor
R1, R2	330-ohm resistor
R3	1-K resistor
R4	220-K resistor

Fig. 4-15. Schematic diagram and parts list for the deluxe logic probe project.

With the component values given in the parts list, the output LED will remain lit for about 3 seconds each time an input pulse is received, no matter how short the incoming pulse's duration may be.

With this deluxe logic probe, you can easily determine the logic state at any point in virtually any digital circuit.

5

Making Successful Substitutions

The best way to be sure that your finished project is going to duplicate the performance of a project described in a magazine or book is to use parts identical to those in the parts list. As you become more comfortable with electronic construction, however, you may become interested in substituting components different from those in the parts list. You may find that it just isn't possible to obtain some parts, or you may want to use some parts that do not cost as much as those in the parts list, or you may have a junkbox which you would prefer to raid before going out to buy new parts. In this final chapter, we will investigate the things you must take into account to insure that the substitute parts will not affect the performance of your project.

RESISTORS

Carbon resistors with a 10-percent tolerance are available in the resistances listed in Table 5-1 and in resistances equal to these values times multiples of 10. For example, consider the fifth value listed, 22. You will find that your radio parts store carries resistors of 22 ohms, 220 ohms, 2.2 K, and so on—all the way to 22 M. You will find 5 percent resistors with values in between these ranges. The tolerances mean that the actual value of the resistor is somewhere between plus and minus so many percent of the resistor's nominal (color coded) value. That is, a so-called 220-ohm, 10-percent resistor will have a value between 198 ohms (220 − 22) and 242 ohms (220 + 22).

You should understand that the resistor manufacturer doesn't have two machines—one making 5 percent, 220-ohm resistors and the other making 10 percent, 220-ohm resistors. Instead, he has several machines, making resistors of unknown values. All of these resistors are funneled to a testing machine

Table 5–1. Ten Percent Resistance and Capacitance Values.

10	18	33	56
12	22	39	68
16	27	47	82

where they are measured and sorted. It's like an electronic sieve separating different sizes of electronic gravel.

The resistors are labeled according to their measured values. All resistors between 176 and 264 ohms are painted with the red-red-brown color bands of the 220-ohm resistor. Those that have a value between 209 and 231 ohms get the additional gold band that signifies a 5-percent tolerance. Some of these may actually be 220-ohm resistors. Of the others, those that measure between 198 and 209 ohms and between 231 and 242 ohms are painted with the silver band that designates a 10-percent tolerance. Note that none of these 10 percenters will ever be exactly 220 ohms. Finally, all the resistors left over, those between 176 and 198 ohms and between 242 and 264 ohms, will be left with no fourth band and will be sold as 20 percent, 220-ohm resistors.

The point of this discussion is to demonstrate the rather wide variation that can exist between two resistors with the same nominal values. This suggests that in most cases we can expect to get away with substituting a resistor of the next higher or lower value for a particular resistor called for in a parts list. The best clue is whether the author has indicated the tolerance on the parts list or not. If he has indicated 10 percent or no tolerance, we can be very confident in substituting the next higher or lower value, if that is convenient. If the author has taken the trouble to specify 5 percent tolerance, he probably has a good reason for it, and it would be well to use the identical part that he has specified.

There is a new development that you will be seeing soon on all resistors made in the U.S., if you have not seen it already. Resistor manufacturers in the U.S. are now making all of their resistors in conformance to a new military specification, called MIL-R39008. These are called ***established reliability resistors*** and they are identified by a fifth color band, following the silver or gold tolerance band. This fifth band indicates the predicted reliability of the resistor in terms of failures per 1000 hours of operation. From your standpoint, as someone who buys one resistor at a time, the fifth band is statistically meaningless—but don't let it throw you the first time you find a resistor with an extra band painted on it.

There are some applications in which you can substitute resistance values quite widely. Figure 5-1 shows the output of a power supply containing a ***bleeder resistor.*** The function of the bleeder is to discharge the power supply filter capacitors when the supply is turned off, to keep them from presenting a shock hazard. Frequently, a value of 1000 ohms is selected for this resistor. It

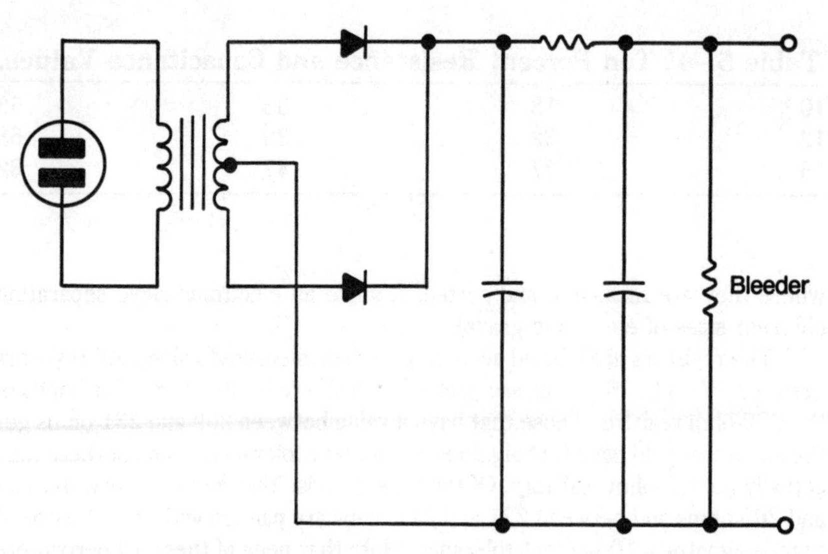

Fig. 5-1. Power supply bleeder resistors may be 1 K to 5 K. Be sure that the power rating (E/R) is adequate.

is quite satisfactory, though, to use a resistor of up to five times this value, if that is easier or cheaper.

Another area in which resistor value is not critical is in the inputs of vacuum-tube or FET amplifiers, such as the ones shown in Fig. 5-2A. In contrast to bipolar transistors, in which current flows in the base circuit, very little, if any, current flows in the control element (grid or gate) circuit of a vacuum tube or FET. Consequently, the only function that these resistors perform is to link the control element to ground and to provide a fairly large resistance at the input of the amplifier. In most amplifiers, this resistance can be anywhere from 10 K to 10 M without affecting performance. Do not confuse this resistor with the voltage divider resistor in Fig. 5-2B, however. In this configuration, the resistor forms part of a bias circuit, and its value is fairly critical.

As important as the actual resistance of a resistor is the ability of the device to dissipate power. It doesn't do you any good to install a resistor that is going to burn up when you apply power to the circuit. You can always substitute a resistor with a higher power rating for one with a lower power rating—for example, a ½-W unit for a ¼-W unit—but unless you calculate the power loss in the resistor and find it to be safely below the resistor's power rating, do not substitute a low-power unit for a high-power one. Another thing to remember is that units of different ratings differ in size. If you are copying a printed circuit pattern published in a magazine, be sure that the resistors you use will fit into the space provided for them on the printed circuit board.

Resistors with power dissipation ratings up to 2 W are available with

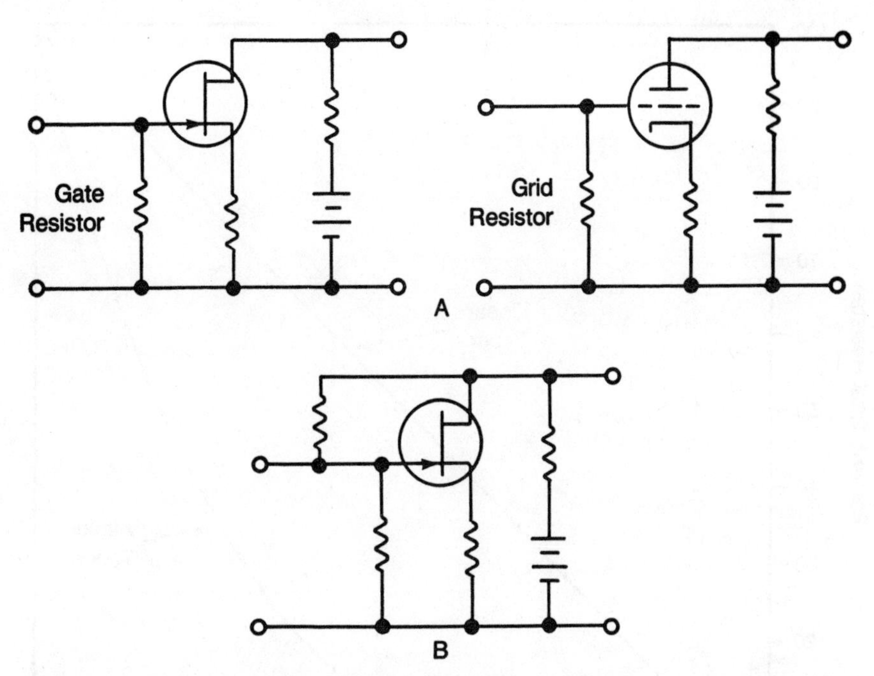

Fig. 5-2. In A, the value of the gate or grid resistor is not critical; it may vary from 10 K to several megohms. In B, the resistor is part of a voltage divider that biases the gate, and its value is more critical.

carbon composition resistive elements. Larger resistors are generally made by wrapping a ceramic core with resistance wire. Most of these resistors are used in power supply applications, and the method of their construction does not often concern us. However, in applications that involve audio or radio frequencies, the inductance created by these windings of resistance wire can create problems. This is not a difficulty encountered very frequently, but when it is, there are two ways to get around it. One way is to use a large number of carbon composition resistors in parallel to achieve the desired resistance at the required power level, and the other way is to wind your own resistor using what is called a *bifilar winding.* This technique involves taking the whole length of resistance wire needed to make your resistance and doubling it back on itself to make a sort of long, narrow hairpin. When this hairpin is wound on the ceramic core of the resistor, the inductance effect is canceled out, because the currents in adjacent turns are flowing in opposite directions.

The considerations outlined in the preceding paragraphs apply to potentiometers and resistors with sliding taps, as well as to fixed resistors. One additional consideration to take into account when substituting potentiometers is the device's *taper,* the shape of the curve of resistance versus shaft position (or slider position, on slider pots). Figure 5-3 compares linear and audio tapers. The audio taper is tailored to the response of the human ear, which is essentially logarithmic. If you substitute a pot with a linear taper in a volume control

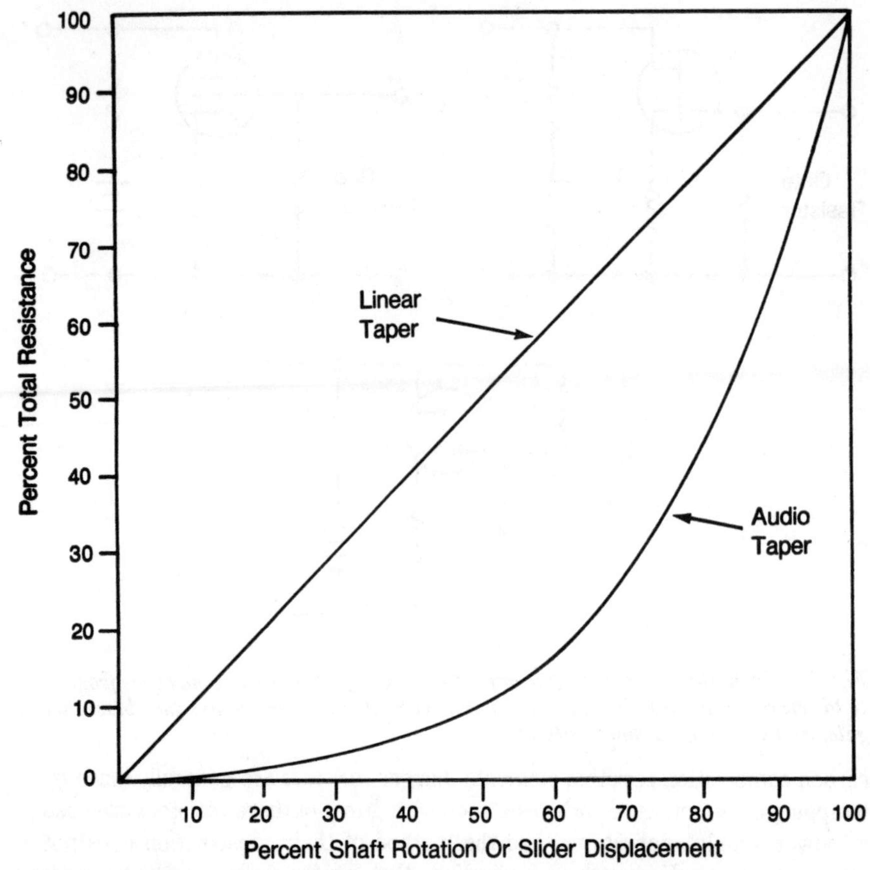

Fig. 5-3. A graph comparing linear and audio taper potentiometers.

application, you'll find it difficult to set a satisfactory level at the low end of its range; it will seem as if a small change in setting will produce an extremely large change in signal level.

CAPACITORS

With respect to choosing the next highest or next lowest value of capacitance instead of the specified value, the remarks under the heading "Resistors" apply equally well to capacitors. For units rated at 10-percent tolerance, the nominal values in Table 5-1 apply, except that appropriate dimensions are picofarads (pF) and microfarads (mF). Incidentally, you may find the expression ***micromicrofarads*** and the abbreviation ***mmf*** used to refer to pf in books written before the IEEE standardized nomenclature in the early 1960s.

In some cases, you have wide latitude in selecting values of capacitance. In power supply filters, for example, anything larger than about 30 μF will work well.

As far as the dielectric material used to make the capacitor is concerned, paper, Mylar, ceramic, and mica capacitors of the same value are all interchangeable. Be careful, though, that there is enough room on a circuit board or inside an enclosure for a substitute capacitor. A paper dielectric capacitor may be several times bigger than a ceramic unit.

In most circuits, you may not substitute an electrolytic capacitor for another type. Electrolytics generally have too much leakage resistance to be useful in applications for which they were not designed. Another precaution is not to use electrolytics with a voltage well below their rated level. The operation of these capacitors involves using the applied voltage to ionize the electrolyte in them, and this requires the application of a certain minimum voltage.

Unlike resistors, capacitors are not rated in terms of their power dissipation (in watts). Instead, they are rated in terms of the maximum DC voltage that can be placed across them. A *DC working voltage* (WVDC) above this level is likely to cause a breakdown of the dielectric and permanent damage to the capacitor.

You can connect several capacitors in series to obtain a voltage rating greater than you could get with a single unit, but you must remember that you decrease the total capacitance as you add capacitors in series. In addition, it is wise to use resistors of 50 K or so in series with each capacitor to equalize the voltage drops.

Generally, it is not practical to change the value of a variable capacitor used for tuning by connecting fixed capacitors in series or parallel with it, because this also diminishes the range of the capacitor and, consequently, the frequency range that can be tuned. If you desire to reduce the capacitance of a variable capacitor, you can remove plates.

COILS AND TRANSFORMERS

When a coil is used in a tuned circuit, its inductance is, of course, critical in determining the resonant frequency of that circuit. In other applications, though, coils may be used to block AC while passing DC. In these ***choke-coil applications,*** the exact value of inductance is not critical, and other chokes of greater inductance may be readily substituted.

A critical consideration in making these substitutions is the current-carrying ability of the coil. Never substitute a coil of lower current capacity than that specified, without checking first to insure that the current actually flowing in the circuit is within the rating of the new coil.

Power transformers are rated in terms of both ***voltamperes*** (VA), which is the product of maximum volts times maximum amperes that can be supplied by the transformer, and in terms of watts, which is voltamperes times the cosine of the phase angle between the voltage and the current. For most electronic projects, loads are nonreactive, voltage and current are in phase, and voltamperes and watts are equal.

If you need a power transformer with a certain power rating for a ham transmitter, you may be able to substitute a TV-type transformer of considerably lower rating. These home entertainment units are rated for continuous duty, while transmitter duty is intermittent. As a rule of thumb, you can increase the power dissipation of home entertainment transformers 40 percent for service in amateur transmitters. Note that this means an increase in ***current capacity***; the voltage is fixed by the ratio of turns in the primary and secondary windings.

Another important rating for power transformers is their design frequency. Conventional transformers used in television sets and common electrical devices are intended to work at 60 Hz. You will occasionally find transformers in surplus stores that were designed for aircraft electrical systems that operate at 400 Hz. There is no way you can use these transformers at 60 Hz. Their reactance at this frequency is so low that they will draw excessive current and burn up. On the other hand, 60 Hz transformers work quite well at 400 Hz, but that isn't likely to do you much good.

As for the actual voltage rating of power transformers, it is best not to deviate too much from the values specified in a project parts list. If you just happen to have a transformer that has an output voltage only 10 percent or so from the specified level, you shouldn't have any trouble substituting it, but voltages that vary widely from the recommended value should be avoided.

Impedance-matching transformers are more critical than power transformers. There are wide gaps between impedance values, so it isn't possible to select the next higher or lower unit. In terms of frequency, it is again not possible to substitute units with vastly different design frequencies for each other. In fact, the only real substitutions you can make between impedance-matching transformers are between units with identical characteristics but made by different manufacturers. Even here, it is wise to make sure that the new unit will fit into the available space before going ahead with the swap.

DIODES

Modern circuits use silicon diodes for all functions. Older power supply circuits used selenium rectifier diodes, but these are completely obsolete, and circuits will operate more efficiently with silicon diodes of the proper current rating. Other circuits used germanium diodes for small-signal applications, but here again, silicon diodes are the modern choice. You may find a project in an old magazine that calls for some other kind of diode than silicon, but you won't find that diode available in the catalogs any more. In general, silicon diodes are superior to all others in terms of heat resistance and current capacity.

The important ratings in selecting diodes and in deciding whether substitutes will work are ***peak inverse voltage*** (PIV) and ***current capacity.*** Current capacity is self-explanatory. The PIV is the maximum voltage that can be applied in the reverse direction across a diode junction without causing breakdown and permanent damage. Note that this is a ***peak rating.*** If your power supply transformer provides an output that is a 200 V ***rms,*** the peak voltage is 1.4 times 200 V, or 280 V.

Older designs use a number of diodes in series to provide increased PIV ratings. These circuits require resistors and capacitors across each diode to equalize the voltage drop across the stack. Presently, it is cheaper to buy rectifier diodes with very high PIV ratings than to buy several diodes and associated resistors and capacitors. Consequently, diode stacks are now used only in very high-voltage power supplies. As semiconductor technology advances, we can expect that even here, diode stacks will be replaced by single high-PIV diodes.

Zener diodes are in a different class than conventional diodes from a design standpoint. However, from a substitution standpoint, the same two considerations—voltage and current—apply. With zeners, the voltage is ***zener breakdown voltage,*** but the only difference is that this reverse voltage does not cause permanent damage to the device. As before, the current rating is self-explanatory.

BIPOLAR TRANSISTORS

When we look at the thousands of types of transistors listed in the various manufacturers' handbooks, we might wonder how one could ever decide which of these are interchangeable. It turns out, though, that the situation isn't as bad as it first appears. Look at the specifications for specific transistors: current gains vary from 100 to 1000 percent. Other ratings change substantially with current gain or temperature. Actually, the chance of finding two perfectly matched transistors (outside an integrated circuit) is about the same as finding two identical snowflakes.

The key to circuit design for bipolar transistors is to bias them so that the variations in their characteristics have little if any effect on circuit performance. This allows you, as an experimenter, some wide latitude in substituting types.

Some manufacturers have simplified your task considerably. Both Motorola and RCA, for example, make lines of transistors specifically for experimenters, and they publish lists cross-referencing their transistors and units for which they may be substituted. The RCA line of replacement transistors is designated "SK," and the Motorola hobbyist transistors are all identified with the prefix "HEP." Substitution guides are available from radio parts stores and mail-order houses.

If you do not want to use one of the devices listed in the substitution guides, you can make a pretty good judgment based on the published transistor parameters. The best way to find out the complete pedigree of a transistor is to obtain the specification sheet published by the manufacturer. You can get sheets for specific transistors by writing to the manufacturer's home office. The addresses of these offices are published in a catalog called the *Electronic Engineers Master* catalog (EEM), which you can find at any good library.

If you are in a hurry or are too lazy to scare up a complete specification sheet, you can get a pretty good notion of the transistor characteristics from the data listed in the manufacturer's transistor handbook.

What data are you interested in? First, you want to know whether the device is PNP or NPN; then whether it is germanium or silicon. Germanium transistors generally drop about 0.7 V between base and emitter at room temperature, while silicons drop only about 0.2 V. This has an influence on the way the engineer designs a circuit and, generally, a germanium transistor cannot be used to replace a silicon device.

The next important consideration is power dissipation. Unless a circuit has been grossly overdesigned, you cannot expect to replace a 20-W transistor with a 10-W transistor. While you are checking to see that your proposed substitute does not have a lower power rating than the transistor it is to replace, take a look at the maximum voltage ratings specified. If they are lower than those listed for the transistor in the parts list, you will have to check the circuit to be sure that they will not be exceeded.

You may be confused by the abbreviations used in the specification sheets and handbooks. Capital letters indicate DC levels; V for voltage, I for current, and so on. Breakdown voltage is indicated in at least two ways, depending on the manufacturer. Sometimes, *BV* is used, and at other times, *BR* is used as a subscript of the letter "V."

Other subscripts refer to the transistor elements between which the voltage is measured or through which the current flows, and to the condition of the remaining element. For example, V_{CBO} refers to the voltage between the transistor collector and base, with the emitter open. And V_{CBS} would be the same voltage measured with the emitter shorted to the base. Other letters used for the third character of the subscript are "R" and "X" or "V." The subscript "R" means that a resistor is connected between the unspecified terminal and one of the others, and "X" and "V" means that a certain voltage is applied between the unspecified terminal and one of the others. The values of resistance or voltage must be indicated adjacent to the listing. If you have to choose between subscripts, remember that the worst case is always the one in which the unspecified terminal is left floating.

If everything matches so far, check the range of values for ***current gain***. This is abbreviated h_{FE} or beta (β). Most circuit designs are unaffected by high values of beta, but beware of replacement transistors with lower minimum beta than the specified transistor. Of course, if you have a transistor tester that measures beta, you may be able to use even these transistors, if their actual beta is greater than the minimum for which the circuit was designed.

If you are dealing with a higher frequency RF application, you will also be interested in the frequency performance of your replacement transistor. The most commonly listed frequency parameter of transistors is the ***gain-bandwidth product*** (f_T). This is the frequency at which the voltage gain of the transistor in a common-emitter amplifier becomes unity. Obviously, you do not want a transistor with a value of f_T near the frequency at which you will be operating.

If you are substituting transistors in a switching circuit, an additional characteristic you will want to look at is the voltage drop from collector to emitter when the transistor is in the ***on***, or ***saturated state***. This is designated "$V_{CE\ (SAT)}$." It should be as low in your replacement, or lower, than it was in the original transistor.

FIELD-EFFECT TRANSISTORS

As with bipolar transistors, RCA and Motorola offer replacement FET designated, respectively, "SK" and "HEP." The same substitutions guides used for bipolars can be used for FETs.

If you are not using a substitution guide, the first thing to check is the ***channel polarity*** of the FET you want to replace. Once you have decided whether it is n-channel or p-channel, you will have to decide whether you have a junction FET or a depletion or enhancement MOSFET. The three types of FETs cannot be interchanged without changing other circuit elements to alter the bias. As with other components we have been discussing, it is important that do you not exceed the power rating or the maximum voltage and current ratings of your replacement FET.

The performance characteristics of FETs vary considerably among devices with the same part number, and circuits that employ FETs are designed with this in mind. This lets you substitute quite widely, within the constraints mentioned above.

While it is possible that some circuits will work with grossly dissimilar transistors (sometimes you can even substitute a dual-gate MOSFET with both gates tied together for a single-gate unit), for best results, you should consult the transistor specification sheet or handbook listings to learn the nominal values of the two most important FET parameters, I_{DSS} and $V_{GS(OFF)}$. The first of these is the current that flows from source to drain when the gate is shorted to the source and the indicated voltage is applied. And $V_{GS(OFF)}$ is the gate-to-source voltage required to pinch off the channel and prevent any current from flowing from drain to source. It is also equal to the minimum drain-to-source voltage that will cause a current equal to I_{DSS} to flow when the gate is shorted to the source. Sometimes, I_{DSS} to flow when the gate is shorted to the source. Sometimes, I_{DSS} and $V_{GS(OFF)}$ are not published in the handbook tables, but ***drain characteristic curves*** are.

Figure 5-4 shows how to find I_{DSS} and $V_{GS(OFF)}$ from the drain characteris-

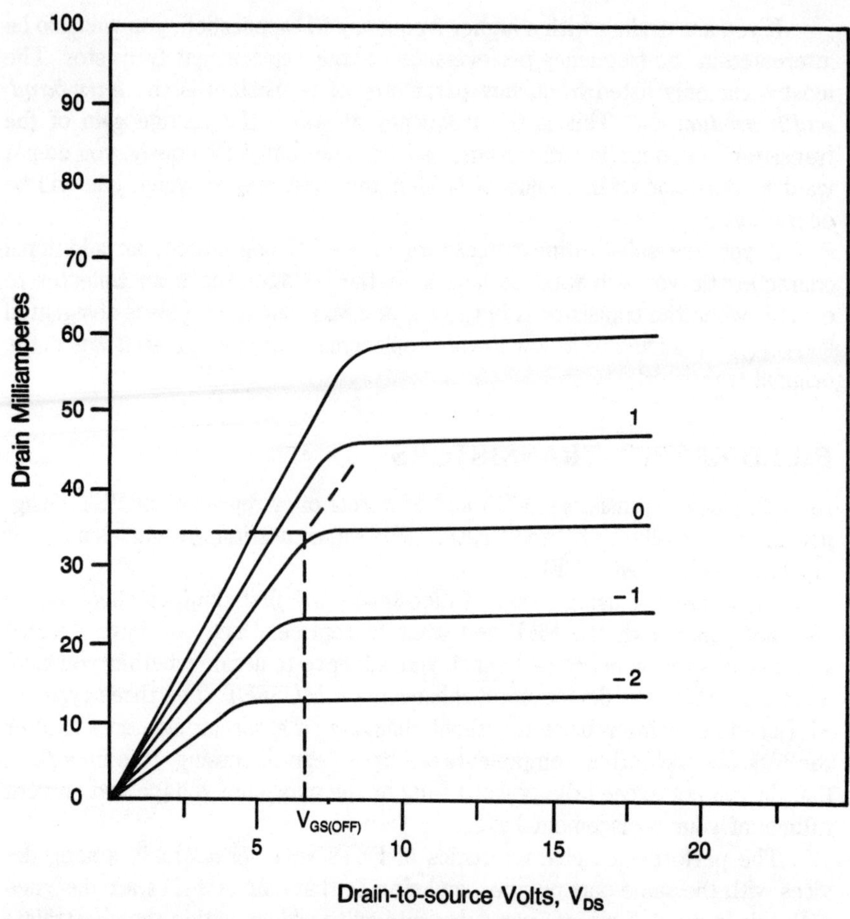

Fig. 5-4. A family of drain characteristic curves.

tic curves. To find I_{DSS} and the drain-to-source voltage equal to $V_{GS(OFF)}$ from the FET drain curves, first find the curve representing $V_{GS} = 0$. Then extend the straight portions of both regions of the curve until they cross. The drain current at this point is I_{DSS}. The drain-to-source voltage equals $-V_{GS(OFF)}$.

UNIJUNCTION TRANSISTORS

Compared to other types of transistors, there are not many types of unijunctions. Most UJT oscillator circuits are not at all critical, and UJTs can be substituted with relative ease.

INTEGRATED CIRCUITS

Substituting ICs is an iffy proposition. Sometimes you can do it, and sometimes you can't. It all depends on the specific IC you need to replace.

Some substitution guides for ICs are available, although they are not as extensive as most transistor substitution guides. Even when a suitable electrical substitute is found, the replacement may not be pin-for-pin compatible with the original chip. That is, the two ICs may function similarly, but the pin numbers for individual functions may be different. The circuit must be mechanically adapted and rewired to suit the replacement device. This can present a problem, particularly when a printed circuit board is used. The original traces leading to certain pins may have to be cut and replaced with jumper wires.

ICs vary widely in their functions, from the relatively simple to the very complex. Generally, it is easier to find a suitable substitute for a low-level (simple function) IC, such as an amplifier or a digital gate, than for a complex device, such as a CPU.

There are two basic types of integrated circuits—digital and analog (or linear). Some specialized function devices are digital/analog hybrids. There are some differences in substituting digital and analog devices.

Digital ICs

Except for some very high-level function chips (such as CPUs), it is generally not too hard to find a substitute for most digital ICs. The majority perform fairly common functions, such as gates, flip-flops, counters, and so forth.

The substitute device must be of the same logic family as the original unit. Different logic families have different power supply requirements and input/output conventions. In most circuits, changing the logic family will require at least moderate alterations in the circuitry.

Linear ICs

Simple linear ICs, such as operational amplifiers (op amps) and differential amplifiers, can often be readily substituted, as long as care is taken to provide the appropriate feedback and frequency compensation connections for each type of amplifier. For example, 741 op amps can be readily substituted for 709 op amps—without the external frequency compensation network required by the 709. The 741 is internally compensated. If you substitute a 709 for a 741, you must remember to add the external frequency compensation network.

Op amps are particularly easy to substitute between type numbers, because most are pin-for-pin compatible with the 741, which has become more or less the "standard reference" op amp IC.

In substituting some differential amplifiers and operational amplifiers, it is possible to simply "float" the inputs; that is, to leave them isolated from ground. However, this can introduce distortion if the signal level is low and the amplifier has a large *offset voltage.* The offset voltage is a DC voltage that appears in series with either input lead of the amplifier, and is an inherent

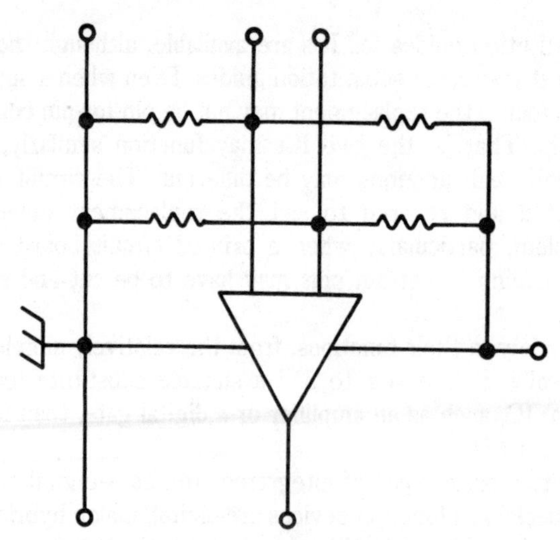

Fig. 5-5. Voltage dividers are used to pull up the DC level at the inputs to this differential amplifier to a potential well above the offset voltage of the IC.

characteristic of the IC. It is usually better to use a voltage divider, as shown in Fig. 5-5, to hold the inputs above ground at some level greater than the offset voltage. This practice is called ***pulling up*** the input, and the voltage divider in this application is called a ***pull-up circuit.***

6

Special Tips for Using ICs

More and more electronic circuits—both on the commercial and the hobbyist level—utilize ICs. These wonderful devices of modern high technology and miniaturization can significantly reduce circuit size and bulk, and even overall expense. In many cases they can simplify design, since much of the work has already been done by the manufacturer of the chip. The IC has also opened up many new projects and areas of experimentation for the hobbyist which would have been impractical, or even impossible, without ICs. In 1965, for example, what hobbyist would have even considered building a computer? Today, thousands have done so.

Essentially, an IC can be thought of as a prefabricated subcircuit which can be used as part of a large system. As long as the proper supply voltages and input signals are fed into the correct pins of the IC, the correct output signals will come out. The hobbyist or circuit designer doesn't have to worry about what goes on in between.

MINIATURIZATION

The big buzzword in electronics over the past couple of decades is unquestionably ***miniaturization***. The IC allows us to make things smaller and lighter. Early computers filled rooms. Today, some models will fit into your coat pocket.

Much of the push for miniaturization came from the space program. In a satellite there is very limited space, and weight must be kept to an absolute minimum. In more earth-bound applications, miniaturization generally isn't quite such a critical issue, although it can still be important. In consumer equipment, portability is always a major selling point. In some cases, ICs are essential to make the product practical as a consumer item. Home computers have already been mentioned. The phenomenally popular VCR machines are another example. Who would buy a VCR they couldn't even fit into the family car to take it home?

But it is possible to take a good thing too far. Many people in electronics —both hobbyists and professionals (who should know better)—place miniaturization above almost everything else.

In designing a project, you should also consider usability. Leave ample space for all controls. Don't pack the front panel so tightly that someone's fingers won't even fit to make adjustments. Every once in awhile, a specific project may need to be made so small that a special tool is required to operate the controls, but unless such extreme miniaturization is absolutely essential for the application, it should be avoided.

You should also consider reliability and serviceability. The more tightly you pack the components together, the greater the chances that you will make an error during construction. Solder bridges become increasingly likely with reduced size.

Thermal conditions must also be considered. All electronic components must dissipate some heat, and that heat must go somewhere. Note that I am not talking about just high-power components which must be heat-sinked. This is true of every component in every circuit. In most practical cases this heat dissipation is no particular problem. The heat is not great and can easily be carried off by the air surrounding the components. But if the components are packed too tightly together their heat is shared and concentrated. It can build up to a harmful level, causing premature failure of one or more of the components.

An extremely miniaturized circuit is also more difficult to service. It is hard to place test probes at the desired test points. Desoldering a faulty component can be an exercise in frustration, especially when there isn't a strong reason for the components to be so tightly packed in the first place.

Don't waste space. Make your circuits as small as practical. Just don't get carried away and forget other practical considerations. This is not just a problem with hobbyists. Commercial manufacturers sometimes make the same sort of mistakes. I remember fixing a function generator from a major manufacturer. I won't mention the brand, but they certainly should have known better. The controls on the front panel required a case of 6 × 5 × 3 inches. Certainly a small instrument, but not too small. The design problems were inside. The circuitry was crammed onto a pc board that measured about 1 × 1.5 inches. The rest of the space in the case was simply wasted. The components were packed extremely tightly. To make matters worse, for some reason 30-gauge wire, which is very fine and fragile, was used to connect the circuit board to the front-panel controls, with a minimum of slack. It should never be used to make any off-board connections. The slightest stress on a wire this thin will cause it to break. In this function generator, several of the wires broke as the case was opened.

Needless to say, the poor design of this instrument made the repair take much longer than it should have. Obviously, the only thing on the designer's mind was "smaller is better," whether smaller made any sense or not. And yes, the original problem that led to the repair in the first place was thermally induced, undoubtedly from the unnecessary crowding of the circuit board. Don't make things harder on yourself by getting carried away with miniaturization.

There is a fairly new construction technique that has been enthusiastically described in some of the hobbyist publications. It is called bricklaying. It makes sense in certain high-density industrial applications, but I think it is ridiculous for a hobbyist. I can't imagine any hobbyist building a product that would require such a technique.

Bricklaying is pretty much the ultimate in miniaturization. ICs are layered like bricks, as shown in Fig. 6-1. Since the individual chips are epoxied in place, the circuit is essentially unrepairable. In industrial applications, the circuit is usually divided into replaceable modules. If one component goes bad, the entire module is replaced. This is faster than detailed troubleshooting. In industry, every second the equipment is down might be costing the company a lot of money. But the module replacement approach would be exceedingly wasteful on the hobbyist level.

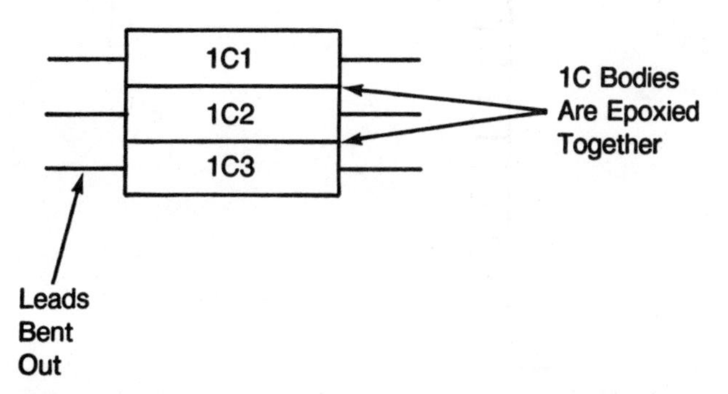

Fig. 6-1. Bricklaying is the ultimate miniaturization construction technique.

Don't over-miniaturize. If you use ICs and reasonably short connecting leads, and pack circuit boards comfortably, your project should be small enough. Don't try for too much of a good thing.

INSTALLING ICS

ICs are relatively easy to work with, but you must use some care when installing them. Most modern ICs are housed in a dual-inline pin (DIP) package, as shown in Fig. 6-2. A small dot or notch indicates the front end of the chip. Use this indicator to identify Pin 1. Because each pin has a different function, an IC must be correctly oriented. Always be careful not to install any IC backwards. It's easier to make this mistake than you might expect. Always double-check before applying power to the circuit. If power is applied to a backwards IC, the odds are very good that the chip will be destroyed. The chief culprit is the supply voltage being fed into an inappropriate input or output.

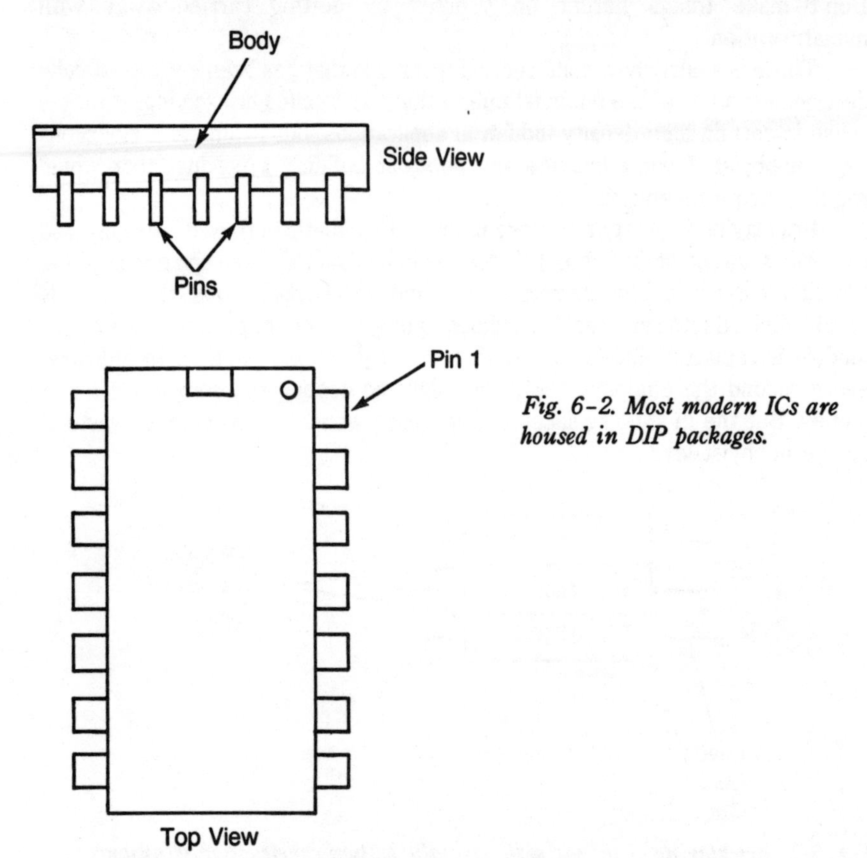

Fig. 6-2. Most modern ICs are housed in DIP packages.

Some ICs are housed in round cans, as shown in Fig. 6-3. These devices resemble somewhat over-sized transistors with eight or more leads. A tab on the case indicates Pin 1. Once again, you must be careful to install the chip correctly. Also, make sure that none of the leads can move and short each other out.

The leads on a DIP IC are rather small. If you are not careful, one or more of the pins can get bent underneath the body of the IC. Of course, this would mean that the appropriate circuit connection will not be made to this pin. The circuit will not function as expected.

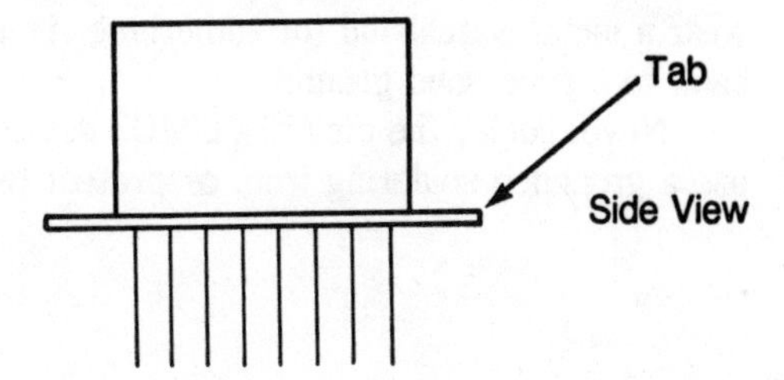

Fig. 6-3. Some ICs are housed in round cans.

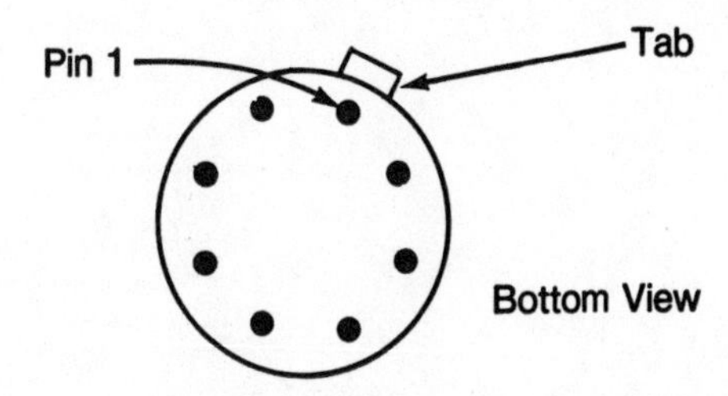

Be careful not to move any of the pins too much. They could break off if you're careless or too rough, which could render the chip useless.

Because the pins are so close to one another, extra care must be taken in soldering. It is all too easy to create a solder bridge between adjacent pins. Use a fine point, low wattage soldering iron, and the narrowest gauge solder you can find.

Never overheat an IC (or any other semiconductor device). The delicate semiconductor crystal can be damaged by excessive heat. If you do not use sockets, some kind of heat-sinking is strongly advised during soldering.

A simple trick is to use an ordinary paper clip as an IC soldering heat sink. It fits neatly over one row of pins, acting as a heat sink for that entire side of the chip. It may be belaboring the obvious to say this, but be sure to remove the clip before applying power to the circuit. If it is not removed, the heat-sink clip will short all of the pins on that side together.

STATIC ELECTRICITY

Some ICs are quite sensitive to static electricity. This is especially true of digital chips of the CMOS family. Earlier devices were more sensitive than modern units, but reasonable care is still advised.

Try not to touch the pins with your fingers, or anything else that might possibly be carrying a static charge. When not actually installed in a circuit, the pins should be shorted together. Mount unused CMOS ICs on a piece of conductive foam, or wrap them in aluminum foil. Sometimes dealers ship CMOS chips in tubes made of a special conductive plastic.

When working with CMOS devices, particularly in a low humidity environment, it is a good idea to ground yourself. The simplest way to do this is to wear a metal watchband (or something similar) with a wire running from the band to a good solid ground.

Never solder the pins of a CMOS device with an ungrounded iron. Either use a grounded soldering iron, or protect the chip with a socket.

A

Color Codes for Electronic Components & Wiring

Color	1st Digit	2nd Digit	Multiplier	Tolerance (percent)
Black	0	0	1	
Brown	1	1	10	
Red	2	2	100	
Orange	3	3	1,000	
Yellow	4	4	10,000	
Green	5	5	100,000	
Blue	6	6	1,000,000	
Violet	7	7	10,000,000	
Gray	8	8	100,000,000	
White	9	9	1,000,000,000	6
Gold			.1	III
Silver			.01	
No Color				2a

A

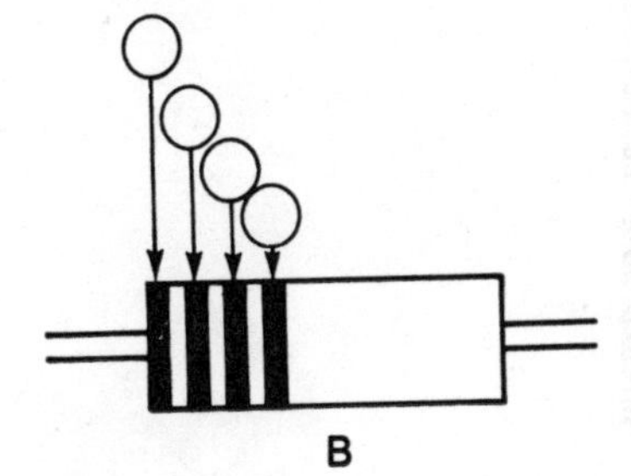

Fig. A1. Resistor color code.

Fig. A2. 6-dot color code for mica and molded paper capacitors.

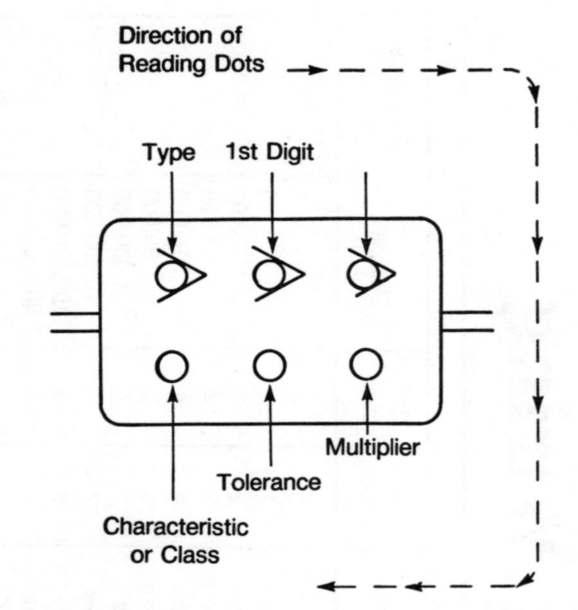

Type	Color	1st Digit	2nd Digit	Multiplier	Tolerance (percent)	Characteristics or Class
Jan. Mica	Black	0	0			Applies to Temperature Coefficient or Methods of Testing
	Brown	1	1	10	±1	
	Red	2	2	100	±2	
	Orange	3	3	1,000	±3	
	Yellow	4	4	10,000	±4	
	Green	5	5	100,000	±5	
	Blue	6	6	1,000,000	±6	
	Violet	7	7	10,000,000	±7	
	Gray	8	8	100,000,000	±8	
Eia. Mica	White	9	9	1,000,000,000	±9	
	Gold			.1	±10	
Molded paper	Silver			.01	±20	
	Body					

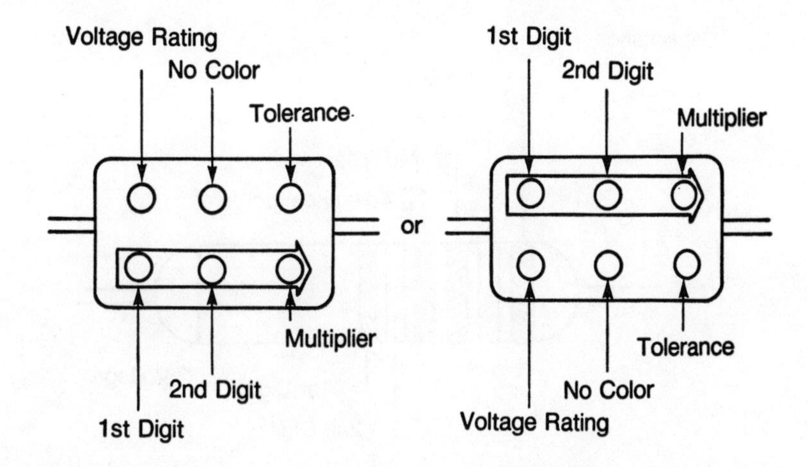

Color	1st Digit	2nd Digit	Multiplier	Tolerance (percent)	Rating
Black	0	0			
Brown	1	1	10	±1	100
Red	2	2	100	±2	200
Orange	3	3	1,000	±3	300
Yellow	4	4	10,000	±4	400
Green	5	5	100,000	±5	500
Blue	6	6	1,000,000	±6	600
Violet	7	7	10,000,000	±7	700
Gray	8	8	100,000,000	±8	800
White	9	9	1,000,000,000	±9	900
Gold					1000
Silver			.1	±10	2000
Body			.01	±20	*

*Where no color is indicated, the voltage rating may be as low as 300 volts.

Fig. A3. 5-dot color code for capacitors (dielectric not specified).

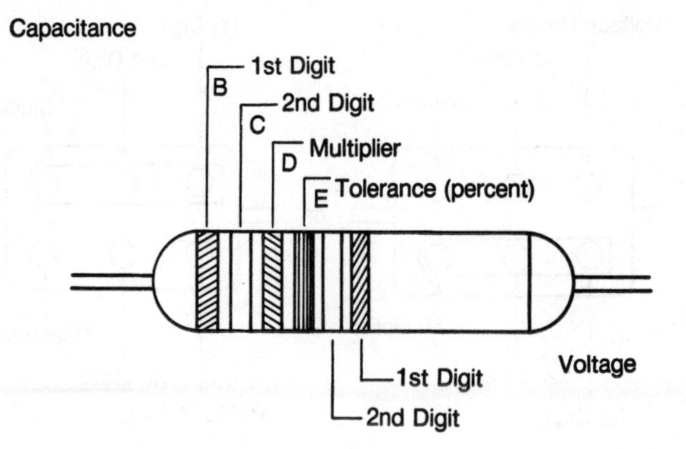

Color	Capacitance			Tolerance (percent)	Voltage Rating	
	1st Digit	2nd Digit	Multiplier		1st Digit	2nd Digit
Black	0	0	1	±20	0	0
Brown	1	1	10		1	1
Red	2	2	100		2	2
Orange	3	3	1,000	±30	3	3
Yellow	4	4	10,000	±40	4	4
Green	5	5	100,000	±5	5	5
Blue	6	6	1,000,000		6	6
Violet	7	7			7	7
Gray	8	8			8	8
White	9	9		±10	9	9

Fig. A4. 6-band color code for tubular paper dielectric capacitors.

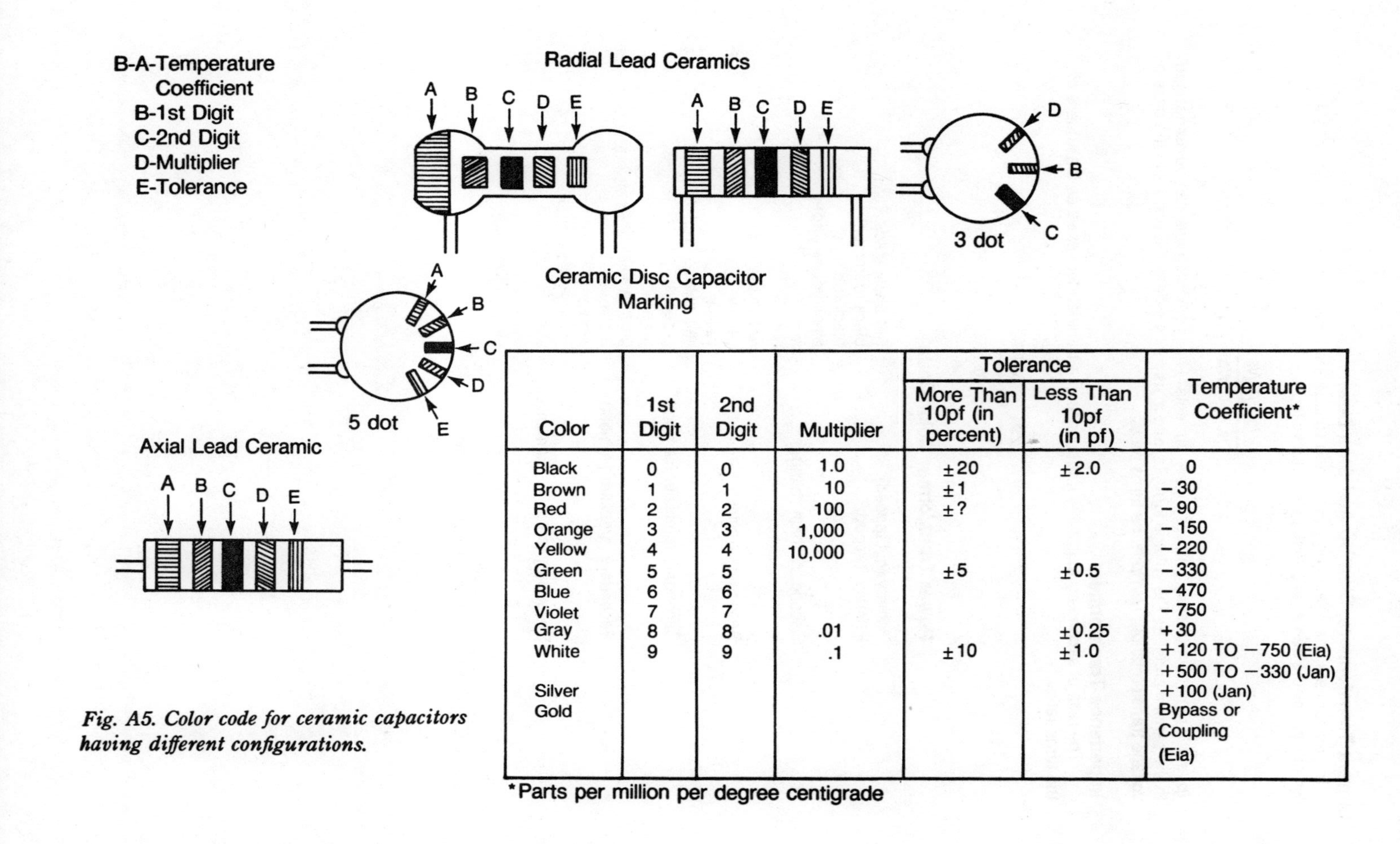

Color	1st Digit	2nd Digit	Multiplier	Tolerance: More Than 10pf (in percent)	Tolerance: Less Than 10pf (in pf)	Temperature Coefficient*
Black	0	0	1.0	±20	±2.0	0
Brown	1	1	10	±1		−30
Red	2	2	100	±?		−90
Orange	3	3	1,000			−150
Yellow	4	4	10,000			−220
Green	5	5		±5	±0.5	−330
Blue	6	6				−470
Violet	7	7				−750
Gray	8	8	.01		±0.25	+30
White	9	9	.1	±10	±1.0	+120 TO −750 (Eia) +500 TO −330 (Jan)
Silver						+100 (Jan)
Gold						Bypass or Coupling (Eia)

*Parts per million per degree centigrade

Fig. A5. Color code for ceramic capacitors having different configurations.

Current Transformers, Step Up And Step Down

In transformers, the effects on current and voltage are inverse. A transformer will step down current by the same ratio that it steps up voltage.

$$\frac{Ip}{Is} = \frac{Ns}{Np}$$

Ip and *Is* represent primary and secondary currents. The current formula can be rearranged in the same manner as the voltage formula, and in as many different ways. As in the case of voltage transformation, *Ns/Np* is the turns ratio.

Impedance Transformers

The ratio of secondary to primary impedance of a transformer varies as the square of the turns ratio.

$$\frac{Zs}{Zp} = \frac{Ns^2}{Np^2}$$

Power Transformer Color Code

Primary (not tapped)	two black leads
Primary (tapped)	black (common)
	black-red
	black-yellow (tap)
Secondary (high voltage)	red
	red
	red-yellow (tap)
Secondary (rectifier filament)	yellow
	yellow
	yellow-blue (tap)
Secondary (amplifier filament)	green
	green
	green-yellow (tap)
Secondary (amplifier filament)	brown
	brown
	brown-yellow (tap)
Secondary (amplifier filament)	slate
	slate
	slate-yellow (tap)

Fig. A6. Color code for power transformers.

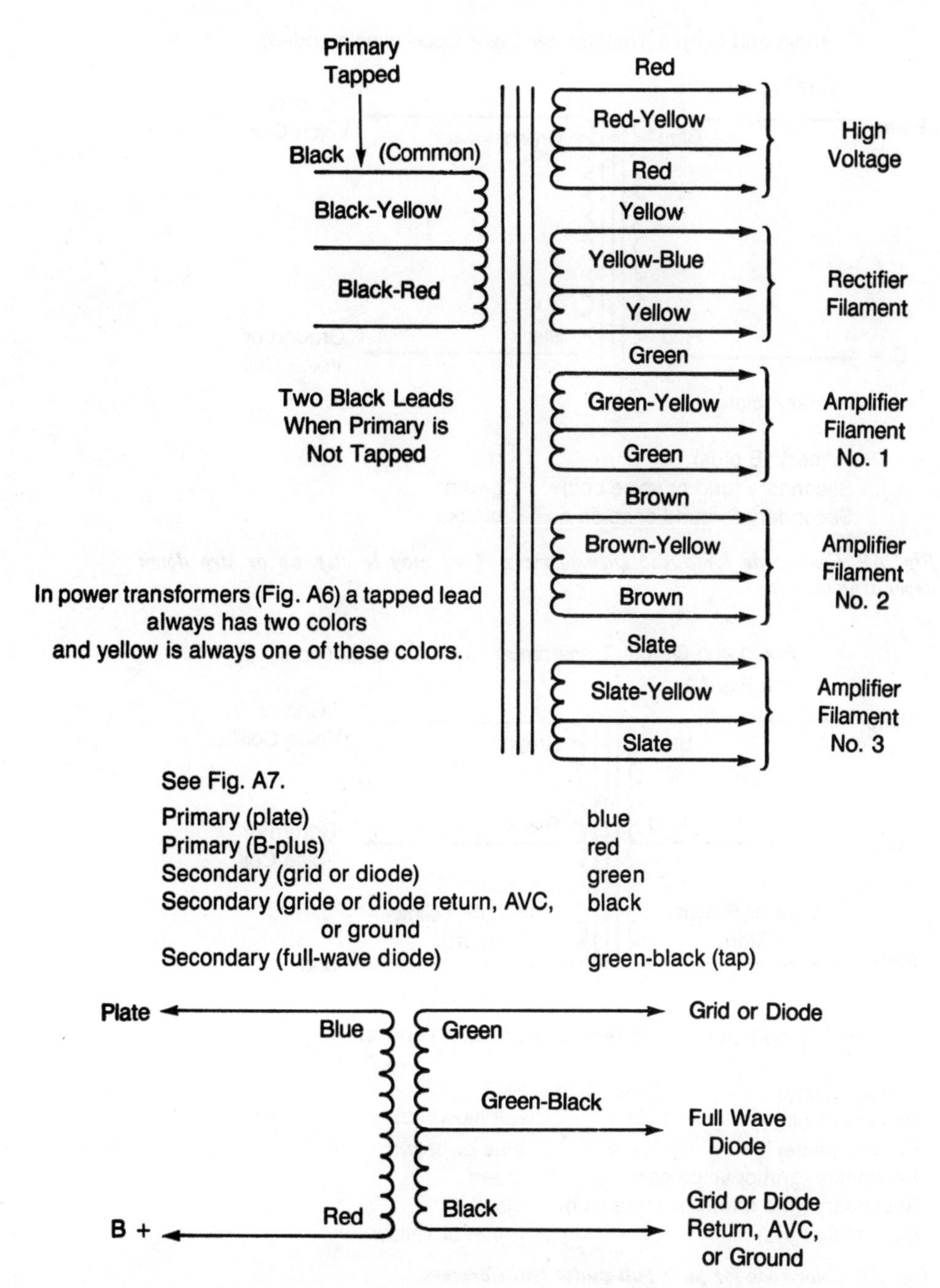

See Fig. A7.

Primary (plate)	blue
Primary (B-plus)	red
Secondary (grid or diode)	green
Secondary (gride or diode return, AVC, or ground	black
Secondary (full-wave diode)	green-black (tap)

Fig. A7. Color code for IF transformers. The center tap on the secondary may or may not be included.

Audio and Output Transformer Color Code (single ended)

See Fig. A8.

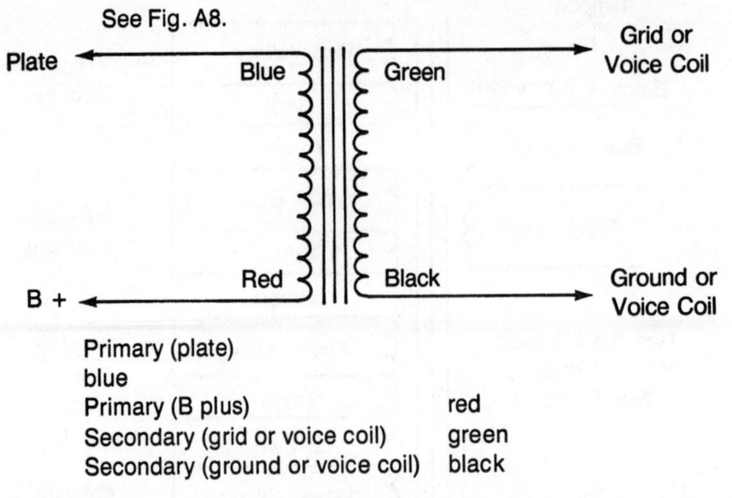

Fig. A8. Color code for audio transformers. They may be step up or step down depending on use.

Audio and Output Transformer Color Code (pushpull)

See Fiq. A9.

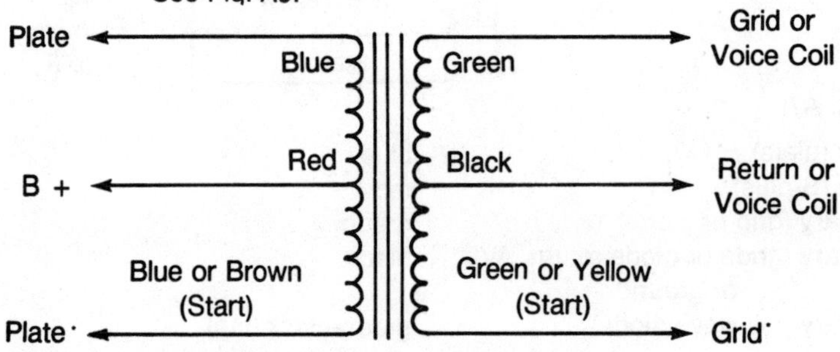

Plate

Blue

Green

Grid or
Voice Coil

Red

Black

B +

Return or
Voice Coil

Blue or Brown
(Start)

Green or Yellow
(Start)

Plate*

Grid*

*Found Only on Push-Pull Primary or Secondary Windings

Primary (plate)	blue
Primary (B plus)	red (tap)
Primary (plate)	blue or brown
Secondary (grid or voice coil)	green
Secondary (grid return or voice coil)	black
Secondary (grid)	green or yellow

Fig. A9. Color code for push-pull audio transformers.

B

Electronic Symbols Used In Schematics

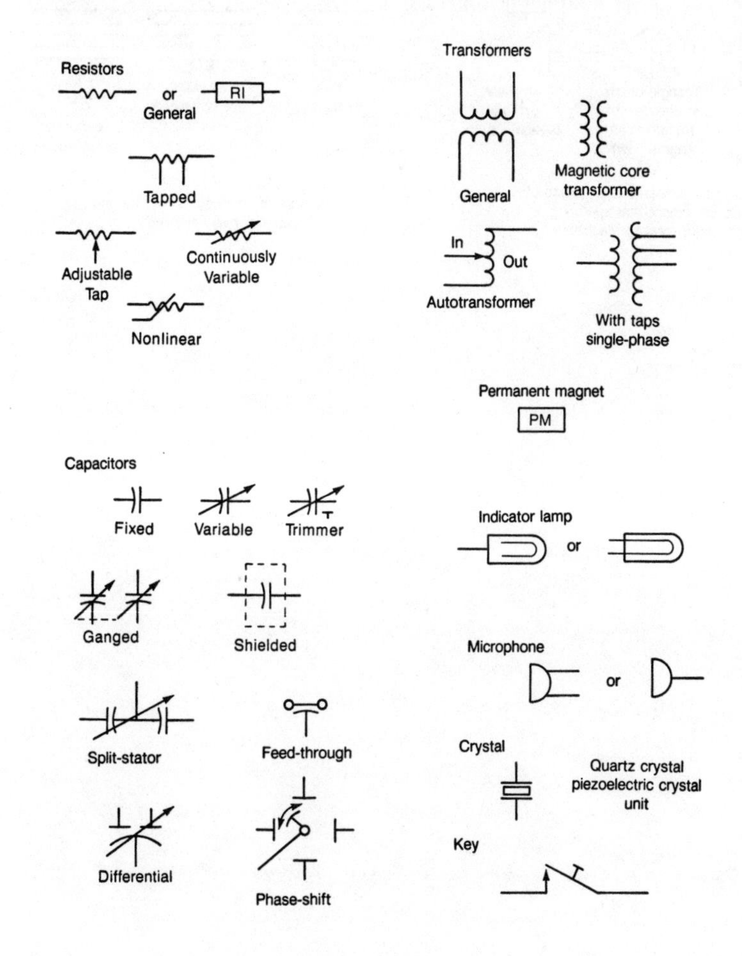

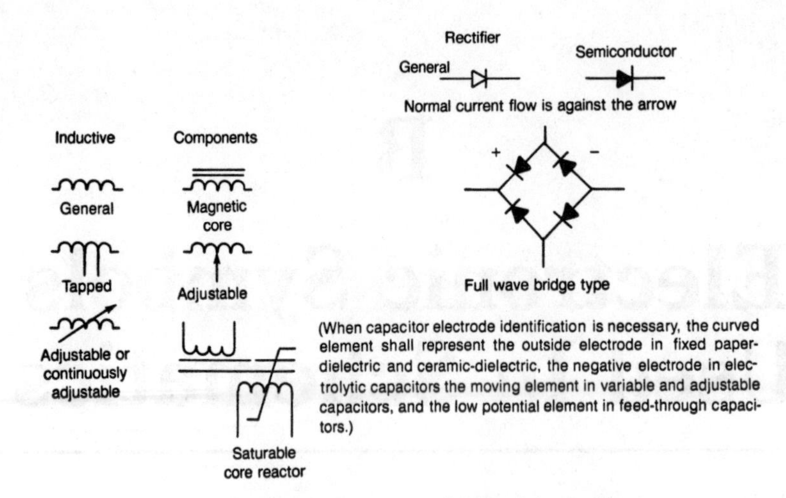

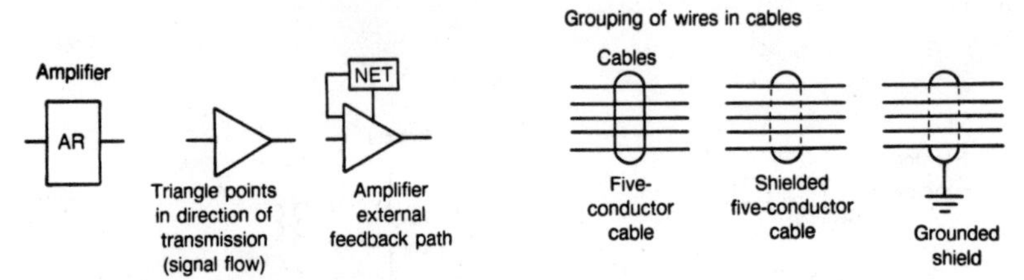

Basic symbol indicates any method of amplification except that operating on the principle of rotating machinery.

Number of conductors may be one or more as necessary

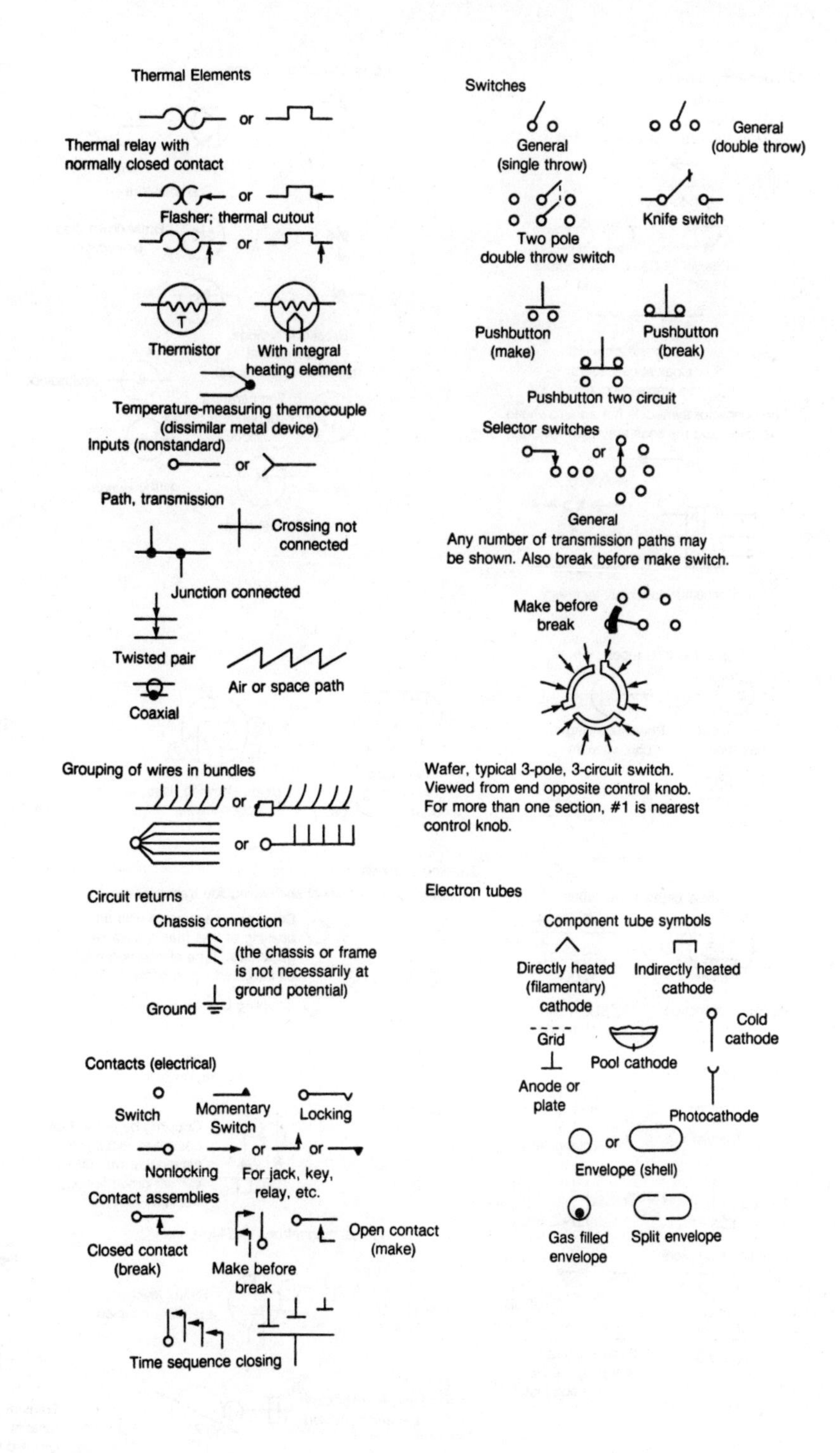
Thermal Elements
or
Thermal relay with normally closed contact
or
Flasher; thermal cutout
or
T
Thermistor
With integral heating element
Temperature-measuring thermocouple (dissimilar metal device)
Inputs (nonstandard)
or
Path, transmission
Crossing not connected
Junction connected
Twisted pair
Air or space path
Coaxial
Grouping of wires in bundles
or
or
Circuit returns
Chassis connection
(the chassis or frame is not necessarily at ground potential)
Ground
Contacts (electrical)
Switch
Momentary Switch
Locking
Nonlocking
or
or
For jack, key, relay, etc.
Contact assemblies
Closed contact (break)
Make before break
Open contact (make)
Time sequence closing
Switches
General (single throw)
General (double throw)
Two pole double throw switch
Knife switch
Pushbutton (make)
Pushbutton (break)
Pushbutton two circuit
Selector switches
or
General
Any number of transmission paths may be shown. Also break before make switch.
Make before break
Wafer, typical 3-pole, 3-circuit switch. Viewed from end opposite control knob. For more than one section, #1 is nearest control knob.
Electron tubes
Component tube symbols
Directly heated (filamentary) cathode
Indirectly heated cathode
Grid
Pool cathode
Cold cathode
Anode or plate
Photocathode
or
Envelope (shell)
Gas filled envelope
Split envelope

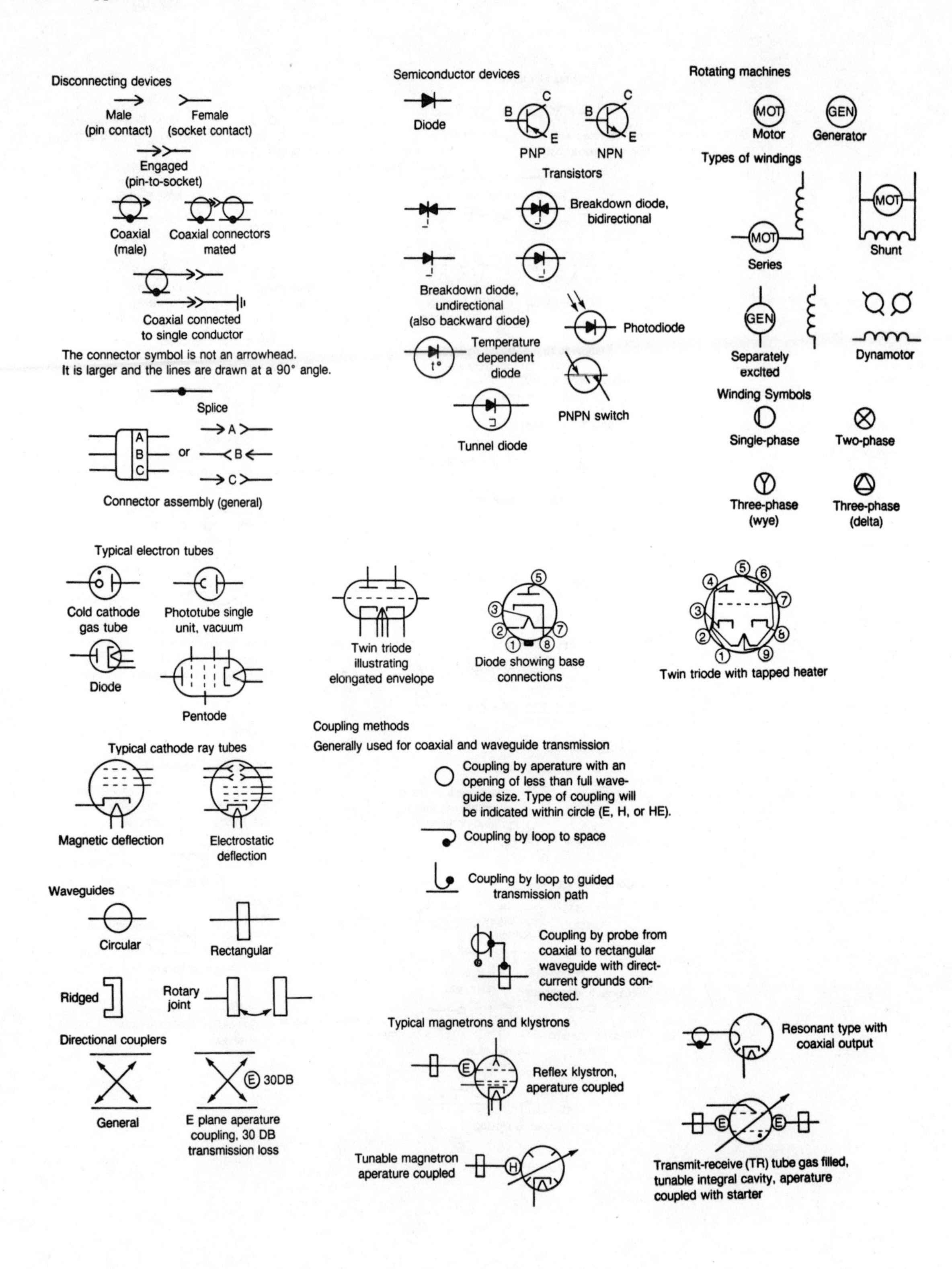
Disconnecting devices
Male
(pin contact)
Female
(socket contact)
Engaged
(pin-to-socket)
Coaxial
(male)
Coaxial connectors
mated
Coaxial connected
to single conductor
The connector symbol is not an arrowhead.
It is larger and the lines are drawn at a 90° angle.
Splice
A
B
C
or
Connector assembly (general)
Semiconductor devices
Diode
C
B
E
PNP
NPN
Transistors
Breakdown diode,
bidirectional
Breakdown diode,
undirectional
(also backward diode)
Photodiode
Temperature
dependent
diode
t°
PNPN switch
Tunnel diode
Rotating machines
MOT
Motor
GEN
Generator
Types of windings
Series
Shunt
Separately
excited
Dynamotor
Winding Symbols
Single-phase
Two-phase
Three-phase
(wye)
Three-phase
(delta)
Typical electron tubes
Cold cathode
gas tube
Phototube single
unit, vacuum
Diode
Pentode
Twin triode
illustrating
elongated envelope
Diode showing base
connections
Twin triode with tapped heater
Typical cathode ray tubes
Magnetic deflection
Electrostatic
deflection
Coupling methods
Generally used for coaxial and waveguide transmission
Coupling by aperture with an
opening of less than full wave-
guide size. Type of coupling will
be indicated within circle (E, H, or HE).
Coupling by loop to space
Coupling by loop to guided
transmission path
Coupling by probe from
coaxial to rectangular
waveguide with direct-
current grounds con-
nected.
Waveguides
Circular
Rectangular
Ridged
Rotary
joint
Directional couplers
General
E 30DB
E plane aperature
coupling, 30 DB
transmission loss
Typical magnetrons and klystrons
Reflex klystron,
aperture coupled
Tunable magnetron
aperature coupled
Resonant type with
coaxial output
Transmit-receive (TR) tube gas filled,
tunable integral cavity, aperature
coupled with starter

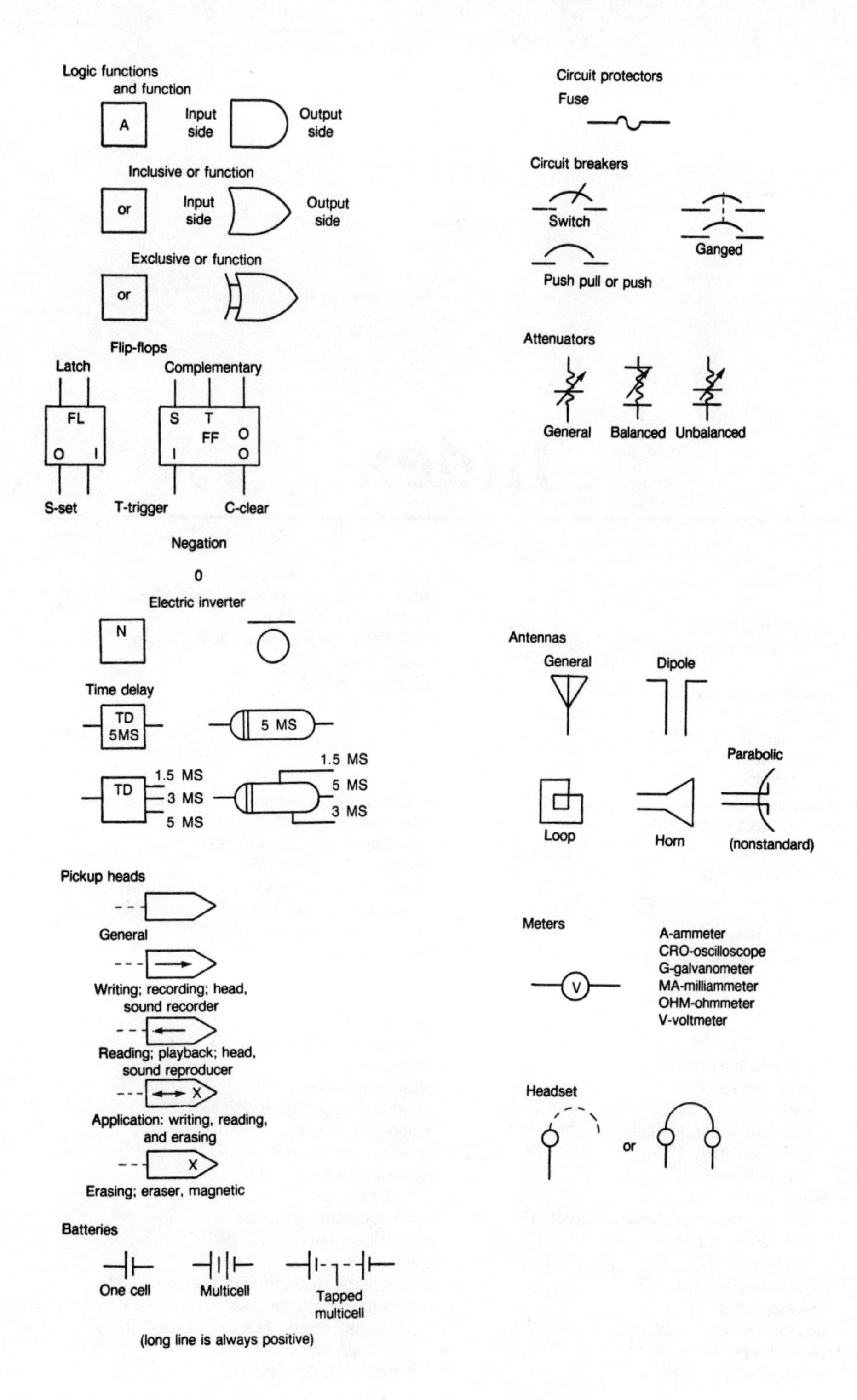
Logic functions
and function
A
Input side
Output side
Inclusive or function
or
Input side
Output side
Exclusive or function
or
Flip-flops
Latch
Complementary
FL
O
I
S
T
FF
O
I
O
S-set
T-trigger
C-clear
Negation
0
Electric inverter
N
Time delay
TD
5MS
5 MS
1.5 MS
5 MS
3 MS
TD
1.5 MS
3 MS
5 MS
Pickup heads
General
Writing; recording; head, sound recorder
Reading; playback; head, sound reproducer
X
Application: writing, reading, and erasing
X
Erasing; eraser, magnetic
Batteries
One cell
Multicell
Tapped multicell
(long line is always positive)
Circuit protectors
Fuse
Circuit breakers
Switch
Ganged
Push pull or push
Attenuators
General
Balanced
Unbalanced
Antennas
General
Dipole
Parabolic
Loop
Horn
(nonstandard)
Meters
V
A-ammeter
CRO-oscilloscope
G-galvanometer
MA-milliammeter
OHM-ohmmeter
V-voltmeter
Headset
or

Index

A

acid core solder, 8
adhesive-backed circuits, 19
Amberlith, 23
ammeters, 89
amplifiers, 146
antenna, 78, 89, 149
attenuators, 149
audio frequencies, 92
audio oscillators, 92
audio output stages, 90

B

Bakelite boxes, 61
balanced modulators, 91
batteries, 149
bench power supply, 83
beta, 102
biasing, FET, 104
bifilar winding resistors, 121
bipolar transistors
  current gain, 126
  gain-bandwidth product, 127
  saturated state, 127
  substitution of, 125
birdies, 90
BJT-FET transistor checker project, 101-107
  BJT tests, 101
  FETs and, 103
blanking circuit, test, 89
blanking input, 73
bleeder resistors, 119
blown fuses, 87
breadboarding, 15-17
break-before-make rotary switch, 65
breakdown diode, 148
breakdown voltage, zener, 125
bricklaying ICs, 133
bridge circuits, 95, 99
bridging, 12

C

calibrating signal generator, 111-113
capaci-bridge project, 95-101
capacitance, reciprocal of, 98
capacitive reactance, 98
capacitors, 87, 93, 145
  color coding, 86, 138-141
  DC working voltages and, 123
  materials used in, 123
  substitution of, 122-123
  tantalum, 41
capillarity, 12
cascaded decade counter, 73
cases, 56
cathode ray tubes, 148
choke-coil applications, 87, 123
chopper circuit, 43
circuit breakers, 149
circuit protectors, 149
Circuit Stik, 19
coils, substitution of, 123
cold solder joints, 8, 11, 86
color codes
  capacitors, dielectric not specified, 139
  ceramic capacitors, 141
  I-F transformers, 143
  mica capacitors, 138
  power transformers, 142

pushpull audio transformers, 144
resistors, 137
tubular paper dielectric capacitor, 140
Colpitts oscillator, 99
commutating voltage, 48
conductance testers, 95
conduction angle, 47
construction methods, 15-55
contacts, 147
continuity checking, 88
control panels, 61-62
controls and shafts, 30
cordwood packaging, 19, 20
couplers, 148
cowled miniboxes, 57
current capacity
coils and transformers, 124
diodes, 125
current divider, 102
current gain, 102
bipolar transistors, 126

D

DC working voltage, capacitors and, 123
decade counter, 72, 73
decoder-driver, 89
depletion MOSFETs, 103
detectors, 91, 92
diacs, 47
dials, 64-66
digital ICs, substitution of, 129
digital multimeters (DMMs), 82
digital output stage, 90
digital universal PC board, 27
diodes, 28, 124, 148
dip meters, 84
divider
current, 102
voltage, 130
drain characteristic curves, FETs, 127
drains, 88, 103, 104
dual in-line packages (DIPs), 66, 133-134
dummy loads, matching, 89

E

edge-card mounting, 26
electric drill and bits, 5
electron tubes, 147, 148
Electronic Engineers Master (EEM) Catalog, 126
electronic symbols, 145
electronic voltmeter (EVM), 81
emission testers, 95
enhancement MOSFETs, 103
established reliability resistors, 119
etching, printed circuits, 25
eutectic solder, 8
exciter, 89

F

facsimile transmissions, 78
FET-set shortwave receiver project, 74-79
FETs, 103-106
biasing, 104
drain characteristic curves, 127
MOSFETs and, 103
n- and p-channels in, 103
substitution of, 127-128
testing, 105
FETVM voltmeters, 81
files, 7
filters, 88
firing circuit, 37
flea clips, 17-18
flip flops, 149
FR signals, 91
fuses, 87, 149

G

gain-bandwidth product, bipolar transistors, 127
gates, 47, 103, 149
grid-dip meters, 84
grounding, 78, 89

H

Hartley oscillator, 108
headsets, 149
heat sinks, 9, 10, 14
hysteresis, 48

I

I-F transformers, color coding, 143
ICs, 95, 131-136
backward installation, 86
bricklaying, 133
digital vs. linear, 129
dual in-line packages (DIPs) of, 133-134
installation of, 133-135
miniaturization and, 131-133
round can-type, 135
static discharge and, 135-136
substitution of, 128-130
indicators, 64-66
inductors, 93, 146
intermediate frequency (IF), 91, 92
inverter circuit, 41, 149
12-v DC to 100-v AC, 39-45

J

junctions, 147

K

keyboard, 53
klystrons, 148
knobs, 30, 64-66

L

labeling, 62-64
lamp test terminal, 89
lamps, 89, 93, 145
layouts, 13
lead dress, 13
LEDs, 66, 89
  troubleshooting, 89
linear ICs, substitution of, 129
litho photography, 23
logic functions, 149
logic probe, 114-117

M

magnetrons, 148
make-before-break rotary switch, 65
metal chassis, 28-29
meter scales, 66-68
meters, 81, 89, 149
microfarads, 122
microphones, 145
mixers, 91, 92
modulators, 91, 92
Morse code, 78
MOSFETs, 127
  depletion, 103
  enhancement, 103
  static discharge from, 14
motherboards, 27
multi-strobe project, 34-39
  power supply for, 35
  timing and firing circuit, 37
  voltage doubler, 35
  xenon flash tube, 35
multivibrators, 92

N

n-channel, 103
NBS frequency and time signals, 79
neutralization, 90
nibbling tool, 6
nonshorting rotary switch, 65
nonsinusoidal oscillators, 92, 93
npn transistors, 84
numerical readouts, 66
nutdrivers, 4

O

offset voltage, linear IC, 129
op amp universal PC board, 27
op amps, substitution of, 129
organ project, 48-55
  keyboard for, 53
  top octave generator, 48
oscillators, 92, 93, 99, 108
oscilloscopes, 85
output devices, troubleshooting, 89, 90

P

p-channel, 103
peak inverse voltage (PIV), 125
perforated boards
  adhesive-backed circuits, 19
  push-in terminals, 17-18
permanent magnet, 145
photodiodes, 148
pickup heads, 149
pliers, 1-3
pnp transistors, 84
polarity
  components, 86
  DC power supplies, 87
potentiometers, 30, 65
power supplies
  AC, 88
  bench, 83
  multi-strobe, 35
  polarity of, 87
power transformers, 87, 90
  color coding, 142
  substitution of, 123
power transistors, 28
printed circuit masters, 19
printed circuits, 21-27
  development, 25
  direct application of resist, 22
  etching, 25
  exposure, 24
  finishing, 26
  layout and planning, 22
  mechanical negatives, 23
  photo negatives, 23
  sensitizing board, 23
  universal PC boards, 27-28
pullup circuits, 130
push pull audio transformer, color coding, 144
push-in terminals, 17-18
pushbuttons, 65, 147

R

randomizer project, 68-74
  blanking input, 73
  decade counter, 72
  ripple-blanking input-output, 73
rat's nest circuits, 32
receiver circuit, 74
rectifiers, 146
regeneration, 75
resist, printed circuit board application, 22
resistors, 87, 88, 93, 145

bifilar winding, 121
bleeder, 119
color coding, 86, 119, 137
established reliability, 119
power dissipation rating, 120
substitutions for, 118-122
taper, 121
RF bridge, 99
RF output stage, 90
ripple-blanking input-output, 73
rosin core solder, 8
rosin joint failure, 11
rotary switches, 65
Rubylith, 23

S

saturated state, bipolar transistors, 127
schematics, electronic symbols for, 145
SCR output stage, 91
semiconductors, 94, 146, 148
shafts, control knobs with, 30
short-circuits, 87
solder-caused, 13
shorting type rotary switch, 65
shortwave receiver, FET-set, 74-79
antenna connection, 78
grounding, 78
receiver circuit, 74
regeneration, 75
tickler, 75
sidebands, 91
signal generator, 83, 89, 107-114
calibration of, 111-113
signal tracers, 83
silicon controlled rectifier (SCR), 47, 93
slide switch, 65
sloping-panel boxes, 58
sockets, 33-34, 148
solderless, 16, 17
solder, 8, 12
solder bridges, 12
soldering, 7-14
adhesive-backed circuits and, 20
aluminum, 9
bridging, 12
cold solder joints, 8, 11
heat sinks for, 9
joint failure in, 11
joint inspection, 86
preparation for, 9
short-circuits caused by, 13
solder flow, 11
solder for, 8
surface tension and capillarity of solder in, 12
tinning soldering gun, 8
tools for, 7
varnishing after, 20
wavesoldering machines for, 26
wires, 9
soldering gun or iron, 7-8
solderless sockets, 16, 17
spectrum analyzer, 92
speed controller circuit, 47
static discharge, 14, 135-136
substitutions, 118-130
superheterodyne receivers, 91
surface tension, 12
switches, 64-66, 147

T

tank circuits, 84, 85
tantalum electrolytic capacitor, 41
taper, resistors, 121
Teletype transmissions, 78
test equipment, 58, 81-86
thermal elements, 147
thermocouples, 147
tickler, 75
time signals, 79
timing circuit, 37
tinning soldering guns, 8
toggle switches, 65
tools, 1-7, 81-86
top octave generator, 48
transformers, 145
color coding, 142-144
substitution of, 123
transistor checker, 84, 101-107
transistors, 28, 86, 87, 88, 148
bipolar (*see* bipolar transistors)
current gain or beta of, 102
field-effect (*see* FETs)
insulating from chassis, 45
unijunction, substitution of, 128
transmitter, 89
triac controller project, 45-48
hysteresis and commutating voltages, 48
speed controller circuit, 47
triac output stage, 91
triacs, 93
triodes, 148
troubleshooting, 80-117
antenna matching, 89
assembly check, 86
audio output stages, 90
birdies or spurious signals, 90
BJT-FET transistor checker for, 103
blown fuses, 87
capaci-bridge for, 95
capacitors, 93
component appearance, 87
detectors, 91
dummy load matching, 89
fastener check, 87
inductors, 93
isolating cause of trouble, 95
LEDs, lamps, meters, 89
logic probe, 114-117
mixers, 91

troubleshooting *(continued)*
modulators, 91
neutralization, 90
oscillators, 92
output devices, 89, 90
polarity of power supply check, 87
power transformers, 90
resistors, 93
RF output stage, 90
semiconductors, 94
signal generator, 111
solder joint inspection, 86
stage-by-stage condition check, 88-93
vacuum tubes, 94
tunnel diode, 148
turns-counting knob, 65

U

unijunction transistors, substitution of, 128
universal PC boards, 27-28

V

vacuum tube voltmeters (VTM), 81
vacuum tubes, 28, 87, 88, 94
variable-frequency oscillator (VFO), 84
varnishing, 20
VFO, 89
volt ohm milliammeter (VOM), 81
voltage divider, 130
voltage doubler, 35
voltage drops, 88
voltage regulator tubes, 87
voltamperes, 123

W

waveguides, 148
wavemeters, 84
wavesoldering machine, 26
windings, 148
wire, 146
oversize, 2
routing of, 31-33
sizes and colors of, 31, 32
soldering, 9
stripping insulation from, 2
wirewrapping technique, 29
wire strippers, 3
wirewrapping, 29
WWW broadcast signals, 79

X

xenon flash tube, multi-strobe, 35

Z

zener breakdown voltage, 125
zener diodes, 125

Getting Started with

VBA

ROBERT T.
GRAUER
UNIVERSITY OF MIAMI

MARYANN
BARBER
UNIVERSITY OF MIAMI

**Upper Saddle River,
New Jersey 07458**

Executive Acquisitions Editor: Jodi McPherson
VP/ Publisher: Natalie E. Anderson
Associate Director of IT Product Development: Melonie Salvati
Senior Project Manager, Editorial: Eileen Clark
Project Manager: Melissa Edwards
Editorial Assistants: Jodi Bolognese and Jasmine Slowik
Media Project Manager: Cathleen Profitko
Marketing Manager: Emily Williams Knight
Marketing Assistant: Nicole Beaudry
Production Manager: Gail Steier de Acevedo
Project Manager, Production: Lynne Breitfeller
Production Editor: Greg Hubit
Associate Director, Manufacturing: Vincent Scelta
Manufacturing Buyer: Lynne Breitfeller
Design Manager: Maria Lange
Interior Design: Michael J. Fruhbeis
Cover Design: Michael J. Fruhbeis
Cover Printer: Phoenix Color
Composition and Project Management: The GTS Companies
Printer/Binder: Banta Menasha

1 2 3 DPC 08 07 06
ISBN 0-13-109038-0

To Marion —
my wife, my lover, and my best friend

Robert Grauer

To Frank —
I love you

To Holly —
for being my friend

Maryann Barber

Contents

Preface vii

GETTING STARTED WITH VBA

Getting Started with VBA: Extending Microsoft Office 2003 1

Objectives 1
Case Study: On-the-Job Training 1
Introduction to VBA 2
The MsgBox Statement 3
The InputBox Function 4
Declaring Variables 5
The VBA Editor 6
HANDS-ON EXERCISE 1: INTRODUCTION TO VBA 7
If . . . Then . . . Else Statement 16
Case Statement 18
Custom Toolbars 19
HANDS-ON EXERCISE 2: DECISION MAKING 20
For . . . Next Statement 28
Do Loops 29
Debugging 30
HANDS-ON EXERCISE 3: LOOPS AND DEBUGGING 32
Putting VBA to Work (Microsoft Excel) 41
HANDS-ON EXERCISE 4: EVENT-DRIVEN PROGRAMMING (MICROSOFT EXCEL) 43
Putting VBA to Work (Microsoft Access) 52
HANDS-ON EXERCISE 5: EVENT-DRIVEN PROGRAMMING (MICROSOFT ACCESS) 54
Summary 62
Key Terms 62
Multiple Choice 63

Preface

THE EXPLORING OFFICE SERIES FOR 2003

Continuing a tradition of excellence, Prentice Hall is proud to announce the new ***Exploring Microsoft Office 2003*** series by Robert T. Grauer and Maryann Barber. The hands-on approach and conceptual framework of this comprehensive series helps students master all aspects of the Microsoft Office 2003 software, while providing the background necessary to transfer and use these skills in their personal and professional lives.

The entire series has been revised to include the new features found in the Office 2003 Suite, which contains Word 2003, Excel 2003, Access 2003, PowerPoint 2003, Publisher 2003, FrontPage 2003, and Outlook 2003.

In addition, this edition includes fully revised end-of-chapter material that provides an extensive review of concepts and techniques discussed in the chapter. Each chapter now begins with an ***introductory case study*** to provide an effective overview of what the reader will be able to accomplish, with additional ***mini cases*** at the end of each chapter for practice and review. The conceptual content within each chapter has been modified as appropriate and numerous end-of-chapter exercises have been added.

The new ***visual design*** introduces the concept of ***perfect pages***, whereby every step in every hands-on exercise, as well as every end-of-chapter exercise, begins at the top of its own page and has its own screen shot. This clean design allows for easy navigation throughout the text.

Continuing the success of the website provided for previous editions of this series, Exploring Office 2003 offers expanded resources that include online, interactive study guides, data file downloads, technology updates, additional case studies and exercises, and other helpful information. Start out at www.prenhall.com/grauer to explore these resources!

Organization of the Exploring Office 2003 Series

The new Exploring Microsoft Office 2003 series includes four combined Office 2003 texts from which to choose:

- ***Volume I*** is Microsoft Office Specialist certified in each of the core applications in the Office suite (Word, Excel, Access, and PowerPoint). Five additional modules (*Essential Computing Concepts*, *Getting Started with Windows XP*, *The Internet and the World Wide Web*, *Getting Started with Outlook*, and *Integrated Case Studies*) are also included.
- ***Volume II*** picks up where Volume I leaves off, covering the advanced topics for the individual applications. A *Getting Started with VBA* module has been added.
- The ***Brief Microsoft Office 2003*** edition provides less coverage of the core applications than Volume I (a total of 10 chapters as opposed to 18). It also includes the *Getting Started with Windows XP* and *Getting Started with Outlook* modules.
- ***Getting Started with Office 2003*** contains the first chapter from each application (Word, Excel, Access, and PowerPoint), plus three additional modules: *Getting Started with Windows XP*, *The Internet and the World Wide Web*, and *Essential Computing Concepts*.

Individual texts for Word 2003, Excel 2003, Access 2003, and PowerPoint 2003 provide complete coverage of the application and are Microsoft Office Specialist certified. For shorter courses, we have created brief versions of the Exploring texts that give students a four-chapter introduction to each application. Each of these volumes is Microsoft Office Specialist certified at the Specialist level.

This series has been approved by Microsoft to be used in preparation for Microsoft Office Specialist exams.

The Microsoft Office Specialist program is globally recognized as the standard for demonstrating desktop skills with the Microsoft Office suite of business productivity applications (Microsoft Word, Microsoft Excel, Microsoft PowerPoint, Microsoft Access, and Microsoft Outlook). With a Microsoft Office Specialist certification, thousands of people have demonstrated increased productivity and have proved their ability to utilize the advanced functionality of these Microsoft applications.

By encouraging individuals to develop advanced skills with Microsoft's leading business desktop software, the Microsoft Office Specialist program helps fill the demand for qualified, knowledgeable people in the modern workplace. At the same time, Microsoft Office Specialist helps satisfy an organization's need for a qualitative assessment of employee skills.

Instructor and Student Resources

The **Instructor's CD** that accompanies the Exploring Office series contains:

- Student data files
- Solutions to all exercises and problems
- PowerPoint lectures
- Instructor's manuals in Word format that enable the instructor to annotate portions of the instructor manuals for distribution to the class

- Instructors may also use our ***test creation software***, TestGen and QuizMaster.

TestGen is a test generator program that lets you view and easily edit testbank questions, transfer them to tests, and print in a variety of formats suitable to your teaching situation. The program also offers many options for organizing and displaying testbanks and tests. A random number test generator enables you to create multiple versions of an exam.

QuizMaster, also included in this package, allows students to take tests created with TestGen on a local area network. The QuizMaster Utility built into TestGen lets instructors view student records and print a variety of reports. Building tests is easy with TestGen, and exams can be easily uploaded into WebCT, BlackBoard, and CourseCompass.

Prentice Hall's Companion Website at www.prenhall.com/grauer offers expanded IT resources and downloadable supplements. This site also includes an online study guide for students containing true/false and multiple choice questions and practice projects.

WebCT www.prenhall.com/webct

Gold level customer support available exclusively to adopters of Prentice Hall courses is provided free-of-charge upon adoption and provides you with priority assistance, training discounts, and dedicated technical support.

Blackboard www.prenhall.com/blackboard

Prentice Hall's abundant online content, combined with Blackboard's popular tools and interface, result in robust Web-based courses that are easy to implement, manage, and use—taking your courses to new heights in student interaction and learning.

CourseCompass www.coursecompass.com

CourseCompass is a dynamic, interactive online course management tool powered by Blackboard. This exciting product allows you to teach with marketing-leading Pearson Education content in an easy-to-use, customizable format.

Training and Assessment www2.phgenit.com/support

Prentice Hall offers Performance Based Training and Assessment in one product, Train&Assess IT. The Training component offers computer-based training that a student can use to preview, learn, and review Microsoft Office application skills. Web or CD-ROM delivered, Train IT offers interactive multimedia, computer-based training to augment classroom learning. Built-in prescriptive testing suggests a study path based not only on student test results but also on the specific textbook chosen for the course.

The Assessment component offers computer-based testing that shares the same user interface as Train IT and is used to evaluate a student's knowledge about specific topics in Word, Excel, Access, PowerPoint, Windows, Outlook, and the Internet. It does this in a task-oriented, performance-based environment to demonstrate proficiency as well as comprehension on the topics by the students. More extensive than the testing in Train IT, Assess IT offers more administrative features for the instructor and additional questions for the student.

Assess IT also allows professors to test students out of a course, place students in appropriate courses, and evaluate skill sets.

Opening Case Study

New! Each chapter now begins with an introductory case study to provide an effective overview of what students will accomplish by completing the chapter.

CHAPTER

1

Getting Started with Microsoft® Windows® XP

OBJECTIVES

After reading this chapter you will:

1. Describe the Windows desktop.
2. Use the Help and Support Center to obtain information.
3. Describe the My Computer and My Documents folders.
4. Differentiate between a program file and a data file.
5. Download a file from the Exploring Office Web site.
6. Copy and/or move a file from one folder to another.
7. Delete a file, and then recover it from the Recycle Bin.
8. Create and arrange shortcuts on the desktop.
9. Use the Search Companion.
10. Use the My Pictures and My Music folders.
11. Use Windows Messenger for instant messaging.

hands-on exercises

1. WELCOME TO WINDOWS XP
   Input: None
   Output: None
2. DOWNLOAD PRACTICE FILES
   Input: Data files from the Web
   Output: Welcome to Windows XP (a Word document)
3. WINDOWS EXPLORER
   Input: Data files from exercise 2
   Output: Screen Capture within a Word document
4. INCREASING PRODUCTIVITY
   Input: Data files from exercise 3
   Output: None
5. FUN WITH WINDOWS XP
   Input: None
   Output: None

CASE STUDY

UNFORESEEN CIRCUMSTANCES

Steve and his wife Shelly have poured their life savings into the dream of owning their own business, a "nanny" service agency. They have spent the last two years building their business and have created a sophisticated database with numerous entries for both families and nannies. The database is the key to their operation. Now that it is up and running, Steve and Shelly are finally at a point where they could hire someone to manage the operation on a part-time basis so that they could take some time off together.

Unfortunately, their process for selecting a person they could trust with their business was not as thorough as it should have been. Nancy, their new employee, assured them that all was well, and the couple left for an extended weekend. The place was in shambles on their return. Nancy could not handle the responsibility, and when Steve gave her two weeks' notice, neither he nor his wife thought that the unimaginable would happen. On her last day in the office Nancy "lost" all of the names in the database—the data was completely gone!

Nancy claimed that a "virus" knocked out the database, but after spending nearly $1,500 with a computer consultant, Steve was told that it had been cleverly deleted from the hard drive and could not be recovered. Of course, the consultant asked Steve and Shelly about their backup strategy, which they sheepishly admitted did not exist. They had never experienced any problems in the past, and simply assumed that their data was safe. Fortunately, they do have hard copy of the data in the form of various reports that were printed throughout the time they were in business. They have no choice but to manually reenter the data.

Your assignment is to read the chapter, paying special attention to the information on file management. Think about how Steve and Shelly could have avoided the disaster if a backup strategy had been in place, then summarize your thoughts in a brief note to your instructor. Describe the elements of a basic backup strategy. Give several other examples of unforeseen circumstances that can cause data to be lost.

1

New! A listing of the input and output files for each hands-on exercise within the chapter. Students will stay on track with what is to be accomplished.

Perfect Pages

hands-on exercise

1 Welcome to Windows XP

Objective To log on to Windows XP and customize the desktop; to open the My Computer folder; to move and size a window; to format a floppy disk and access the Help and Support Center. Use Figure 7 as a guide.

Step 1: Log On to Windows XP

- Turn on the computer and all of the peripheral devices. The floppy drive should be empty prior to starting your machine.
- Windows XP will load automatically, and you should see a login screen similar to Figure 7a. (It does not matter which version of Windows XP you are using.) The number and names of the potential users and their associated icons will be different on your system.
- Click the icon for the user account you want to access. You may be prompted for a password, depending on the security options in effect.

(a) Log On to Windows XP (step 1)

FIGURE 7 Hands-on Exercise 1

USER ACCOUNTS

The available user names are cr Windows XP, but you can add or d click Control Panel, switch to the Ca the desired task, such as creating then supply the necessary informati user accounts in a school setting.

10 GETTING STARTED WITH MICROSOFT WINDOWS XP

Each step in the hands-on exercises begins at the top of the page to ensure that students can easily navigate through the text.

Step 2: Choose the Theme and Start Menu

- Check with your instructor to see if you are able to modify the desktop and other settings at your school or university. If your network administrator has disabled these commands, skip this step and go to step 3.
- Point to a blank area on the desktop, click the **right mouse button** to display a context-sensitive menu, then click the **Properties command** to open the Display Properties dialog box. Click the **Themes tab** and select the **Windows XP theme** if it is not already selected. Click **OK**.
- We prefer to work without any wallpaper (background picture) on the desktop. **Right click** the desktop, click **Properties**, then click the **Desktop tab** in the Display Properties dialog box. Click **None** as shown in Figure 7b, then click **OK**. The background disappears.
- The Start menu is modified independently of the theme. **Right click** a blank area of the taskbar, click the **Properties command** to display the Taskbar and Start Menu Properties dialog box, then click the **Start Menu tab**.
- Click the **Start Menu option button**. Click **OK**.

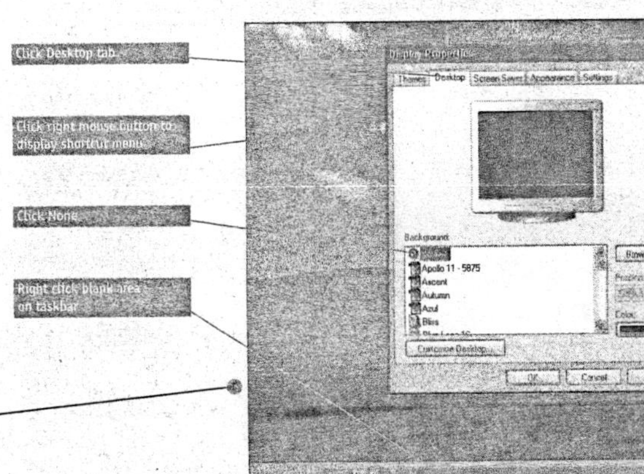

(b) Choose the Theme and Start Menu (step 2)

FIGURE 7 Hands-on Exercise 1 *(continued)*

IMPLEMENT A SCREEN SAVER

A screen saver is a delightful way to personalize your computer and a good way to practice with basic commands in Windows XP. Right click a blank area of the desktop, click the Properties command to open the Display Properties dialog box, then click the Screen Saver tab. Click the down arrow in the Screen Saver list box, choose the desired screen saver, then set the option to wait an appropriate amount of time before the screen saver appears. Click OK to accept the settings and close the dialog box.

GETTING STARTED WITH MICROSOFT WINDOWS XP 11

New! Larger screen shots with clear callouts.

Boxed tips provide students with additional information.

MINI CASES AND PRACTICE EXERCISES

New!
We've added mini cases at the end of each chapter for expanded practice and review.

MINI CASES

The Financial Consultant

A friend of yours is in the process of buying a home and has asked you to compare the payments and total interest on a 15- and 30-year loan at varying interest rates. You have decided to analyze the loans in Excel, and then incorporate the results into a memo written in Microsoft Word. As of now, the principal is $150,000, but it is very likely that your friend will change his mind several times, and so you want to use the linking and embedding capability within Windows to dynamically link the worksheet to the word processing document. Your memo should include a letterhead that takes advantage of the formatting capabilities within Word; a graphic logo would be a nice touch.

Fun with the If Statement

Open the *Chapter 4 Mini Case—Fun with the If Statement* workbook in the Exploring Excel folder, then follow the directions in the worksheet to view a hidden message. The message is displayed by various If statements scattered throughout the worksheet, but the worksheet is protected so that you cannot see these formulas. (Use help to see how to protect a worksheet.) We made it easy for you, however, because you can unprotect the worksheet since a password is not required. Once the worksheet is unprotected, pull down the Format menu, click the Cells command, click the Protection tab, and clear the Hidden check box. Prove to your professor that you have done this successfully, by changing the text of our message. Print the completed worksheet to show both displayed values and cell formulas.

The Lottery

Many states raise money through lotteries that advertise prizes of several million dollars. In reality, however, the actual value of the prize is considerably less than the advertised value, although the winners almost certainly do not care. One state, for example, recently offered a twenty million dollar prize that was to be distributed in twenty annual payments of one million dollars each. How much was the prize actually worth, assuming a long-term interest rate of five percent? Use the PV (Present Value) function to determine the answer. What is the effect on the answer if payments to the recipient are made at the beginning of each year, rather than at the end of each year?

A Penny a Day

What if you had a rich u
salary each day for the n
prised at how quickly th
use the Goal Seek comm
(if any) will your uncle p
uncle pay you on the 31

The Rule of 72

Delaying your IRA for o
on when you begin. Tha
a calculator, using the "R
long it takes money to
money earning 8% ann
money doubles again in
your IRA at age 21, rath
initial contribution. Use
lose, assuming an 8% ra
determine the exact am

New!
Each project in the end-of-chapter material begins at the top of a page—now students can easily see where their assignments begin and end.

PRACTICE WITH EXCEL

1. **Theme Park Admissions:** A partially completed version of the worksheet in Figure 3.13 is available in the Exploring Excel folder as *Chapter 3 Practice 1*. Follow the directions in parts (a) and (b) to compute the totals and format the worksheet, then create each of the charts listed below.
   a. Use the AutoSum command to enter the formulas to compute the total number of admissions for each region and each quarter.
   b. Select the entire worksheet (cells A1 through F8), then use the AutoFormat command to format the worksheet. You do not have to accept the entire design, nor do you have to use the design we selected. You can also modify the design after it has been applied to the worksheet by changing the font size of selected cells and/or changing boldface and italics.
   c. Create a column chart showing the total number of admissions in each quarter as shown in Figure 3.13. Add the graphic shown in the figure for emphasis.
   d. Create a pie chart that shows the percentage of the total number of admissions in each region. Create this chart in its own chart sheet with an appropriate name.
   e. Create a stacked column chart that shows the total number of admissions for each region and the contribution of each quarter within each region. Create this chart in its own chart sheet with an appropriate name.
   f. Create a stacked column chart showing the total number of admissions for each quarter and the contribution of each region within each quarter. Create this chart in its own chart sheet with an appropriate name.
   g. Change the color of each of the worksheet tabs.
   h. Print the entire workbook, consisting of the worksheet in Figure 3.13 plus the three additional sheets that you create. Use portrait orientation for the Sales Data worksheet and landscape orientation for the other worksheets. Create a custom header for each worksheet that includes your name, your course, and your instructor's name. Create a custom footer for each worksheet that includes the name of the worksheet. Submit the completed assignment to your instructor.

AutoSum button

Enter formula to compute total admissions for region

Enter formula to compute total admissions for quarter

Add arrow and text box

Create column chart

Change color of worksheet tabs

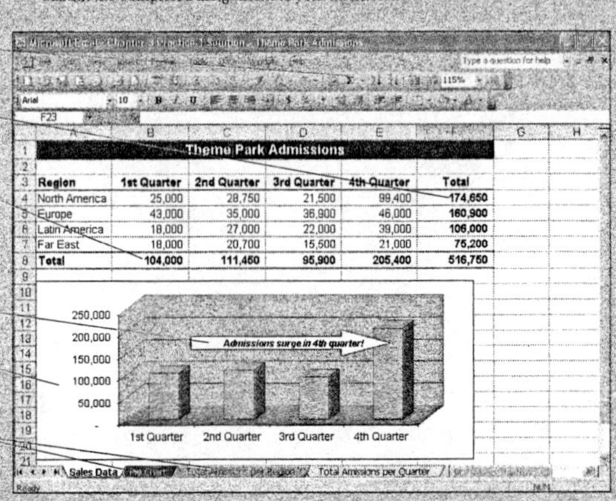

FIGURE 3.13 Theme Park Admissions (exercise 1)

Integrated Case Studies

New!

Each case study contains multiple exercises that use Microsoft Office applications in conjunction with one another.

CASE STUDY 1

Integrated Case Study: The Totally Fit Wellness Center

OBJECTIVES

1. Use the core applications in Office 2003 individually or in conjunction with one another.
2. Link an Excel chart to a Word document and a PowerPoint presentation.
3. Import an Excel worksheet into an Access database.
4. Create Access tables, queries, reports, and forms.
5. Use an Access table as the basis for a mail merge.
6. Use Word to create a Web page for the organization.

integrated exercises

1. Totally Fit Revenue Workbook and Chart (Excel)
2. Report to the President (Word and Excel)
3. Presentation to the President (PowerPoint and Excel)
4. Last-minute Change (Word, Excel, and PowerPoint)
5. Importing Data (Access and Excel)
6. A Relational Database (Access)
7. Access Objects: Forms, Queries, and Reports (Access)
8. An Access Switchboard (Access)
9. Mail Merge (Word and Access)
10. Worksheet References (Excel)
11. Presentation to the Board (PowerPoint and Excel)
12. Letter to the Board (Word and Excel)
13. Create a Home Page (Word)
14. Submission Checklist (Word)

This document provides a series of 13 exercises in Microsoft Office that relate to the Totally Fit Wellness Center, a health center such as might exist at your school or university. Each exercise describes a specific task for the organization to accomplish and typically requires the use of multiple applications within Microsoft Office for solution. Many of the exercises are cumulative in nature. You may, for example, be asked to create an Excel chart in one exercise, and then incorporate that chart into a Word memo and a PowerPoint presentation in subsequent exercises. Other exercises will ask you to create or modify an Access database, to exchange data between Excel and Access, and to create a Web page for the Wellness Center. Collectively, you will gain considerable practice in the core applications of Microsoft Office 2003, and especially in how they work together to accomplish complex tasks.

All of the exercises are based on material from the core application chapters in *Exploring Office 2003 Volume I.* Specific chapter references are deliberately not provided so that the case study is as realistic as possible. Think of this assignment as your first day on the job; you were hired by the Wellness Center because you are proficient in Microsoft Office. This is your chance to show what you know.

Your assignment is to follow the directions of your instructor with respect to which of the exercises you are to complete within the case description. You should begin by downloading the practice files from our Web site at www.prenhall.com/grauer. Go to the site, click the book icon for Office 2003, and then click the Students Download tab. Scroll until you can select the link to Integrated Case Studies, and then download the file to your desktop. Double click the file icon and follow the onscreen instructions to install the practice files on your computer. The files you need will be in the Totally Fit folder within the Exploring Integrated Cases folder.

1

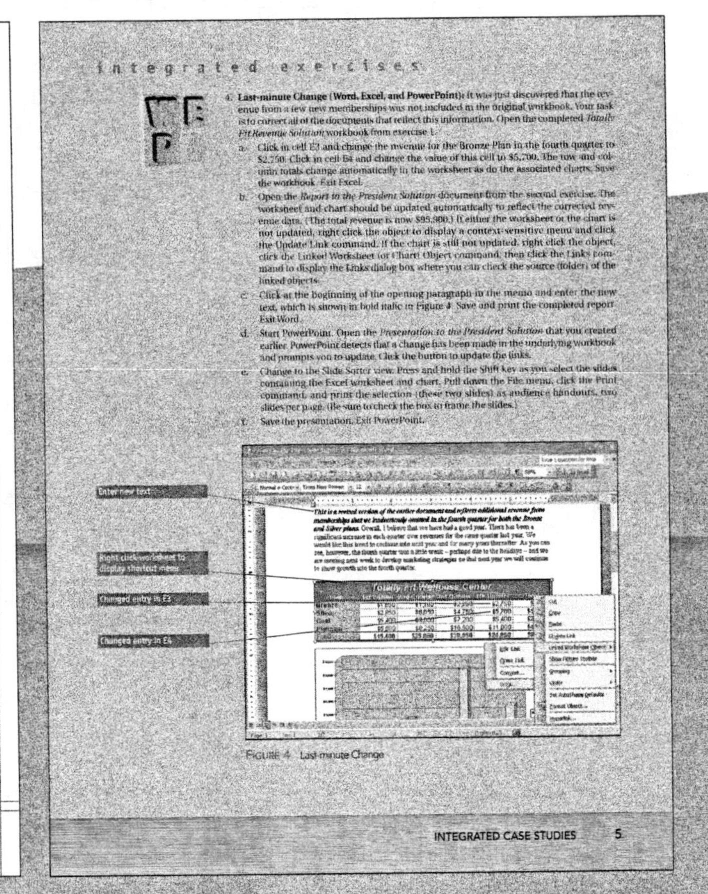
integrated exercises

4. **Last-minute Change (Word, Excel, and PowerPoint):** It was just discovered that the revenue from a few new memberships was not included in the original workbook. Your task is to correct all of the documents that reflect this information. Open the completed *Totally Fit Revenue Solution* workbook from exercise 1.
   a. Click in cell E3 and change the revenue for the Bronze Plan in the fourth quarter to $2,750. Click in cell E4 and change the value of this cell to $5,700. The row and column totals change automatically in the worksheet as do the associated charts. Save the workbook. Exit Excel.
   b. Open the *Report to the President Solution* document from the second exercise. The worksheet and chart should be updated automatically to reflect the corrected revenue data. (The total revenue is now $95,800.) If either the worksheet or the chart is not updated, right click the object to display a context-sensitive menu and click the Update Link command. If the chart is still not updated, right click the object, click the Linked Worksheet (or Chart) Object command, then click the Links command to display the Links dialog box where you can check the source folder of the linked objects.
   c. Click at the beginning of the opening paragraph in the memo and enter the new text, which is shown in bold italic in Figure 4. Save and print the completed report. Exit Word.
   d. Start PowerPoint. Open the *Presentation to the President Solution* that you created earlier. PowerPoint detects that a change has been made in the underlying workbook and prompts you to update. Click the button to update the links.
   e. Change to the Slide Sorter view. Press and hold the Shift key as you select the slides containing the Excel worksheet and chart. Pull down the File menu, click the Print command, and print the selection (these two slides) as audience handouts, two slides per page. (Be sure to check the box to frame the slides.)
   f. Save the presentation. Exit PowerPoint.

FIGURE 4 Last-minute Change

INTEGRATED CASE STUDIES 5

Companion Web site

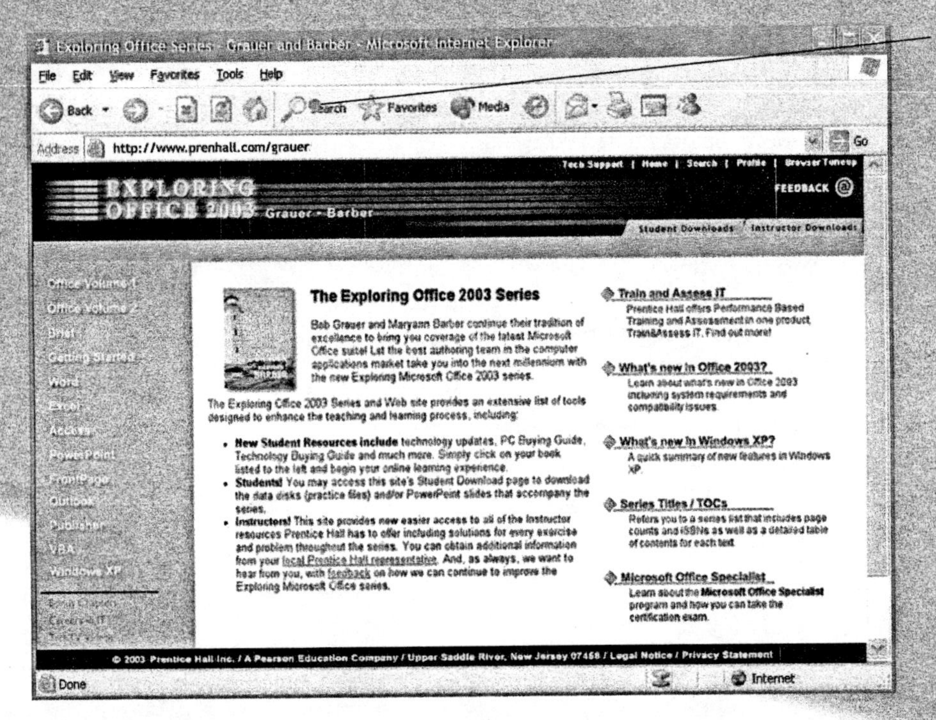

New!

Updated and enhanced Companion Web site. Find everything you need—student practice files, PowerPoint lectures, online study guides, and instructor support (solutions)!

www.prenhall.com/grauer

Acknowledgments

We want to thank the many individuals who have helped to bring this project to fruition. Jodi McPherson, executive acquisitions editor at Prentice Hall, has provided new leadership in extending the series to Office 2003. Cathi Profitko did an absolutely incredible job on our Web site. Shelly Martin was the creative force behind the chapter-opening case studies. Emily Knight coordinated the marketing and continues to inspire us with suggestions for improving the series. Greg Hubit has been masterful as the external production editor for every book in the series from its inception. Eileen Clark coordinated the myriad details of production and the certification process. Lynne Breitfeller was the project manager and manufacturing buyer. Lori Johnson was the project manager at The GTS Companies and in charge of composition. Chuck Cox did his usual fine work as copyeditor. Melissa Edwards was the supplements editor. Cindy Stevens, Tom McKenzie, and Michael Olmstead wrote the instructor manuals. Michael Fruhbeis developed the innovative and attractive design. We also want to acknowledge our reviewers who, through their comments and constructive criticism, greatly improved the series.

Lynne Band, Middlesex Community College
Don Belle, Central Piedmont Community College
Stuart P. Brian, Holy Family College
Carl M. Briggs, Indiana University School of Business
Kimberly Chambers, Scottsdale Community College
Jill Chapnick, Florida International University
Alok Charturvedi, Purdue University
Jerry Chin, Southwest Missouri State University
Dean Combellick, Scottsdale Community College
Cody Copeland, Johnson County Community College
Larry S. Corman, Fort Lewis College
Janis Cox, Tri-County Technical College
Douglass Cross, Clackamas Community College
Martin Crossland, Southwest Missouri State University
Bill Daley, University of Oregon
Paul E. Daurelle, Western Piedmont Community College
Shawna DePlonty, Sault College of Applied Arts and Technology
Carolyn DiLeo, Westchester Community College
Judy Dolan, Palomar College
David Douglas, University of Arkansas
Carlotta Eaton, Radford University
Judith M. Fitspatrick, Gulf Coast Community College
James Franck, College of St. Scholastica
Raymond Frost, Central Connecticut State University
Susan Fry, Boise State University
Midge Gerber, Southwestern Oklahoma State University
James Gips, Boston College
Vernon Griffin, Austin Community College
Ranette Halverson, Midwestern State University
Michael Hassett, Fort Hays State University
Mike Hearn, Community College of Philadelphia
Wanda D. Heller, Seminole Community College

Bonnie Homan, San Francisco State University
Ernie Ivey, Polk Community College
Walter Johnson, Community College of Philadelphia
Mike Kelly, Community College of Rhode Island
Jane King, Everett Community College
Rose M. Laird, Northern Virginia Community College
David Langley, University of Oregon
John Lesson, University of Central Florida
Maurie Lockley, University of North Carolina at Greensboro
Daniela Marghitu, Auburn University
David B. Meinert, Southwest Missouri State University
Alan Moltz, Naugatuck Valley Technical Community College
Kim Montney, Kellogg Community College
Bill Morse, DeVry Institute of Technology
Kevin Pauli, University of Nebraska
Mary McKenry Percival, University of Miami
Marguerite Nedreberg, Youngstown State University
Jim Pruitt, Central Washington University
Delores Pusins, Hillsborough Community College
Gale E. Rand, College Misericordia
Judith Rice, Santa Fe Community College
David Rinehard, Lansing Community College
Marilyn Salas, Scottsdale Community College
Herach Safarian, College of the Canyons
John Shepherd, Duquesne University
Barbara Sherman, Buffalo State College
Robert Spear, Prince George's Community College
Michael Stewardson, San Jacinto College—North
Helen Stoloff, Hudson Valley Community College
Margaret Thomas, Ohio University
Mike Thomas, Indiana University School of Business
Suzanne Tomlinson, Iowa State University
Karen Tracey, Central Connecticut State University
Antonio Vargas, El Paso Community College
Sally Visci, Lorain County Community College
David Weiner, University of San Francisco
Connie Wells, Georgia State University
Wallace John Whistance-Smith, Ryerson Polytechnic University
Jack Zeller, Kirkwood Community College

A final word of thanks to the unnamed students at the University of Miami who make it all worthwhile. Most of all, thanks to you, our readers, for choosing this book. Please feel free to contact us with any comments and suggestions.

Robert T. Grauer
rgrauer@miami.edu
www.prenhall.com/grauer

Maryann Barber
mbarber@miami.edu

CHAPTER 1

Getting Started with VBA: Extending Microsoft Office 2003

OBJECTIVES

After reading this chapter you will:

1. Describe the relationship of VBA to Microsoft Office 2003.
2. Explain how to create, edit, and run a VBA procedure.
3. Use the MsgBox statement and InputBox function.
4. Explain how to debug a procedure by stepping through its statements.
5. Use the If . . . Then . . . Else statement to implement a decision.
6. Explain the Case statement.
7. Create a custom toolbar.
8. Describe several statements used to implement a loop.
9. Describe event-driven programming.

hands-on exercises

1. INTRODUCTION TO VBA
   Input: None
   Output: VBA workbook
2. DECISION MAKING
   Input: VBA workbook
   Output: VBA workbook
3. LOOPS AND DEBUGGING
   Input: VBA workbook
   Output: VBA workbook
4. EVENT-DRIVEN PROGRAMMING
   Input: VBA workbook; Financial Consultant workbook
   Output: VBA workbook; Financial Consultant workbook
5. EVENT-DRIVEN PROGRAMMING
   Input: VBA Switchboard and Security database
   Output: VBA Switchboard and Security database

CASE STUDY

ON-THE-JOB TRAINING

Your first job is going exceedingly well. The work is very challenging and your new manager, Phyllis Simon, is impressed with the Excel workbooks that you have developed thus far. Phyllis has asked you to take it to the next level by incorporating VBA procedures into future projects. You have some knowledge of Excel macros and have already used the macro recorder to record basic macros. You are able to make inferences about the resulting code, but you will need additional proficiency in VBA to become a true expert in Excel.

The good news is that you work for a company that believes in continuing education and promotes from within. Phyllis has assigned you to a new interdepartmental team responsible for creating high-level Excel applications that will be enhanced through VBA. Moreover, you have been selected to attend a week-long seminar to learn VBA so that you can become a valued member of the team. The seminar will be held in San Diego, California, where there is a strong temptation to study sand and surf rather than VBA. Thus, Phyllis expects you to complete a series of VBA procedures upon your return—just to be sure that you were not tempted to skip class and dip your toes in the water. ■

Your assignment is to read the VBA primer at the end of the text and focus on the first three hands-on exercises that develop the syntax for basic VBA statements—MsgBox, InputBox, decision making through If/Else and Case statements, and iteration through the For . . . Next and Do Until statements. You will then open the partially completed *VBA Case Study—On-the-Job Training*, start the VBA editor, and then complete the tasks presented in the procedures in Module1. (The requirements for each procedure appear as comments within the procedure.) Add a command button for each macro to the Excel workbook, and then print the worksheet and a copy of the completed module for your instructor. Last, but not least, create a suitable event procedure for closing the workbook.

INTRODUCTION TO VBA

Visual Basic for Applications (VBA) is a powerful programming language that is accessible from all major applications in Microsoft Office XP. You do not have to know VBA to use Office effectively, but even a basic understanding will help you to create more powerful documents. Indeed, you may already have been exposed to VBA through the creation of simple macros in Word or Excel. A ***macro*** is a set of instructions (i.e., a program) that simplifies the execution of repetitive tasks. It is created through the ***macro recorder*** that captures commands as they are executed, then converts those commands into a VBA program. (The macro recorder is present in Word, Excel, and PowerPoint, but not in Access.) You can create and execute macros without ever looking at the underlying VBA, but you gain an appreciation for the language when you do.

The macro recorder is limited, however, in that it captures only commands, mouse clicks, and/or keystrokes. As you will see, VBA is much more than just recorded keystrokes. It is a language unto itself, and thus, it contains all of the statements you would expect to find in any programming language. This lets you enhance the functionality of any macro by adding extra statements as necessary—for example, an InputBox function to accept data from the user, followed by an If . . . Then . . . Else statement to take different actions based on the information supplied by the user.

This supplement presents the rudiments of VBA and is suitable for use with any Office application. We begin by describing the VBA editor and how to create, edit, and run simple procedures. The examples are completely general and demonstrate the basic capabilities of VBA that are found in any programming language. We illustrate the MsgBox statement to display output to the user and the InputBox function to accept input from the user. We describe the For . . . Next statement to implement a loop and the If . . . Then . . . Else and Case statements for decision making. We also describe several debugging techniques to help you correct the errors that invariably occur. The last two exercises introduce the concept of event-driven programming, in which a procedure is executed in response to an action taken by the user. The material here is application-specific in conjunction with Excel and Access, but it can be easily extended to Word or PowerPoint.

One last point before we begin is that this supplement assumes no previous knowledge on the part of the reader. It is suitable for someone who has never been exposed to a programming language or written an Office macro. If, on the other hand, you have a background in programming or macros, you will readily appreciate the power inherent in VBA. VBA is an incredibly rich language that can be daunting to the novice. Stick with us, however, and we will show you that it is a flexible and powerful tool with consistent rules that can be easily understood and applied. You will be pleased at what you will be able to accomplish.

VBA is a programming language, and like any other programming language its programs (or procedures, as they are called) are made up of individual statements. Each ***statement*** accomplishes a specific task such as displaying a message to the user or accepting input from the user. Statements are grouped into ***procedures***, and procedures, in turn, are grouped into ***modules***. Every VBA procedure is classified as either public or private. A ***private procedure*** is accessible only from within the module in which it is contained. A ***public procedure***, on the other hand, can be accessed from any module.

The statement, however, is the basic unit of the language. Our approach throughout this supplement will be to present individual statements, then to develop simple procedures using those statements in a hands-on exercise. As you read the discussion, you will see that every statement has a precise ***syntax*** that describes how the statement is to be used. The syntax also determines the ***arguments*** (or parameters) associated with that statement, and whether those arguments are required or optional.

THE MSGBOX STATEMENT

The ***MsgBox statement*** displays information to the user. It is one of the most basic statements in VBA, but we use it to illustrate several concepts in VBA programming. Figure 1a contains a simple procedure called MsgBoxExamples, consisting of four individual MsgBox statements. All procedures begin with a ***procedure header*** and end with the ***End Sub statement.***

The MsgBox statement has one required argument, which is the message (or prompt) that is displayed to the user. All other arguments are optional, but if they are used, they must be entered in a specified sequence. The simplest form of the MsgBox statement is shown in example 1, which specifies a single argument that contains the text (or prompt) to be displayed. The resulting message box is shown in Figure 1b. The message is displayed to the user, who responds accordingly, in this case by clicking the OK button.

Example 2 extends the MsgBox statement to include a second parameter that displays an icon within the resulting dialog box as shown in Figure 1c. The type of icon is determined by a VBA ***intrinsic*** (or predefined) ***constant*** such as vbExclamation, which displays an exclamation point in a yellow triangle. VBA has many such constants that enable you to simplify your code, while at the same time achieving some impressive results.

Example 3 uses a different intrinsic constant, vbInformation, to display a different icon. It also extends the MsgBox statement to include a third parameter that is displayed on the title bar of the resulting dialog box. Look closely, for example, at Figures 1c and 1d, whose title bars contain "Microsoft Excel" and "Grauer/Barber", respectively. The first is the default entry (given that we are executing the procedure from within Microsoft Excel). You can, however, give your procedures a customized look by displaying your own text in the title bar.

Procedure header

End Sub statement

```
Public Sub MsgBoxExamples()
'This procedure was written by John Doe on 6/10/2003

    MsgBox "Example 1 - VBA is not difficult"
    MsgBox "Example 2 - VBA is not difficult", vbExclamation
    MsgBox "Example 3 - VBA is not difficult", vbInformation, "Grauer/Barber"
    MsgBox "Example 4 - VBA is not difficult", , "Your name goes here"

End Sub
```

(a) VBA Code

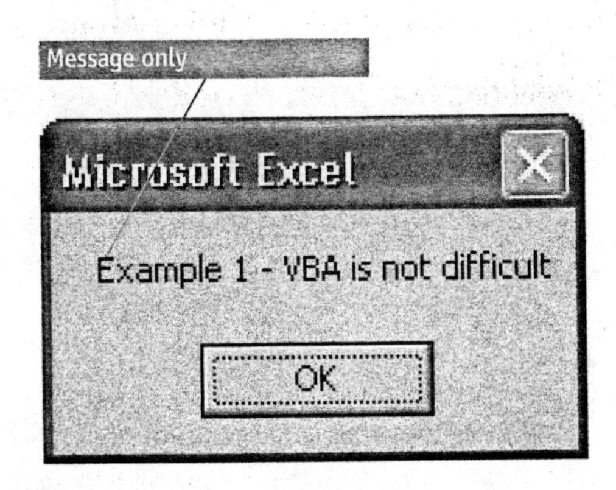

(b) Example 1—One Argument

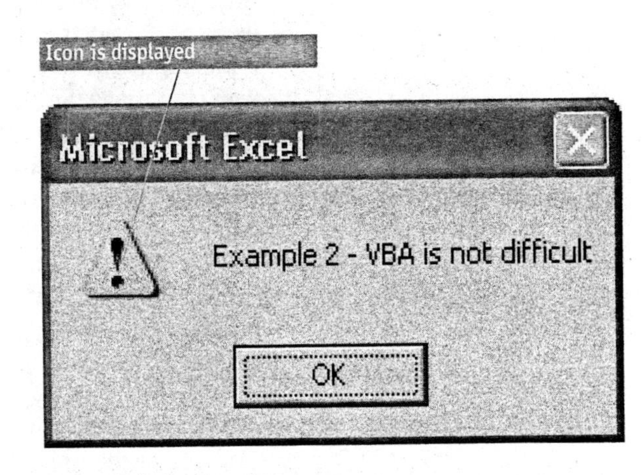

(c) Example 2—Two Arguments

FIGURE 1 The MsgBox Statement

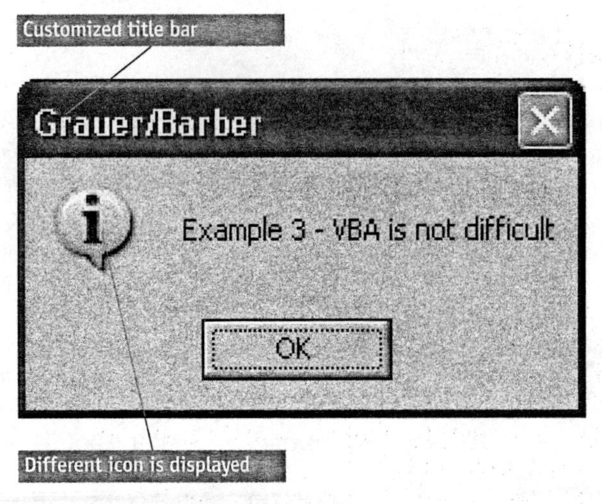

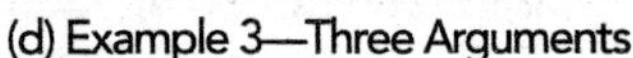

(d) Example 3—Three Arguments

(e) Example 4—Omitted Parameter

FIGURE 1 The MsgBox Statement (*continued*)

Example 4 omits the second parameter (the icon), but includes the third parameter (the entry for the title bar). The parameters are positional, however, and thus the MsgBox statement contains two commas after the message to indicate that the second parameter has been omitted.

THE INPUTBOX FUNCTION

The MsgBox statement displays a prompt to the user, but what if you want the user to respond to the prompt by entering a value such as his or her name? This is accomplished using the ***InputBox function***. Note the subtle change in terminology in that we refer to the InputBox *function*, but the MsgBox *statement*. That is because a function returns a value, in this case the user's name, which is subsequently used in the procedure. In other words, the InputBox function asks the user for information, then it stores that information (the value returned by the user) for use in the procedure.

Figure 2 displays a procedure that prompts the user for a first and last name, after which it displays the information using the MsgBox statement. (The Dim statement at the beginning of the procedure is explained shortly.) Let's look at the first InputBox function, and the associated dialog box in Figure 2b. The InputBox function displays a prompt on the screen, the user enters a value ("Bob" in this example), and that value is stored in the variable that appears to the left of the equal sign (strFirstName). The concept of a variable is critical to every programming language. Simply stated, a ***variable*** is a named storage location that contains data that can be modified during program execution.

The MsgBox statement then uses the value of strFirstName to greet the user by name as shown in Figure 2c. This statement also introduces the ampersand to ***concatenate*** (join together) two different character strings, the literal "Good morning", followed by the value within the variable strFirstName.

The second InputBox function prompts the user for his or her last name. In addition, it uses a second argument to customize the contents of the title bar (VBA Primer in this example) as can be seen in Figure 2d. Finally, the MsgBox statement in Figure 2e displays both the first and last name through concatenation of multiple strings. This statement also uses the ***underscore*** to continue a statement from one line to the next.

VBA is not difficult, and you can use the MsgBox statement and InputBox function in conjunction with one another as the basis for several meaningful procedures. You will get a chance to practice in the hands-on exercise that follows shortly.

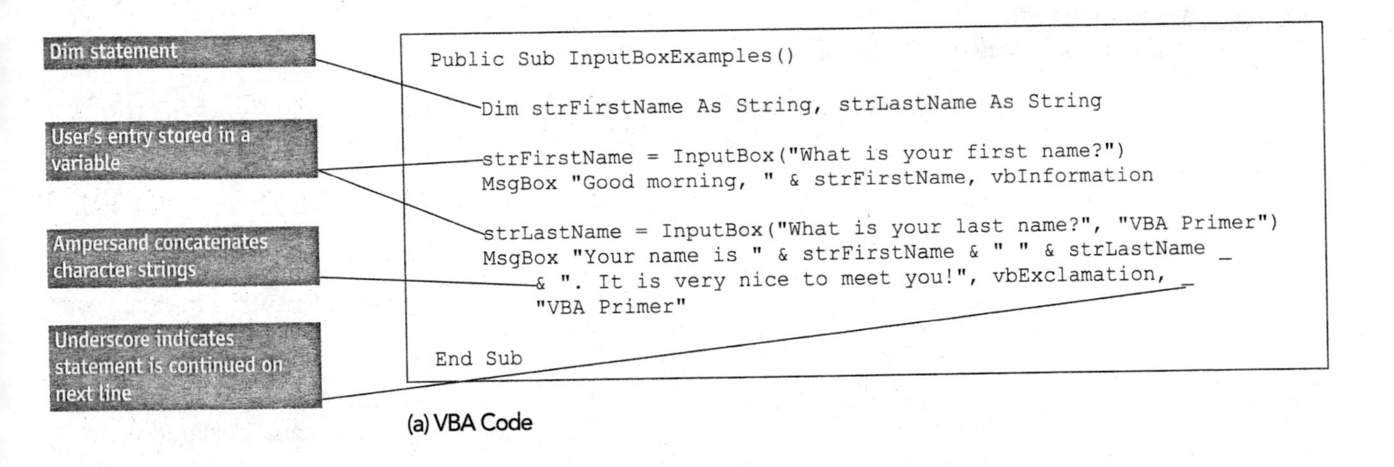


```
Public Sub InputBoxExamples()

    Dim strFirstName As String, strLastName As String

    strFirstName = InputBox("What is your first name?")
    MsgBox "Good morning, " & strFirstName, vbInformation

    strLastName = InputBox("What is your last name?", "VBA Primer")
    MsgBox "Your name is " & strFirstName & " " & strLastName _
        & ". It is very nice to meet you!", vbExclamation, _
        "VBA Primer"

End Sub
```

(a) VBA Code

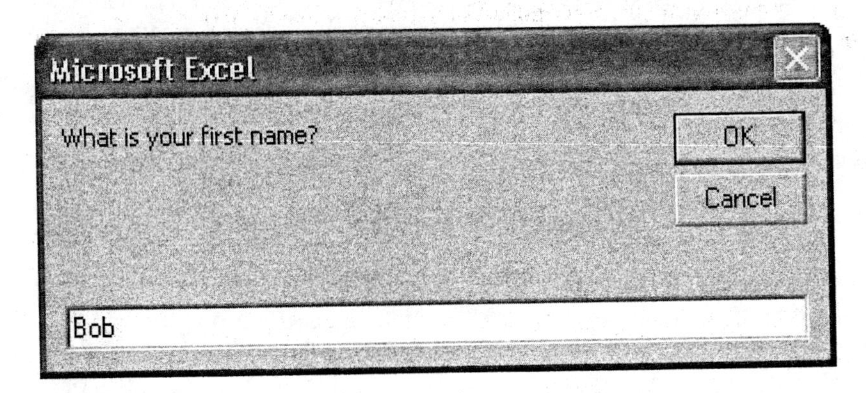

(b) InputBox

(c) Concatenation

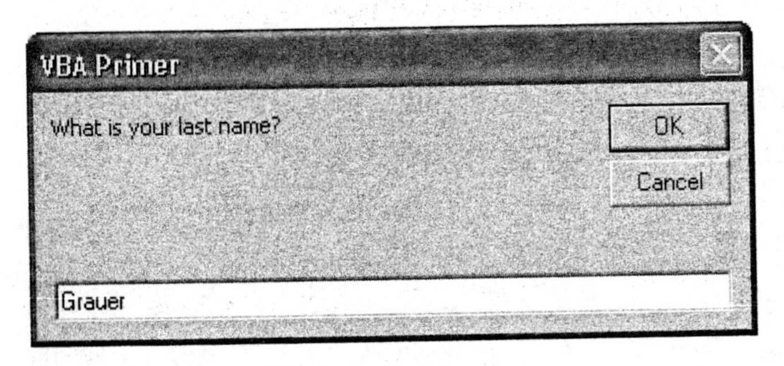

(d) Input Box Includes Argument for Title Bar

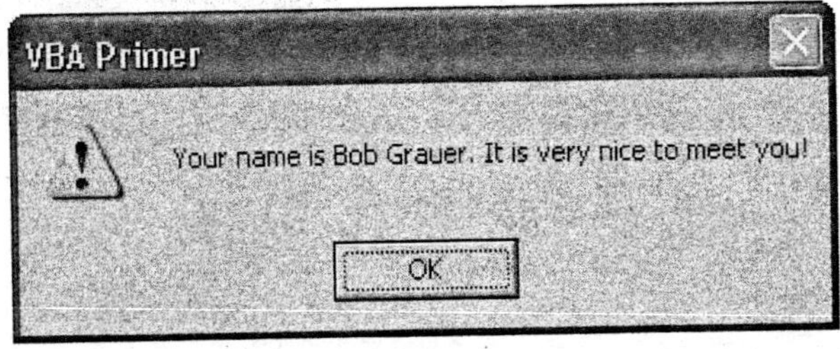

(e) Concatenation and Continuation

FIGURE 2 The InputBox Function

Declaring Variables

Every variable must be declared (defined) before it can be used. This is accomplished through the ***Dim*** (short for Dimension) ***statement*** that appears at the beginning of a procedure. The Dim statement indicates the name of the variable and its type (for example, whether it will hold characters or numbers), which in turn reserves the appropriate amount of memory for that variable.

A variable name must begin with a letter and cannot exceed 255 characters. It can contain letters, numbers, and various special characters such as an underscore, but it cannot contain a space or the special symbols !, @, &, $, or #. Variable names typically begin with a prefix to indicate the type of data that is stored within the variable such as "str" for a character string or "int" for integers. The use of a prefix is optional with respect to the rules of VBA, but it is followed almost universally.

THE VBA EDITOR

All VBA procedures are created using the ***Visual Basic editor*** as shown in Figure 3. You may already be familiar with the editor, perhaps in conjunction with creating and/or editing macros in Word or Excel, or event procedures in Microsoft Access. Let's take a moment, however, to review its essential components.

The left side of the editor displays the ***Project Explorer***, which is similar in concept and appearance to the Windows Explorer, except that it displays the objects associated with the open document. If, for example, you are working in Excel, you will see the various sheets in a workbook, whereas in an Access database you will see forms and reports.

The VBA statements for the selected module (Module1 in Figure 3) appear in the code window in the right pane. The module, in turn, contains declarations and procedures that are separated by horizontal lines. There are two procedures, MsgBoxExamples and InputBoxExamples, each of which was explained previously. A ***comment*** (nonexecutable) statement has been added to each procedure and appears in green. It is the apostrophe at the beginning of the line, rather than the color, that denotes a comment.

The ***Declarations section*** appears at the beginning of the module and contains a single statement, ***Option Explicit***. This option requires every variable in a procedure to be explicitly defined (e.g., in a Dim statement) before it can be used elsewhere in the module. It is an important option and should appear in every module you write.

The remainder of the window should look reasonably familiar in that it is similar to any other Office application. The title bar appears at the top of the window and identifies the application (Microsoft Visual Basic) and the current document (VBA Examples.xls). The right side of the title bar contains the Minimize, Restore, and Close buttons. A menu bar appears under the title bar. Toolbars are displayed under the menu bar. Commands are executed by pulling down the appropriate menu, via buttons on the toolbar, or by keyboard shortcuts.

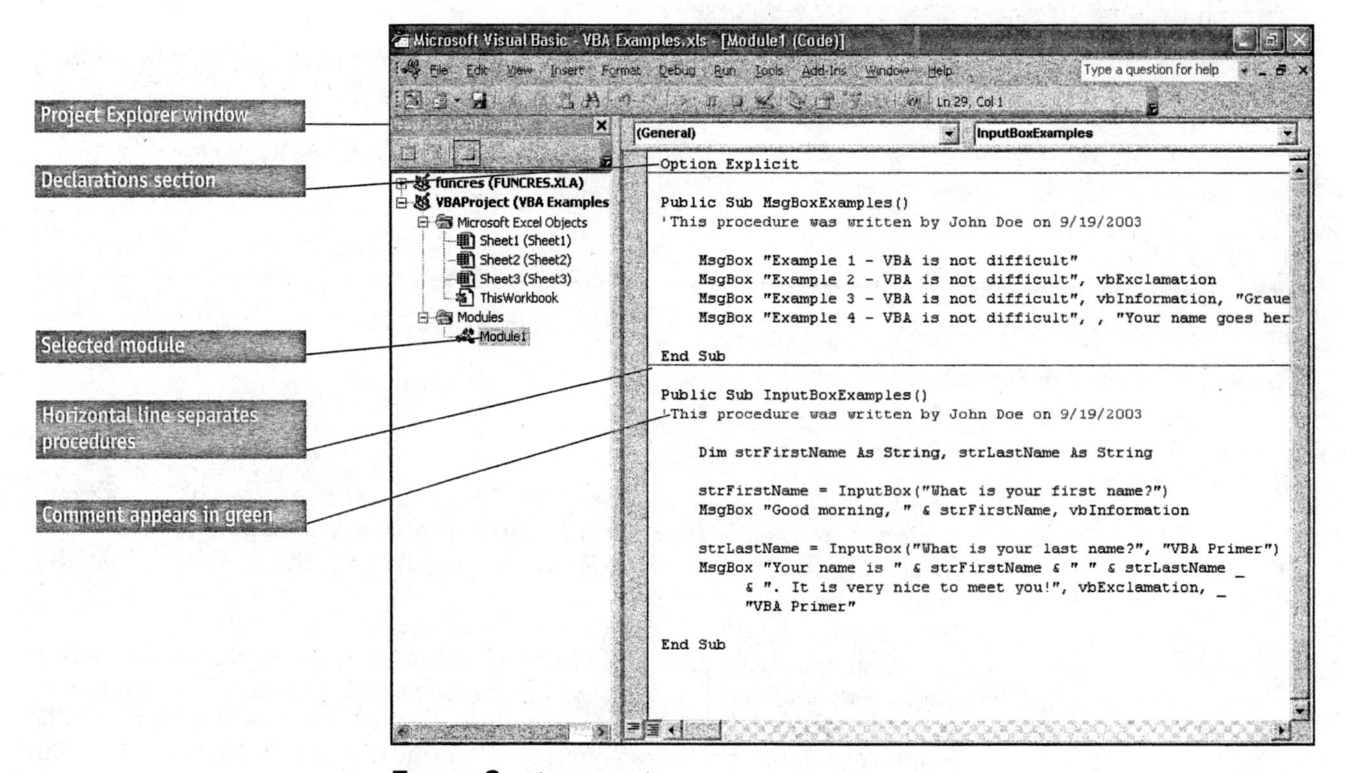

FIGURE 3 The VBA Editor

hands-on exercise

1 Introduction to VBA

Objective To create and test VBA procedures using the MsgBox and InputBox statements. Use Figure 4 as a guide in the exercise. You can do the exercise in any Office application.

Step 1a: Start Microsoft Excel

- We suggest you do the exercise in either Excel or Access (although you could use Word or PowerPoint just as easily). Go to step 1b for Access.
- Start **Microsoft Excel** and open a new workbook. Pull down the **File menu** and click the **Save command** (or click the **Save button** on the Standard toolbar) to display the Save As dialog box. Choose an appropriate drive and folder, then save the workbook as **VBA Examples**.
- Pull down the **Tools menu**, click the **Macro command**, then click the **Visual Basic Editor command** as shown in Figure 4a. Go to step 2.

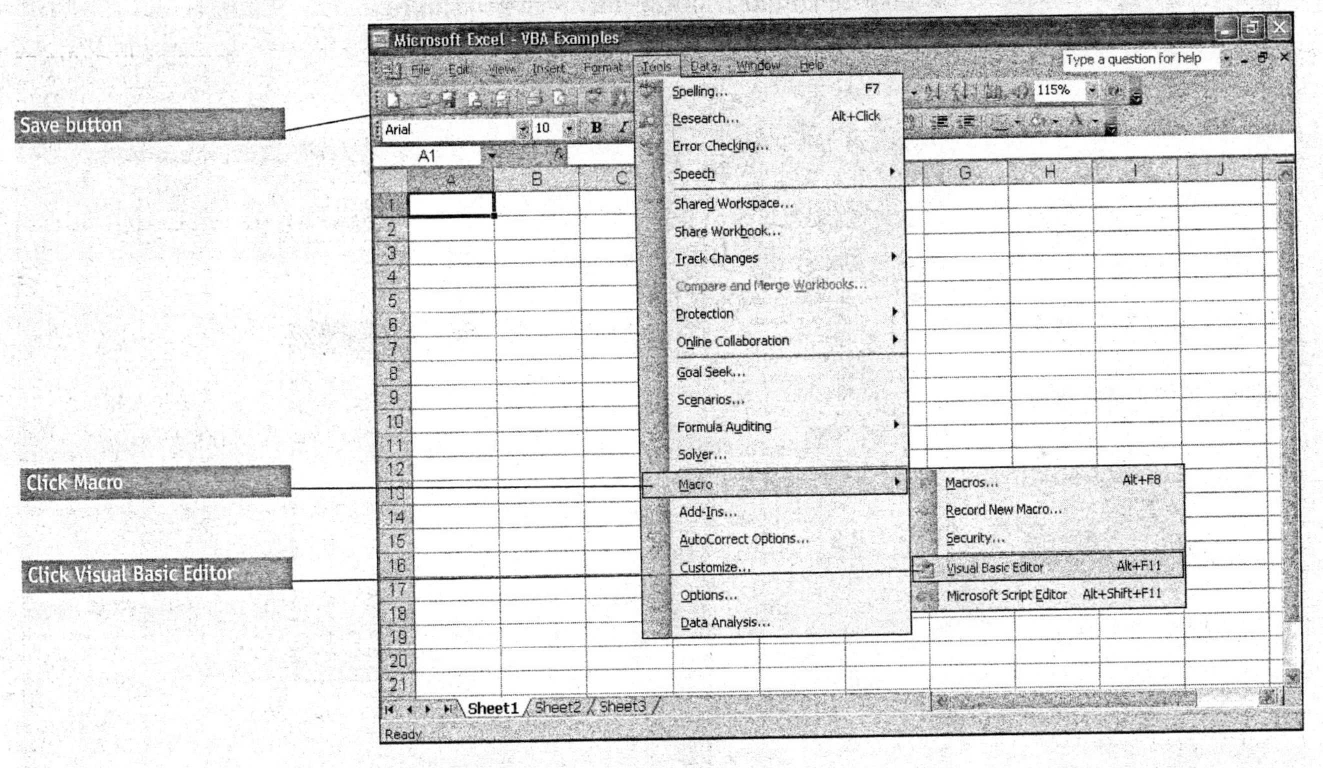

(a) Start Microsoft Excel (step 1a)

FIGURE 4 Hands-on Exercise 1

Step 1b: Start Microsoft Access

- Start **Microsoft Access** and choose the option to create a **Blank Access database**. Save the database as **VBA Examples**.
- Pull down the **Tools menu**, click the **Macro command**, then click the **Visual Basic Editor command.** (You can also use the **Alt+F11** keyboard shortcut to open the VBA editor without going through the Tools menu.)

Step 2: Insert a Module

- You should see a window similar to Figure 4b, but Module1 is not yet visible. Close the Properties window if it appears.
- If necessary, pull down the **View menu** and click **Project Explorer** to display the Project Explorer pane at the left of the window. Our figure shows Excel objects, but you will see the "same" window in Microsoft Access.
- Pull down the **Insert menu** and click **Module** to insert Module1 into the current project. The name of the module, Module1 in this example, appears in the Project Explorer pane.
- The Option Explicit statement may be entered automatically, but if not, click in the code window and type the statement **Option Explicit**.
- Pull down the **Insert menu** a second time, but this time select **Procedure** to display the Add Procedure dialog box in Figure 4b. Click in the **Name** text box and enter **MsgBoxExamples** as the name of the procedure. (Spaces are not allowed in a procedure name.)
- Click the option buttons for a **Sub procedure** and for **Public scope**. Click **OK**. The sub procedure should appear within the module and consist of the Sub and End Sub statements.

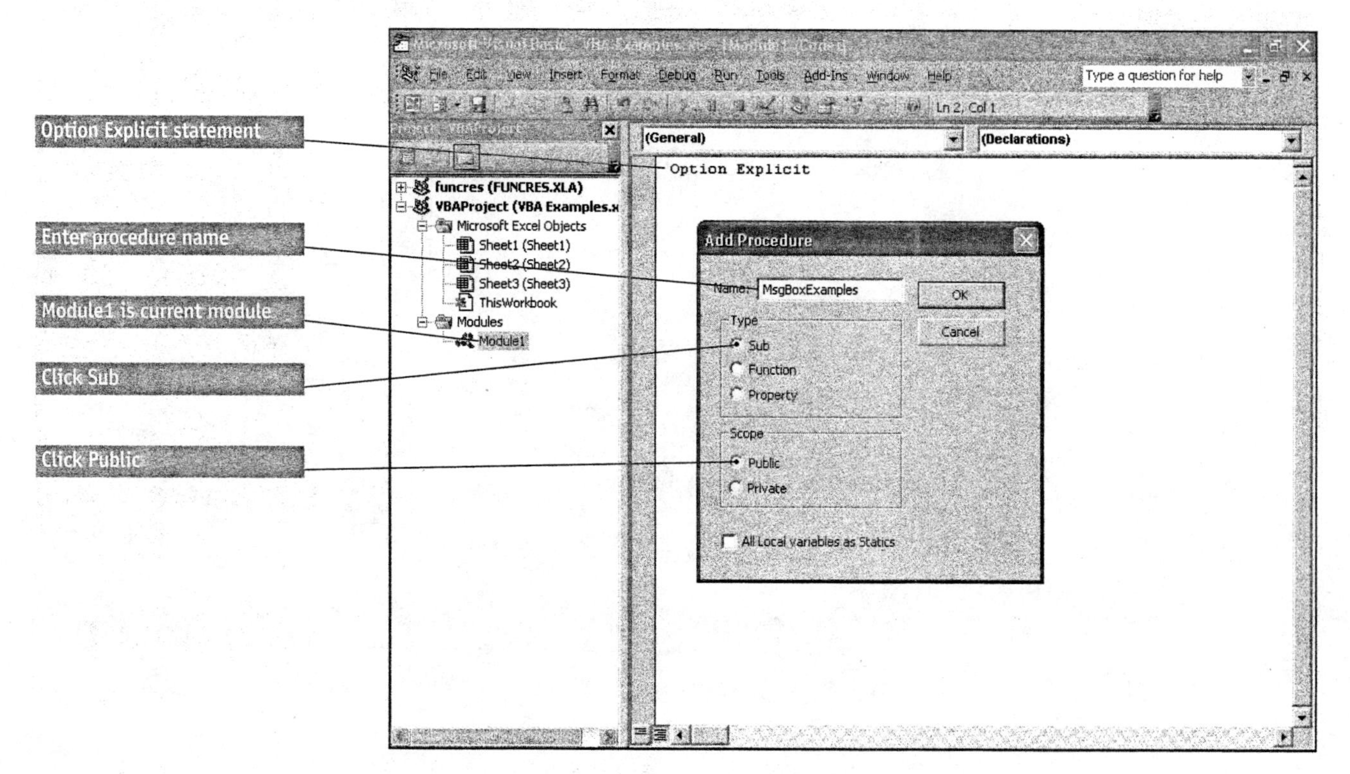

(b) Insert a Module (step 2)

FIGURE 4 Hands-on Exercise 1 (*continued*)

THE OPTION EXPLICIT STATEMENT

The Option Explicit statement is optional, but if it is used it must appear in a module before any procedures. The statement requires that all variables in the module be declared explicitly by the programmer (typically with a Dim, Public, or Private statement), as opposed to VBA making an implicit assumption about the variable. It is good programming practice and it should be used every time.

Step 3: The MsgBox Statement

- The insertion point (the flashing cursor) appears below the first statement. Press the **Tab key** to indent the next statement. (Indentation is not a VBA requirement, but is used to increase the readability of the statement.)
- Type the keyword **MsgBox**, then press the **space bar**. VBA responds with Quick Info that displays the syntax of the statement as shown in Figure 4c.
- Type a **quotation mark** to begin the literal, enter the text of your message, **This is my first VBA procedure**, then type the closing **quotation mark**.
- Click the **Run Sub button** on the Standard toolbar (or pull down the **Run menu** and click the **Run Sub command**) to execute the procedure.
- You should see a dialog box, containing the text you entered, within the Excel workbook (or other Office document) on which you are working.
- After you have read the message, click **OK** to return to the VBA editor.

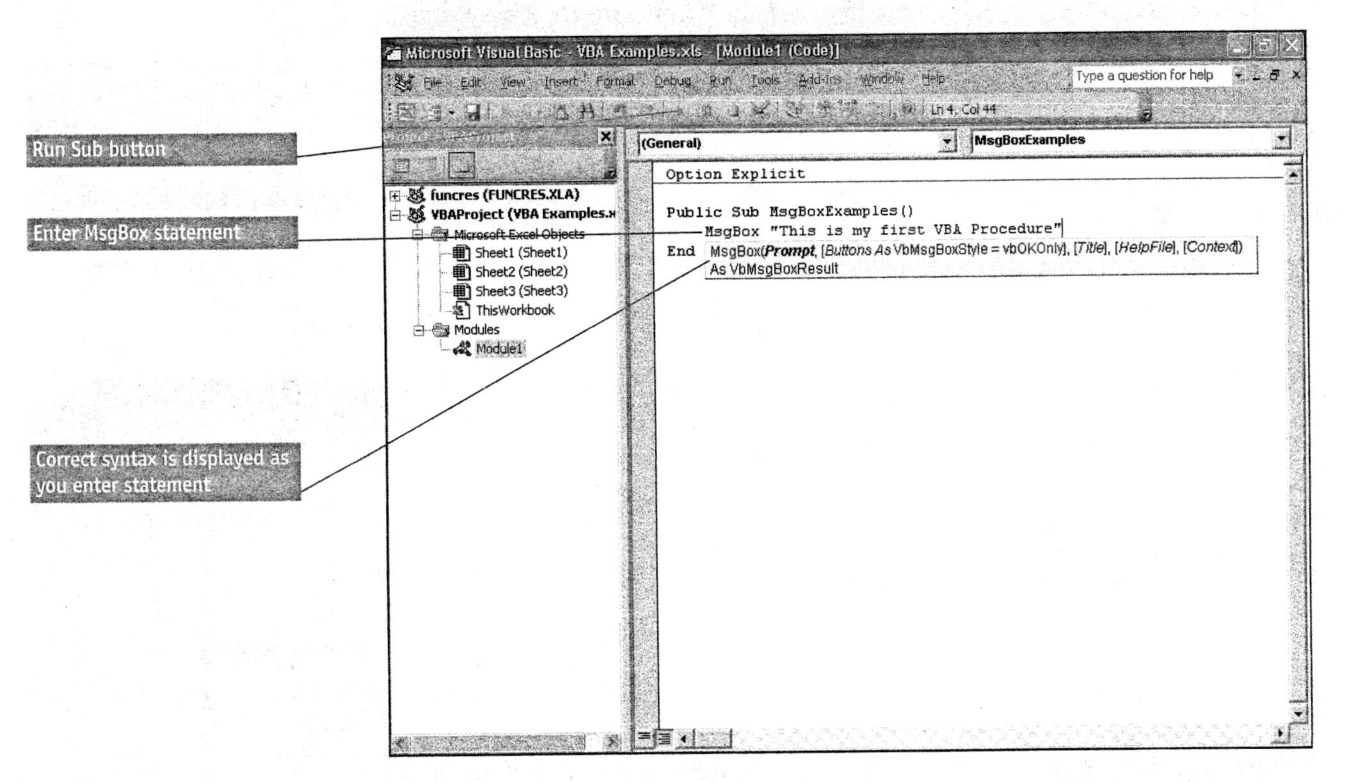

(c) The MsgBox Statement (step 3)

FIGURE 4 Hands-on Exercise 1 (*continued*)

QUICK INFO—HELP WITH VBA SYNTAX

Press the space bar after entering the name of a statement (e.g., MsgBox), and VBA responds with a Quick Info box that displays the syntax of the statement. You see the arguments in the statement and the order in which those arguments appear. Any argument in brackets is optional. If you do not see this information, pull down the Tools menu, click the Options command, then click the Editor tab. Check the box for Auto Quick Info and click OK.

Step 4: **Complete the Procedure**

- You should be back within the MsgBoxExamples procedure. If necessary, click at the end of the MsgBox statement, then press **Enter** to begin a new line. Type **MsgBox** and press the **space bar** to begin entering the statement.
- The syntax of the MsgBox statement will appear on the screen. Type a **quotation mark** to begin the message, type **Add an icon** as the text of this message, then type the closing **quotation mark**. Type a **comma**, then press the **space bar** to enter the next parameter.
- VBA automatically displays a list of appropriate parameters, in this case a series of intrinsic constants that define the icon or command button that is to appear in the statement.
- You can type the first several letters (e.g., **vbi**, for vbInformation), then press the **space bar**, or you can use the **down arrow** to select **vbInformation** and then press the **space bar**. Either way you should complete the second MsgBox statement as shown in Figure 4d. Press **Enter**.
- Enter the third MsgBox statement as shown in Figure 4d. Note the presence of the two consecutive commas to indicate that we omitted the second parameter within the MsgBox statement. Enter your name instead of John Doe where appropriate. Press **Enter**.
- Enter the fourth (and last) MsgBox statement following our figure. Select **vbExclamation** as the second parameter, type a **comma**, then enter the text of the title bar, as you did for the previous statement.
- Click the **Save button** to save the changes to the module.

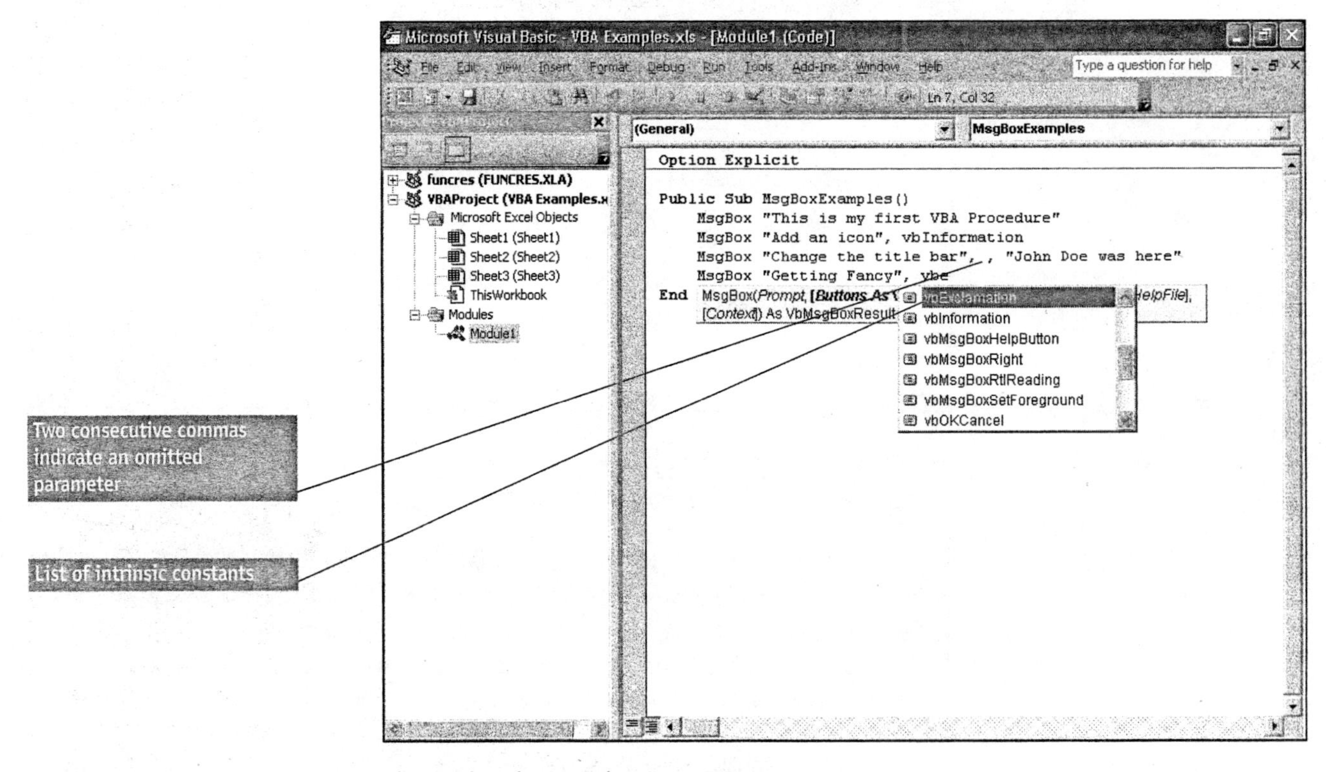

(d) Complete the Procedure (step 4)

FIGURE 4 Hands-on Exercise 1 (*continued*)

Step 5: Test the Procedure

- It's convenient if you can see the statements in the VBA procedure at the same time you see the output of those statements. Thus we suggest that you tile the VBA editor and the associated Office application.
- Minimize all applications except the VBA editor and the Office application (e.g., Excel).
- Right click the taskbar and click **Tile Windows Horizontally** to tile the windows as shown in Figure 4e. (It does not matter which window is on top. (If you see more than these two windows, minimize the other open window, then right click the taskbar and retile the windows.)
- Click anywhere in the VBA procedure, then click the **Run Sub button** on the Standard toolbar.
- The four messages will be displayed one after the other. Click **OK** after each message.
- Maximize the VBA window to continue working.

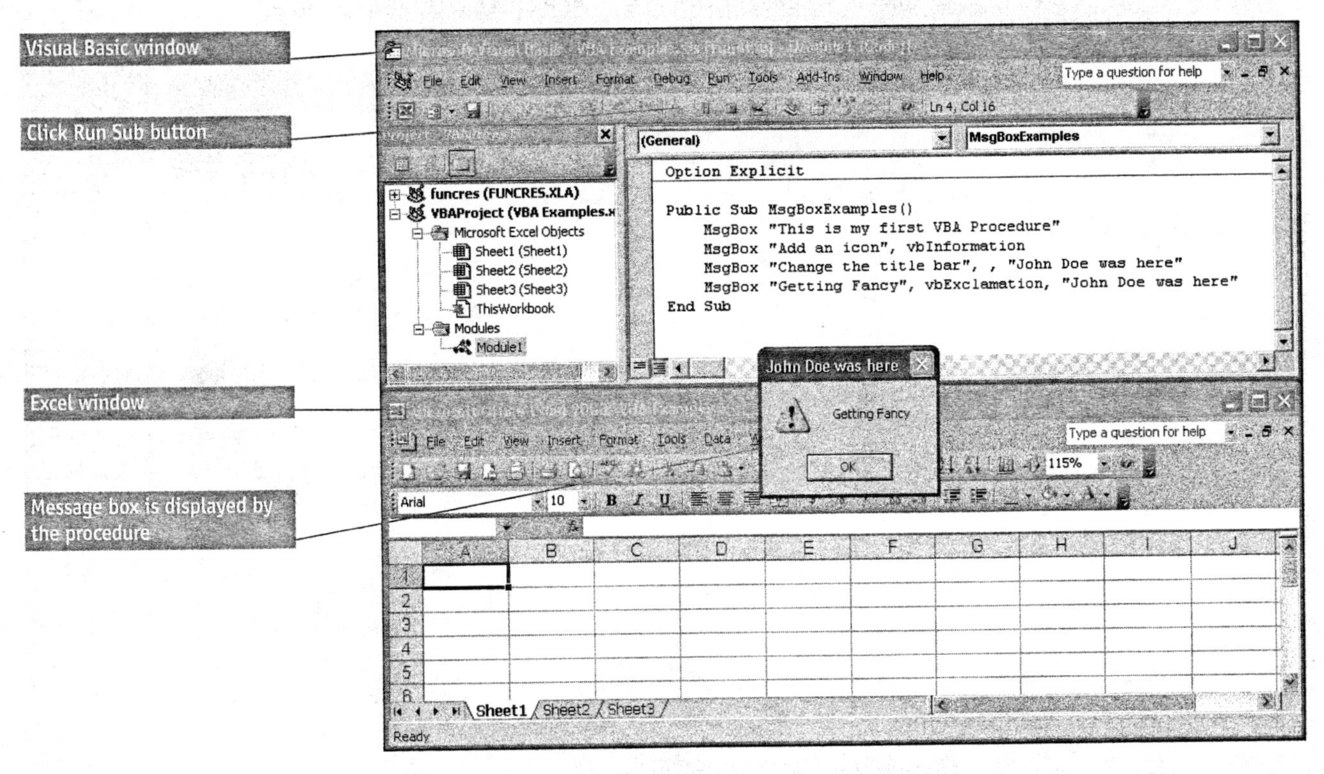

(e) Test the Procedure (step 5)

FIGURE 4 Hands-on Exercise 1 (*continued*)

HIDE THE WINDOWS TASKBAR

You can hide the Windows taskbar to gain additional space on the desktop. Right click any empty area of the taskbar to display a context-sensitive menu, click Properties to display the Taskbar properties dialog box, and if necessary click the Taskbar tab. Check the box to Auto Hide the taskbar, then click OK. The taskbar disappears from the screen but will reappear as you point to the bottom edge of the desktop.

Step 6: Comments and Corrections

- All VBA procedures should be documented with the author's name, date, and other comments as necessary to explain the procedure. Click after the procedure header. Press the **Enter key** to leave a blank line.
- Press **Enter** a second time. Type an **apostrophe** to begin the comment, then enter a descriptive statement similar to Figure 4f. Press **Enter** when you have completed the comment. The line turns green to indicate it is a comment.
- The best time to experiment with debugging is when you know your procedure is correct. Go to the last MsgBox statement and delete the quotation mark in front of your name. Move to the end of the line and press **Enter**.
- You should see the error message in Figure 4f. Unfortunately, the message is not as explicit as it could be; VBA cannot tell that you left out a quotation mark, but it does detect an error in syntax.
- Click **OK** in response to the error. Click the **Undo button** twice, to restore the quotation mark, which in turn corrects the statement.
- Click the **Save button** to save the changes to the module.

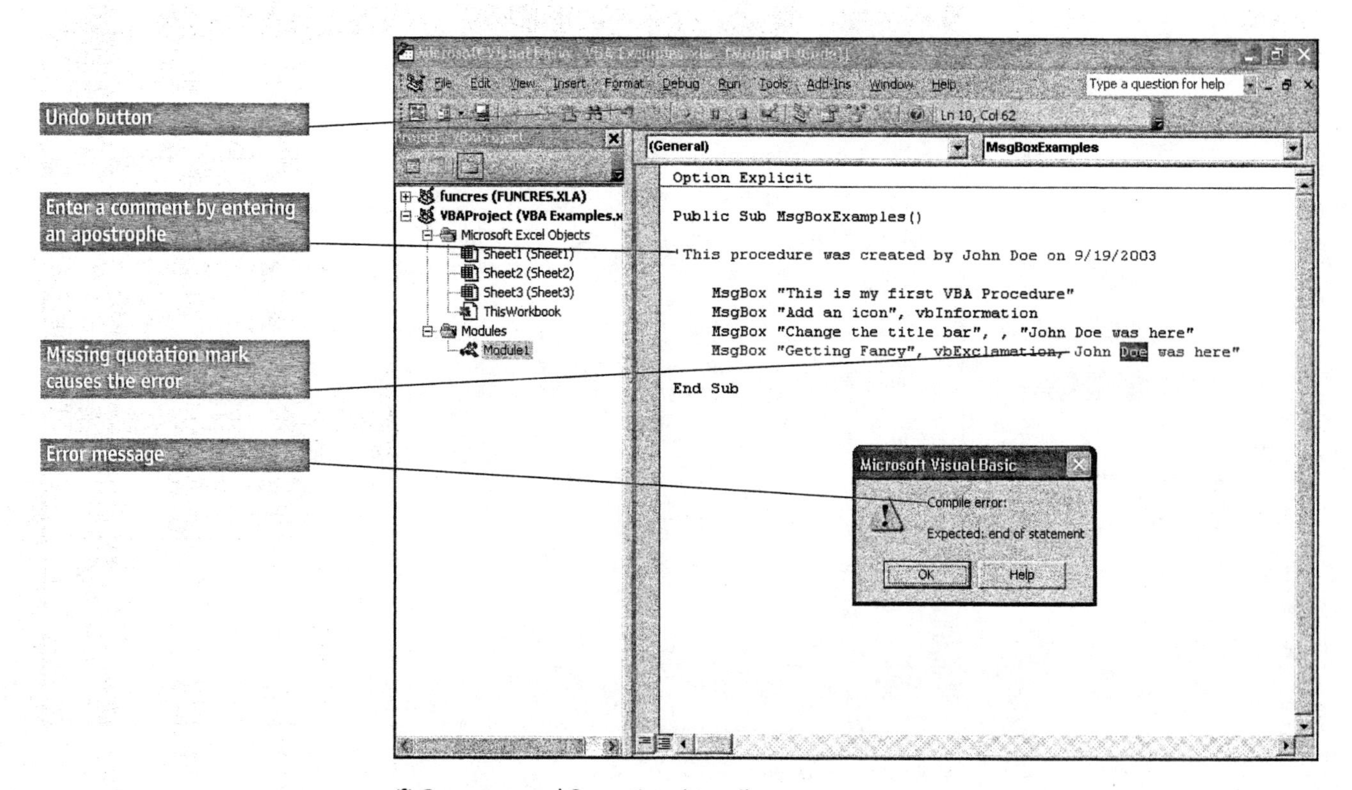

(f) Comments and Corrections (step 6)

FIGURE 4 Hands-on Exercise 1 (*continued*)

RED, GREEN, AND BLUE

Visual Basic for Applications uses different colors for different types of statements (or a portion of those statements). Any statement containing a syntax error appears in red. Comments appear in green. Keywords, such as Sub and End Sub, appear in blue.

Step 7: Create a Second Procedure

- Pull down the **Insert menu** and click **Procedure** to display the Add Procedure dialog box. Enter **InputBoxExamples** as the name of the procedure. (Spaces are not allowed in a procedure name.)
- Click the option buttons for a **Sub procedure** and for **Public scope**. Click **OK**. The new sub procedure will appear within the existing module below the existing MsgBoxExamples procedure.
- Enter the statements in the procedure as they appear in Figure 4g. Be sure to type a space between the ampersand and the underscore in the second MsgBox statement. Click the **Save button** to save the procedure before testing it.
- You can display the output of the procedure directly in the VBA window if you minimize the Excel window. Thus, **right click** the Excel button on the taskbar to display a context-sensitive menu, then click the **Minimize command**. There is no visible change on your monitor.
- Click the **Run Sub button** to test the procedure. This time you see the Input box displayed on top of the VBA window because the Excel window has been minimized.
- Enter your first name in response to the initial prompt, then click **OK**. Click **OK** when you see the message box that says "Hello".
- Enter your last name in response to the second prompt and click **OK**. You should see a message box similar to the one in Figure 4g. Click **OK**.
- Return to the VBA procedure to correct any mistakes that might occur. Save the module.

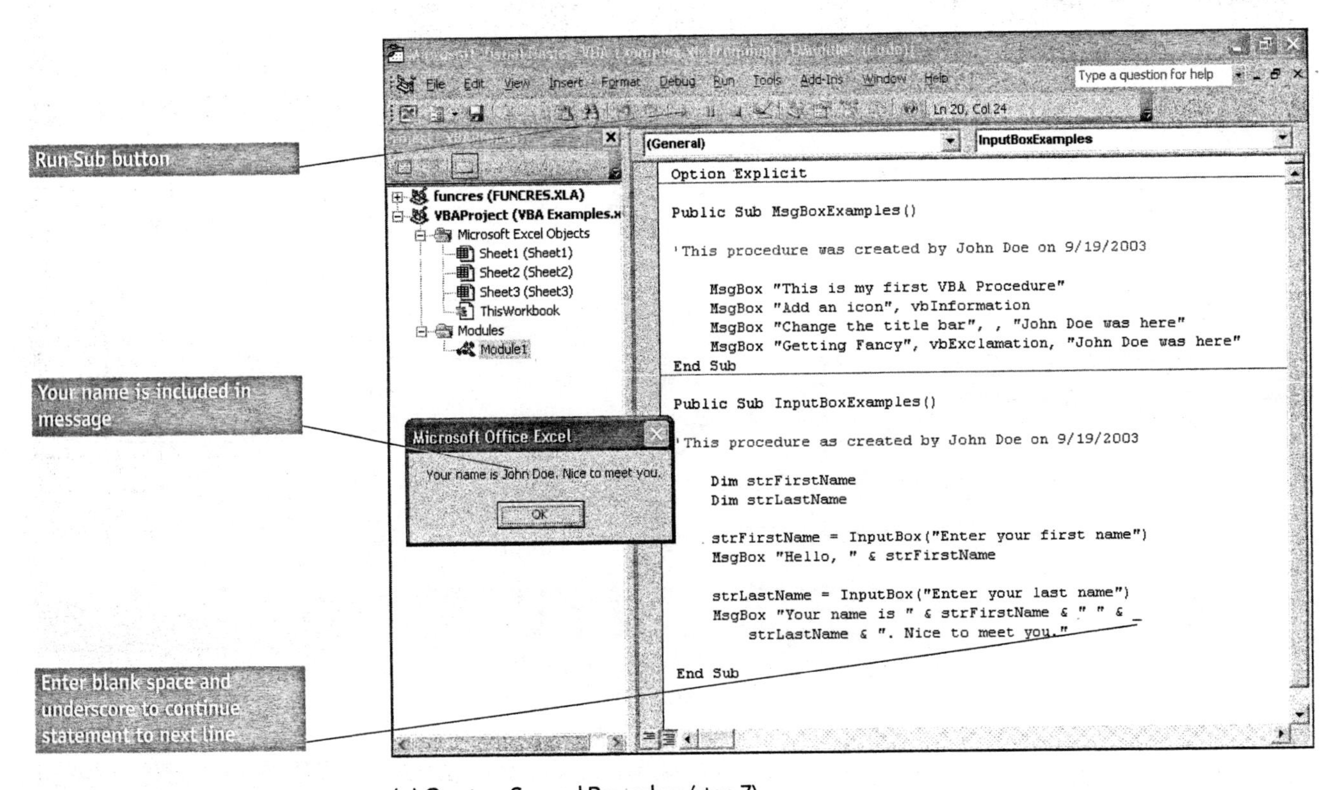

(g) Create a Second Procedure (step 7)

FIGURE 4 Hands-on Exercise 1 (*continued*)

Step 8: Create a Public Constant

- Click after the Options Explicit statement and press **Enter** to move to a new line. Type the statement to define the constant, **ApplicationTitle**, as shown in Figure 4h, and press **Enter**.
- Click anywhere in the MsgBoxExamples procedure, then change the third argument in the last MsgBox statement to ApplicationTitle. Make the four modifications in the InputBoxExamples procedure as shown in Figure 4h.
- Click anywhere in the InputBoxExamples procedure, then click the **Run Sub button** to test the procedure. The title bar of each dialog box will contain a descriptive title corresponding to the value of the ApplicationTitle constant.
- Change the value of the ApplicationTitle constant in the General Declarations section, then rerun the InputBoxExamples procedure. The title of every dialog box changes to reflect the new value.
- Save the procedure. Do you see the advantage of defining a title in the General Declarations section?

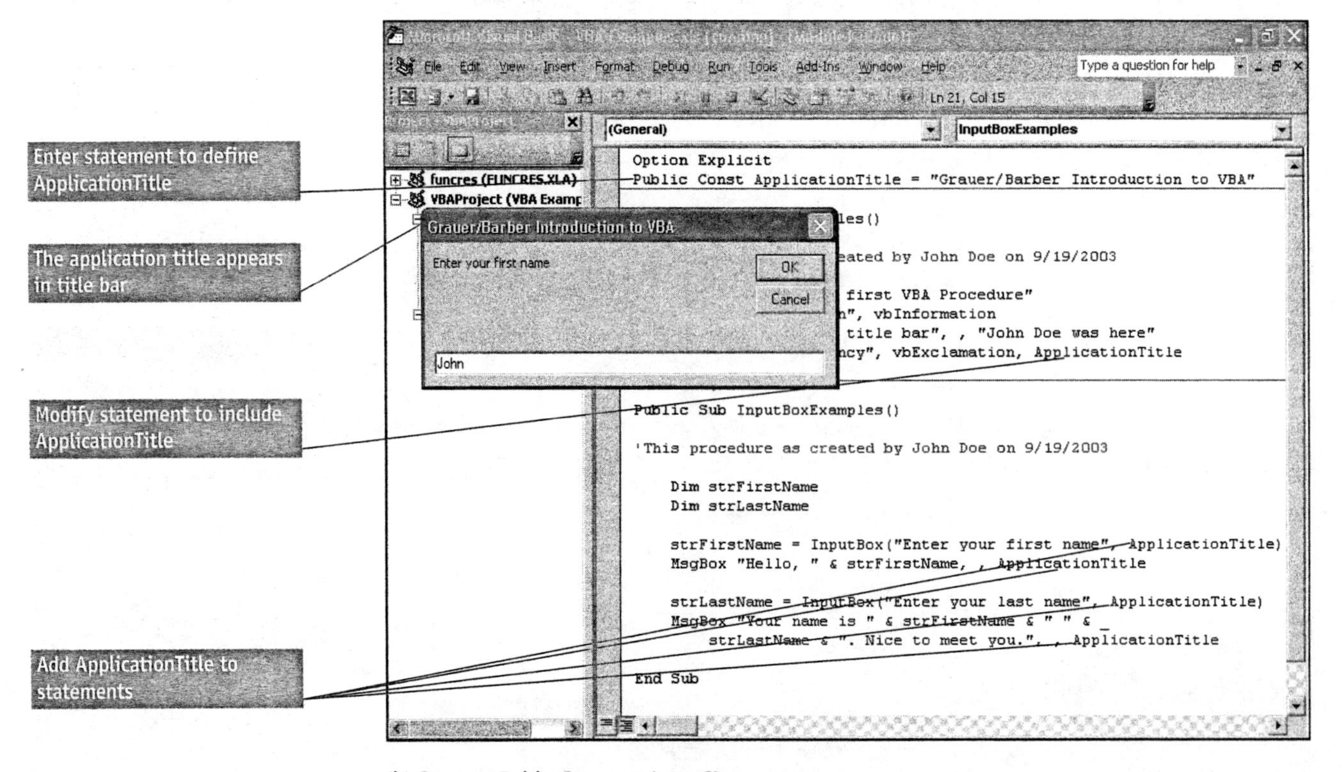

(h) Create a Public Constant (step 8)

FIGURE 4 Hands-on Exercise 1 (*continued*)

CONTINUING A VBA STATEMENT—THE & AND THE UNDERSCORE

A VBA statement can be continued from one line to the next by typing a space at the end of the line to be continued, typing the underscore character, then continuing on the next line. You may not, however, break a line in the middle of a literal (character string). Thus, you need to complete the character string with a closing quotation mark, add an ampersand (as the concatenation operator to display this string with the character string on the next line), then leave a space followed by the underscore to indicate continuation.

Step 9: Help with VBA

- You should be in the VBA editor. Pull down the **Help menu** and click the **Microsoft Visual Basic Help command** to open the Help pane.
- Type **Input Box function** in the Search box, then click the arrow to initiate the search. The results should include a hyperlink to InputBox function. Click the **hyperlink** to display the Help screen in Figure 4i.
- Maximize the Help window, then explore the information on the InputBox function to reinforce your knowledge of this statement.
  - Click the **Print button** to print this page for your instructor.
  - Click the link to **Example** within the Help window to see actual code.
  - Click the link to **See Also**, which displays information about the MsgBox statement.
- Close the Help window, but leave the task pane open. Click the **green** (back) **arrow** within the task pane to display the Table of Contents for Visual Basic Help, then explore the table of contents.
  - Click any closed book to open the book and "drill down" within the list of topics. The book remains open until you click the icon a second time to close it.
  - Click any question mark icon to display the associated help topic.
- Close the task pane. Pull down the **File menu** and click the **Close and Return to Microsoft Excel command** (or click the **Close button** on the VBA title bar) to close the VBA window and return to the application. Click **Yes** if asked whether to save the changes to Module1.
- You should be back in the Excel (or Access) application window. Close the application if you do not want to continue with the next exercise at this time.
- Congratulations. You have just completed your first VBA procedure. Remember to use Help any time you have a question.

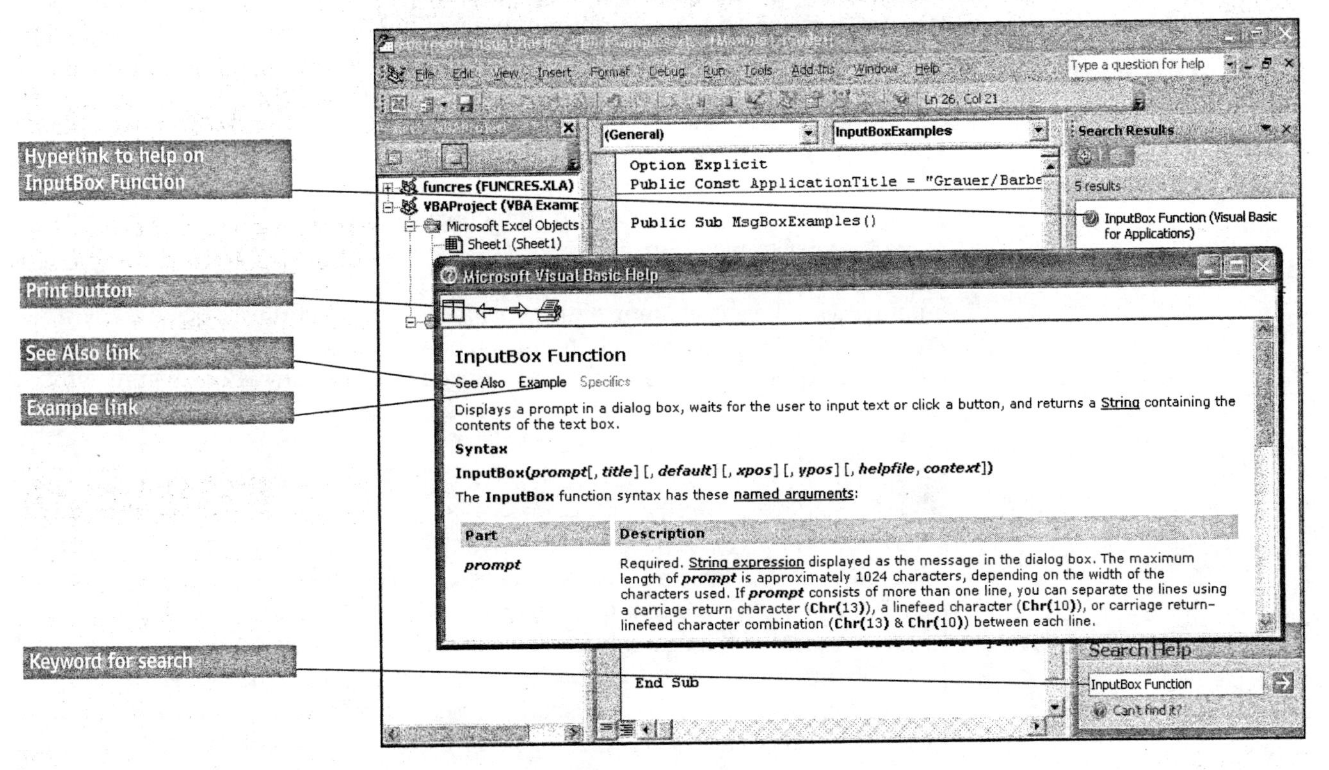

(i) Help with VBA (step 9)

FIGURE 4 Hands-on Exercise 1 (*continued*)

IF . . . THEN . . . ELSE STATEMENT

The ability to make decisions within a program, and then execute alternative sets of statements based on the results of those decisions, is crucial to any programming language. This is typically accomplished through an ***If statement***, which evaluates a condition as either true or false, then branches accordingly. The If statement is not used in isolation, however, but is incorporated into a procedure to accomplish a specific task as shown in Figure 5a. This procedure contains two separate If statements, and the results are displayed in the message boxes shown in the remainder of the figure.

The InputBox statement associated with Figure 5b prompts the user for the name of his or her instructor, then it stores the answer in the variable strInstructorName. The subsequent If statement then compares the user's answer to the literal "Grauer". If the condition is true (i.e., Grauer was entered into the input box), then the message in Figure 5c is displayed. If, however, the user entered any other value, then the condition is evaluated as false, the MsgBox is not displayed, and processing continues with the next statement in the procedure.

The second If statement includes an optional ***Else clause***. Again, the user is asked for a value, and the response is compared to the number 50. If the condition is true (i.e., the value of intUserStates equals 50), the message in Figure 5d is displayed to indicate that the response is correct. If, however, the condition is false (i.e., the user entered a number other than 50), the user sees the message in Figure 5e. Either way, true or false, processing continues with the next statement in the procedure. That's it—it's simple and it's powerful, and we will use the statement in the next hands-on exercise.

You can learn a good deal about VBA by looking at existing code and making inferences. Consider, for example, the difference between literals and numbers. ***Literals*** (also known as ***character strings***) are stored differently from numbers, and this is manifested in the way that comparisons are entered into a VBA statement. Look closely at the condition that references a literal (strInstructorName = "Grauer") compared to the condition that includes a number (intUserStates = 50). The literal ("Grauer") is enclosed in quotation marks, whereas the number (50) is not. (The prefix used in front of each variable, "str" and "int", is a common VBA convention to indicate the variable type—a string and an integer, respectively. Both variables are declared in the Dim statements at the beginning of the procedure.)

Note, too, that indentation and spacing are used throughout a procedure to make it easier to read. This is for the convenience of the programmer and not a requirement for VBA. The If, Else, and End If keywords are aligned under one another, with the subsequent statements indented under the associated keyword. We also indent a continued statement, such as a MsgBox statement, which is typically coded over multiple lines. Blank lines can be added anywhere within a procedure to separate blocks of statements from one another.

THE MSGBOX FUNCTION—YES OR NO

A simple MsgBox statement merely displays information to the user. MsgBox can also be used as a function, however, to accept information from the user such as clicking a Yes or No button, then combined with an If statement to take different actions based on the user's input. In essence, you enclose the arguments of the MsgBox function in parentheses (similar to what is done with the InputBox function), then test for the user response using the intrinsic constants vbYes and vbNo. The statement, If MsgBox("Are you having fun?", vbYesNo)=vbYes asks the user a question, displays Yes and No command buttons, then tests to see if the user clicked the Yes button.

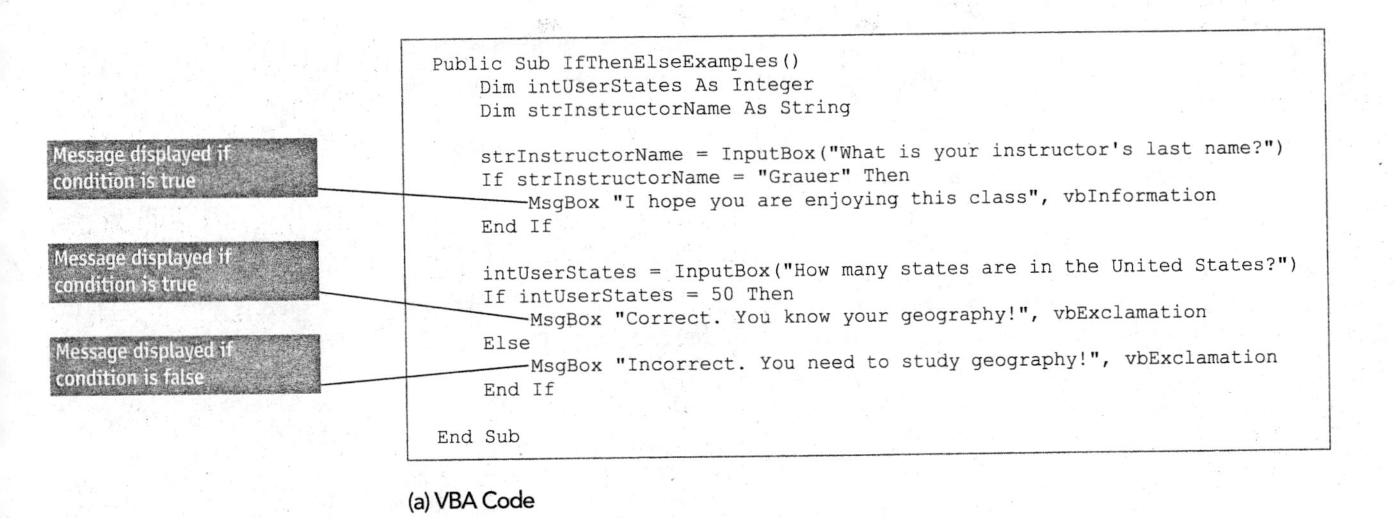

(a) VBA Code

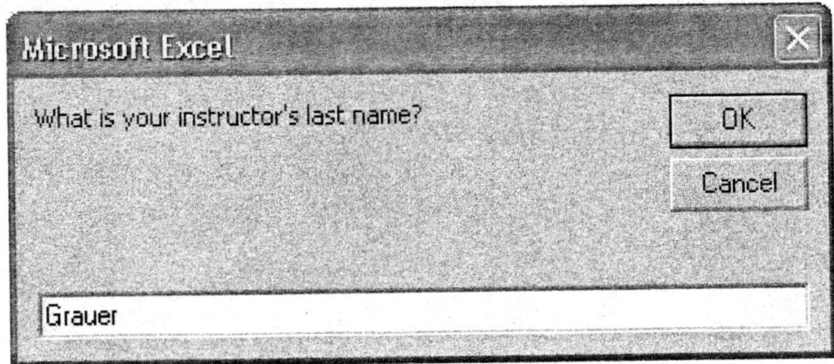

(b) InputBox Prompts for User Response

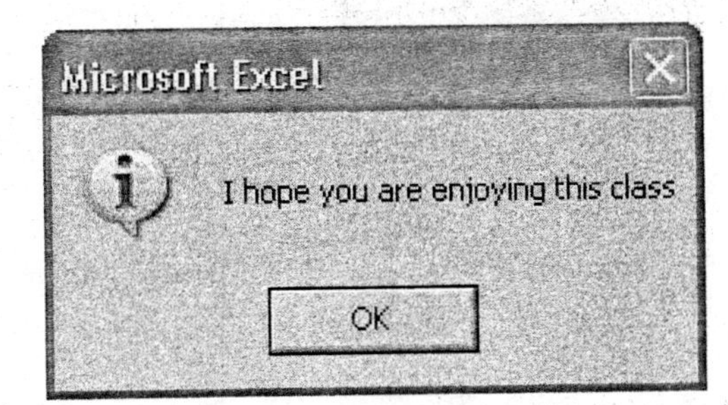

(c) Condition Is True

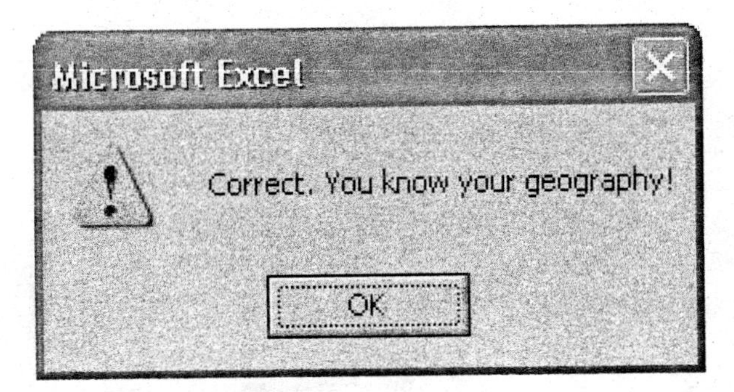

(d) Answer Is Correct (condition is true)

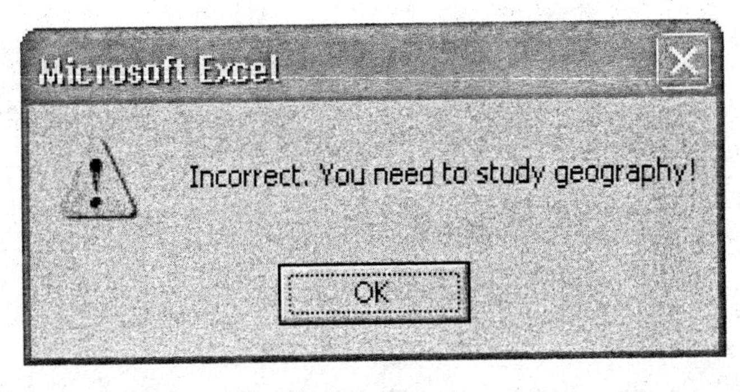

(e) Answer Is Wrong (condition is false)

FIGURE 5 The If Statement

CASE STATEMENT

The If statement is ideal for testing simple conditions and taking one of two actions. Although it can be extended to include additional actions by including one or more ElseIf clauses (If . . . Then . . . ElseIf . . . ElseIf . . .), this type of construction is often difficult to follow. Hence the ***Case statement*** is used when multiple branches are possible.

The procedure in Figure 6a accepts a student's GPA, then displays one of several messages, depending on the value of the GPA. The individual cases are evaluated in sequence. (The GPAs must be evaluated in descending order if the statement is to work correctly.) Thus, we check first to see if the GPA is greater than or equal to 3.9, then 3.75, then 3.5, and so on. If none of the cases is true, the statement following the Else clause is executed.

Note, too, the format of the comparison in that numbers (such as 3.9 or 3.75) are not enclosed in quotation marks because the associated variable (sngUserGPA) was declared as numeric. If, however, we had been evaluating a string variable (such as, strUserMajor), quotation marks would have been required around the literal values (e.g., Case Is = "Business", Case Is = "Liberal Arts", and so on.) The distinction between numeric and character (string) variables is important.

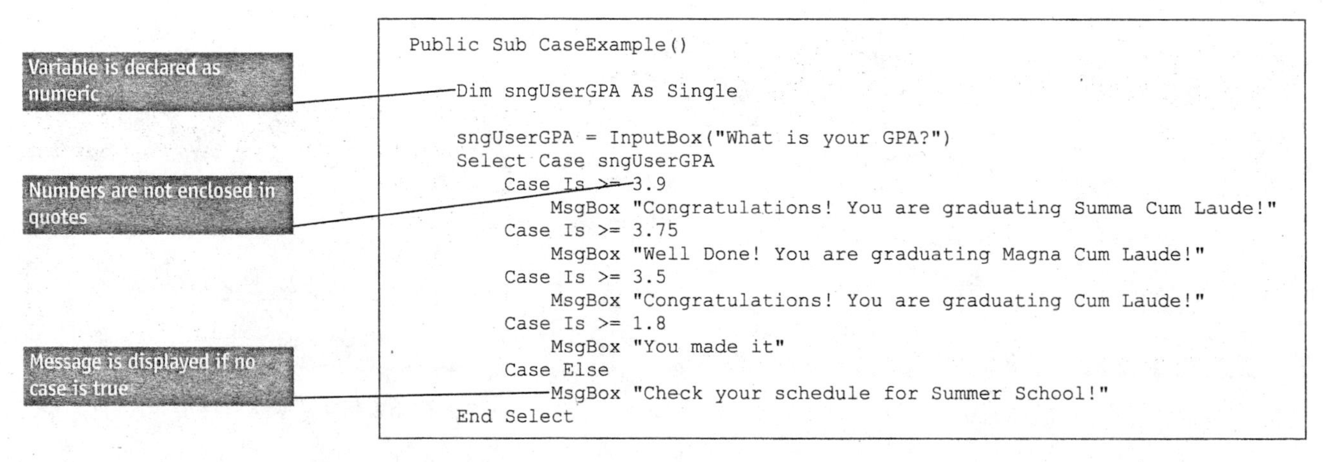


```
Public Sub CaseExample()

    Dim sngUserGPA As Single

    sngUserGPA = InputBox("What is your GPA?")
    Select Case sngUserGPA
        Case Is >= 3.9
            MsgBox "Congratulations! You are graduating Summa Cum Laude!"
        Case Is >= 3.75
            MsgBox "Well Done! You are graduating Magna Cum Laude!"
        Case Is >= 3.5
            MsgBox "Congratulations! You are graduating Cum Laude!"
        Case Is >= 1.8
            MsgBox "You made it"
        Case Else
            MsgBox "Check your schedule for Summer School!"
    End Select
```

(a) VBA Code

(b) Enter the GPA

(c) Third Option Is Selected

FIGURE 6 The Case Statement

CUSTOM TOOLBARS

A VBA procedure can be executed in several different ways. It can be run from the Visual Basic editor by pulling down the Run menu and clicking the Run Sub button on the Standard toolbar, or using the F5 function key. It can also be run from within the Office application (Word, Excel, or PowerPoint, but not Access), by pulling down the Tools menu, clicking the Macro command, then choosing the name of the macro that corresponds to the name of the procedure.

Perhaps the best way, however, is to create a ***custom toolbar*** that is displayed within the application as shown in Figure 7. (A custom menu can also be created that contains the same commands as the custom toolbar.) The toolbar has its own name (Bob's Toolbar), yet it functions identically to any other Office toolbar. You have your choice of displaying buttons only, text only, or both buttons and text. Our toolbar provides access to four commands, each corresponding to a procedure that was discussed earlier. Click the Case Example button, for example, and the associated procedure is executed, starting with the InputBox statement asking for the user's GPA.

A custom toolbar is created via the Toolbars command within the View menu. The new toolbar is initially big enough to hold only a single button, but you can add, move, and delete buttons following the same procedure as for any other Office toolbar. You can add any command at all to the toolbar; that is, you can add existing commands from within the Office application, or you can add commands that correspond to VBA procedures that you have created. Remember, too, that you can add more buttons to existing office toolbars.

Once the toolbar has been created, it is displayed or hidden just like any other Office toolbar. It can also be docked along any edge of the application window or left floating as shown in Figure 7. It's fun, it's easy, and as you may have guessed, it's time for the next hands-on exercise.

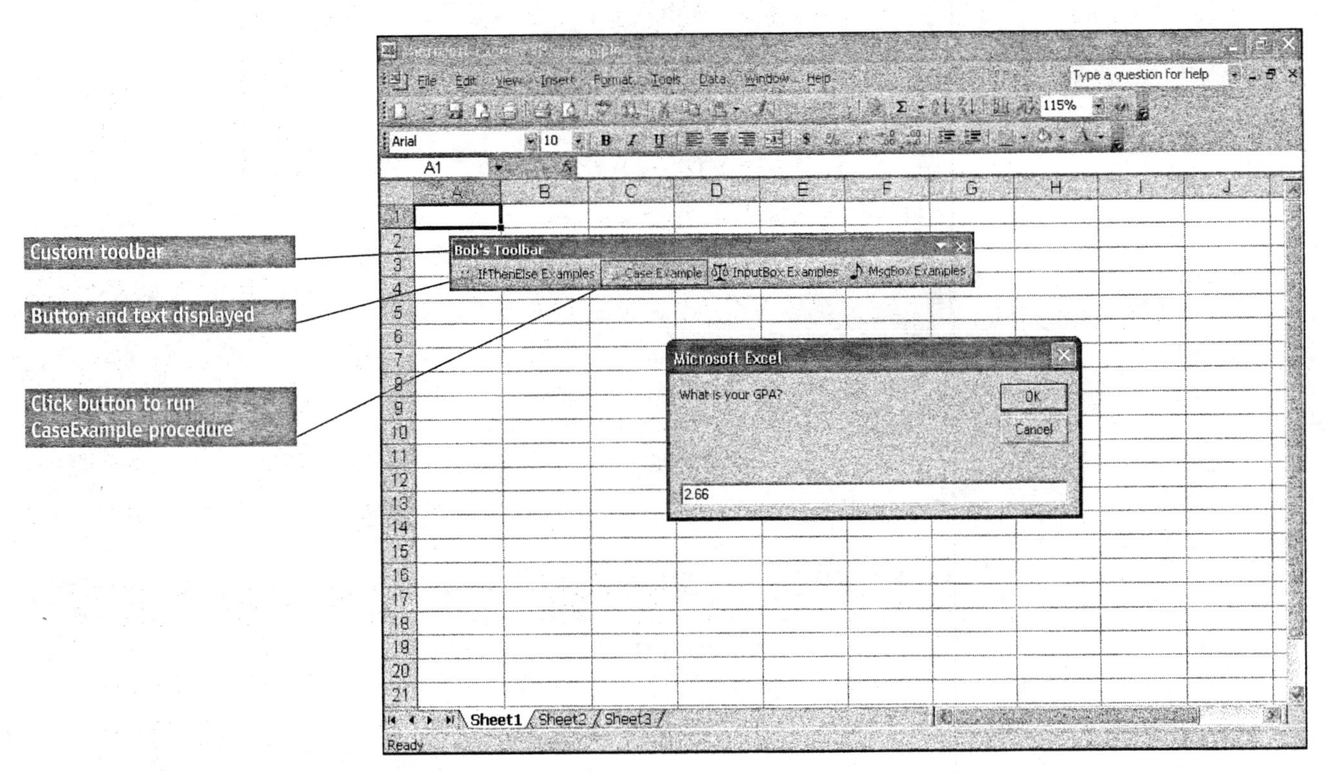

FIGURE 7 Custom Toolbars

hands-on exercise

2 Decision Making

Objective To create procedures with If . . . Then . . . Else and Case statements, then create a custom toolbar to execute those procedures. Use Figure 8 as a guide in the exercise.

Step 1: Open the Office Document

- Open the **VBA Examples workbook** or Access database from the previous exercise. The procedure differs slightly, depending on whether you are using Access or Excel.
  - In Access, you simply open the database.
  - In Excel you will be warned that the workbook contains a macro as shown in Figure 8a. Click the button to **Enable Macros.**
- Pull down the **Tools menu**, click the **Macro command**, then click the **Visual Basic Editor command**. You can also use the **Alt+F11** keyboard shortcut to open the VBA editor without going through the Tools menu.

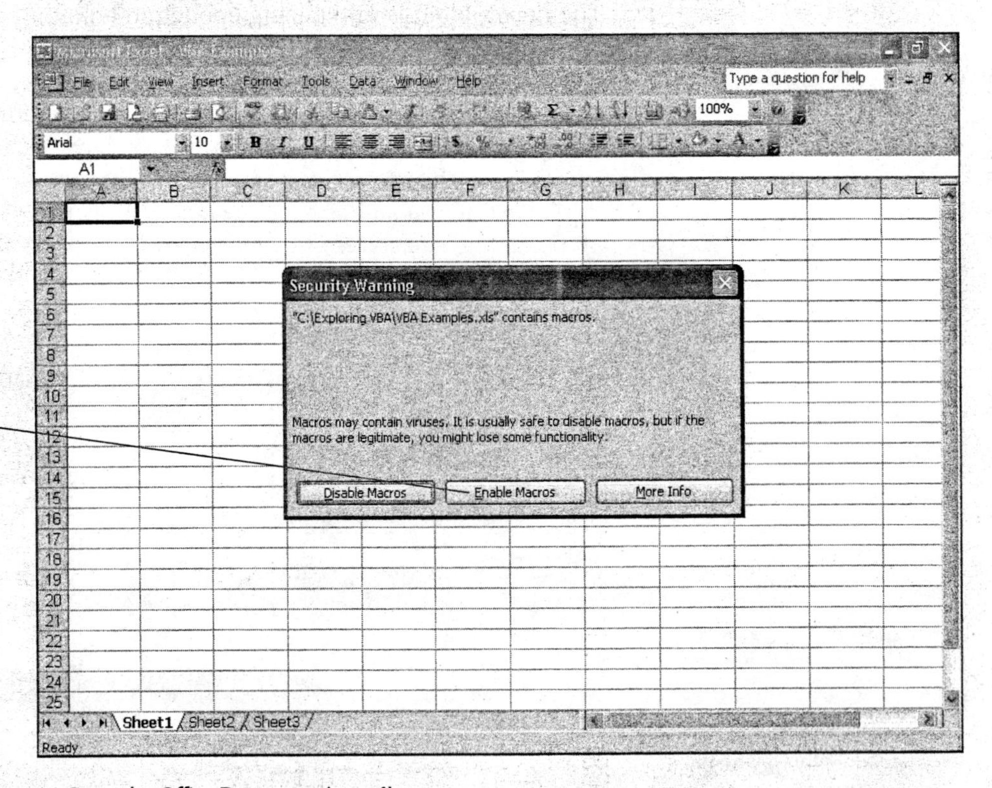

(a) Open the Office Document (step 1)

FIGURE 8 Hands-on Exercise 2

MACRO SECURITY

A computer virus could take the form of an Excel macro; thus, Excel will warn you that a workbook contains a macro, provided the security option is set appropriately. Pull down the Tools menu, click the Options command, click the Security tab, and then set the Macro Security to either High or Medium. High security disables all macros except those from a trusted source. Medium security gives you the option to enable macros. Click the button only if you are sure the macro is from a trusted source.

Step 2: Insert a New Procedure

- You should be in the Visual Basic editor as shown in Figure 8b. If necessary, double click **Module1** in the Explorer Window to open this module. Pull down the **Insert menu** and click the **Procedure command** to display the Add Procedure dialog box.
- Click in the **Name** text box and enter **IfThenElseExamples** as the name of the procedure. Click the option buttons for a **Sub procedure** and for **Public scope**. Click **OK** to create the procedure.
- The Sub procedure should appear within the module and consist of the Sub and End Sub statements as shown in Figure 8b.
- Click within the newly created procedure, then click the **Procedure View button** at the bottom of the window. The display changes to show just the current procedure.
- Click the **Save button** to save the module with the new procedure.

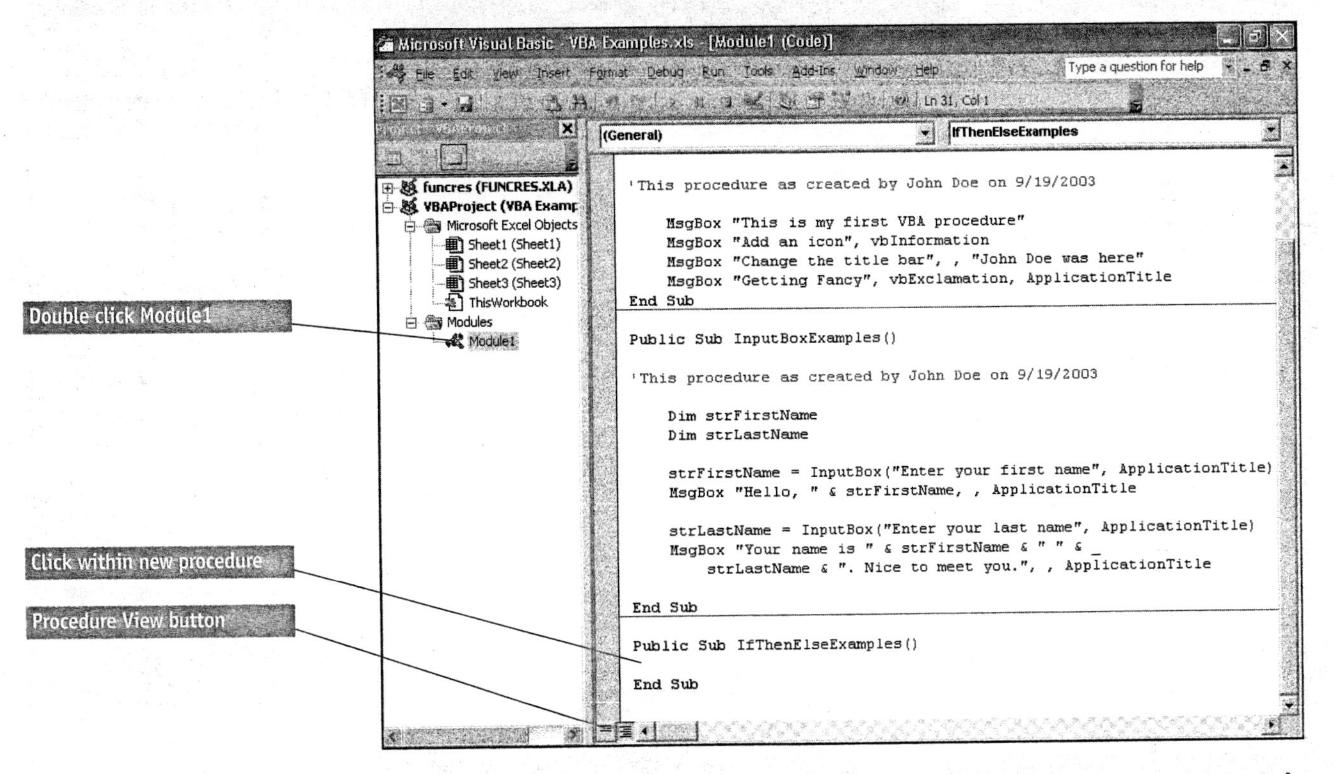

(b) Insert a New Procedure (step 2)

FIGURE 8 Hands-on Exercise 2 (*continued*)

PROCEDURE VIEW VERSUS FULL MODULE VIEW

The procedures within a module can be displayed individually, or alternatively, multiple procedures can be viewed simultaneously. To go from one view to the other, click the Procedure View button at the bottom of the window to display just the procedure you are working on, or click the Full Module View button to display multiple procedures. You can press Ctrl+PgDn and Ctrl+PgUp to move between procedures in either view.

Step 3: Create the If... Then... Else Procedure

- Enter the IfThenElseExamples procedure as it appears in Figure 8c, but use your instructor's name instead of Bob's. Note the following:
  - The Dim statements at the beginning of the procedure are required to define the two variables that are used elsewhere in the procedure.
  - The syntax of the comparison is different for string variables versus numeric variables. String variables require quotation marks around the comparison value (e.g., strInstructorName = "Grauer"). Numeric variables (e.g., intUserStates = 50) do not.
  - Indentation and blank lines are used within a procedure to make the code easier to read, as distinct from a VBA requirement. Press the **Tab key** to indent one level to the right.
  - Comments can be added to a procedure at any time.
- Save the procedure.

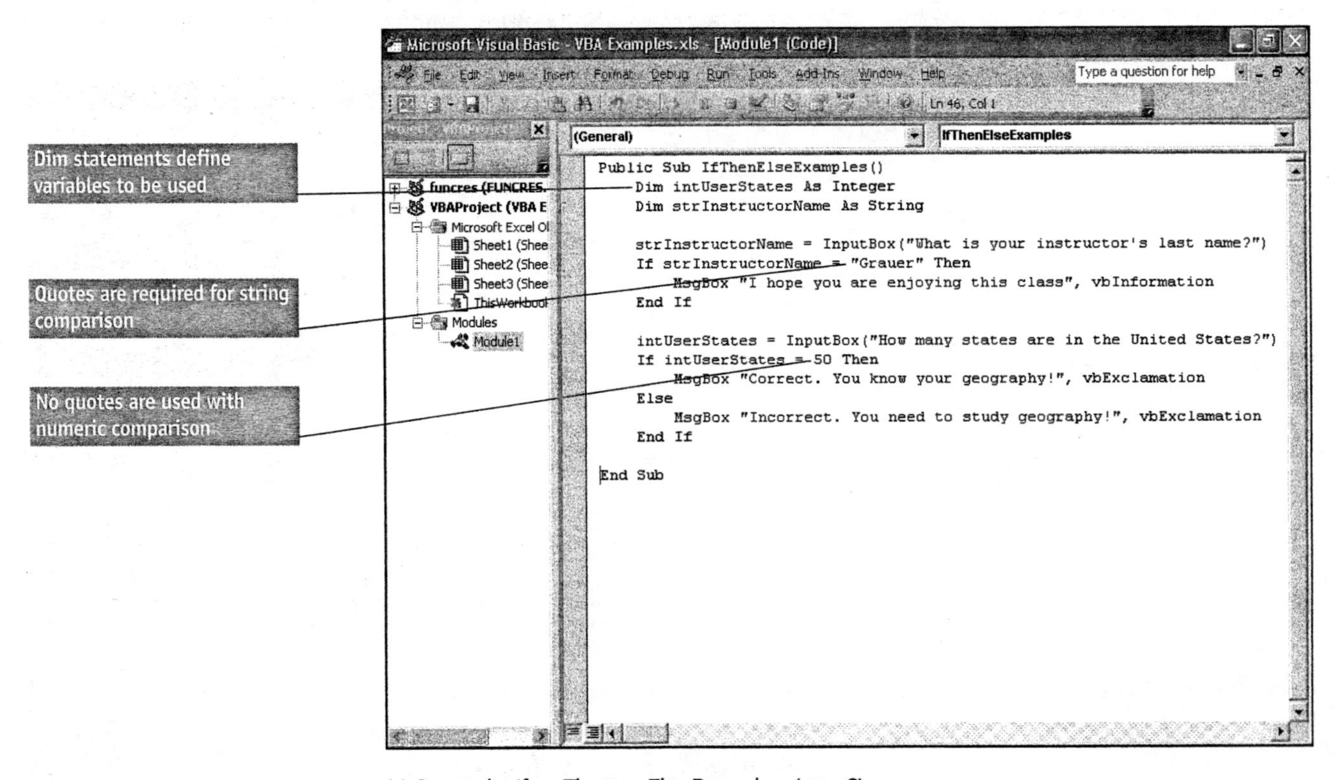

(c) Create the If... Then... Else Procedure (step 3)

FIGURE 8 Hands-on Exercise 2 (*continued*)

THE COMPLETE WORD TOOL

It's easy to misspell a variable name within a procedure, which is why the Complete Word tool is so useful. Type the first several characters in a variable name (such as "intU" or "strI" in the current procedure), then press Ctrl+Space. VBA will complete the variable name for you, provided that you have already entered a sufficient number of letters for a unique reference. Alternatively, it will display all of the elements that begin with the letters you have entered. Use the down arrow to scroll through the list until you find the item, then press the space bar to complete the entry.

Step 4: Test the Procedure

- The best way to test a procedure is to display its output directly in the VBA window (without having to switch back and forth between that and the application window). Thus, right click the Excel button on the taskbar to display a context-sensitive menu, then click the **Minimize command**.
- There is no visible change on your monitor. Click anywhere within the procedure, then click the **Run Sub button**. You should see the dialog box in Figure 8d.
- Enter your instructor's name, exactly as it was spelled within the VBA procedure. Click **OK**.
- You should see a second message box that hopes you are enjoying the class. This box will be displayed only if you spell the instructor's name correctly. Click **OK**.
- You should see a second input box that asks how many states are in the United States. Enter **50** and click **OK**. You should see a message indicating that you know your geography. Click **OK** to close the dialog box.
- Click the **Run Sub button** a second time, but enter a different set of values in response to the prompts. Misspell your instructor's name, and you will not see the associated message box.
- Enter any number other than 50, and you will be told to study geography.
- Continue to test the procedure until you are satisfied it works under all conditions. We cannot overemphasize the importance of thorough testing!

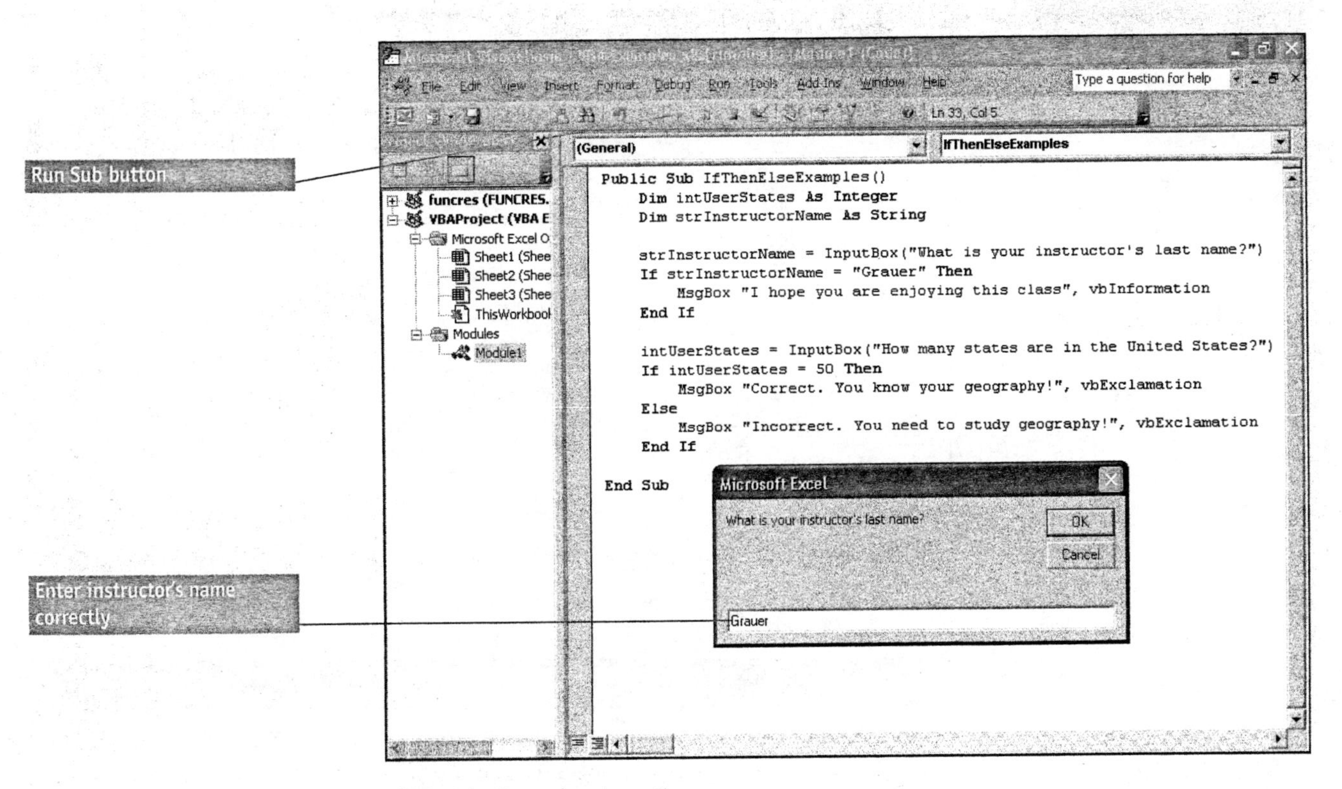

(d) Test the Procedure (step 4)

FIGURE 8 Hands-on Exercise 2 (*continued*)

Step 5: Create and Test the CaseExample Procedure

- Pull down the **Insert menu** and create a new procedure called **CaseExample**, then enter the statements exactly as they appear in Figure 8e. Note:
  - The variable sngUserGPA is declared to be a single-precision floating-point number (as distinct from the integer type that was used previously). A floating-point number is required in order to maintain a decimal point.
  - The GPA must be tested in descending order if the statement is to work correctly.
  - You may use any editing technique with which you are comfortable. You could, for example, enter the first case, copy it four times in the procedure, then modify the copied text as necessary.
  - The use of indentation and blank lines is for the convenience of the programmer and not a requirement of VBA.
- Click the **Run Sub button**, then test the procedure. Be sure to test it under all conditions; that is, you need to run it several times and enter a different GPA each time to be sure that all of the cases are working correctly.
- Save the procedure.

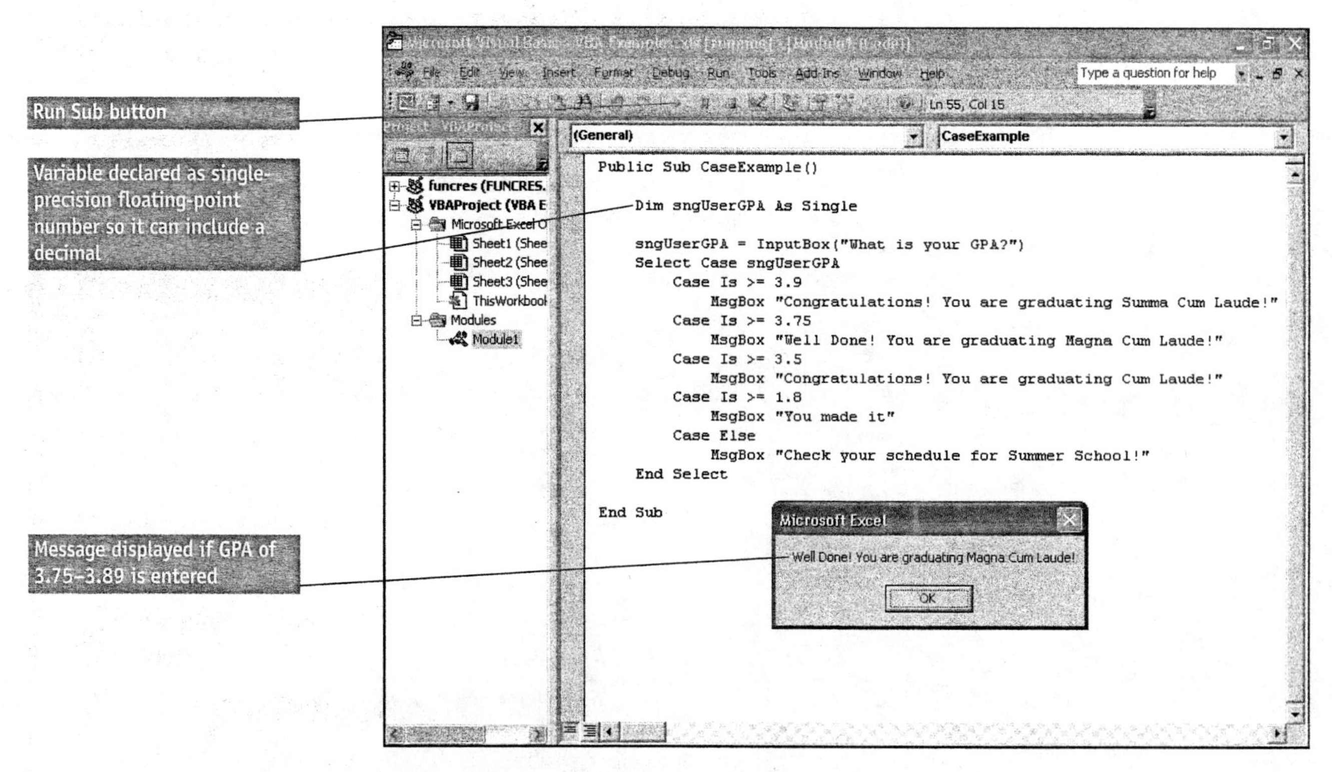

(e) Create and Test the CaseExample Procedure (step 5)

FIGURE 8 Hands-on Exercise 2 (*continued*)

RELATIONAL OPERATORS

The condition portion of an If or Case statement uses one of several relational operators. These include =, <, and > for equal to, less than, or greater than, respectively. You can also use >=, <=, or <> for greater than or equal to, less than or equal to, or not equal. This is basic, but very important, information if you are to code these statements correctly.

Step 6: Create a Custom Toolbar

- Click the **View Microsoft Excel** (or **Access**) **button** to display the associated application window. Pull down the **View menu**, click (or point to) the **Toolbars command**, then click **Customize** to display the Customize dialog box in Figure 8f. (Bob's toolbar is not yet visible.) Click the **Toolbars tab**.
- Click the **New button** to display the New Toolbar dialog box. Enter the name of your toolbar—e.g., **Bob's toolbar**—then click **OK** to create the toolbar and close the New Toolbar dialog box.
- Your toolbar should appear on the screen, but it does not yet contain any buttons. If necessary, click and drag the title bar of your toolbar to move the toolbar within the application window.
- Toggle the check box that appears next to your toolbar within the Customize dialog box on and off to display or hide your toolbar. Leave the box checked to display the toolbar and continue with this exercise.

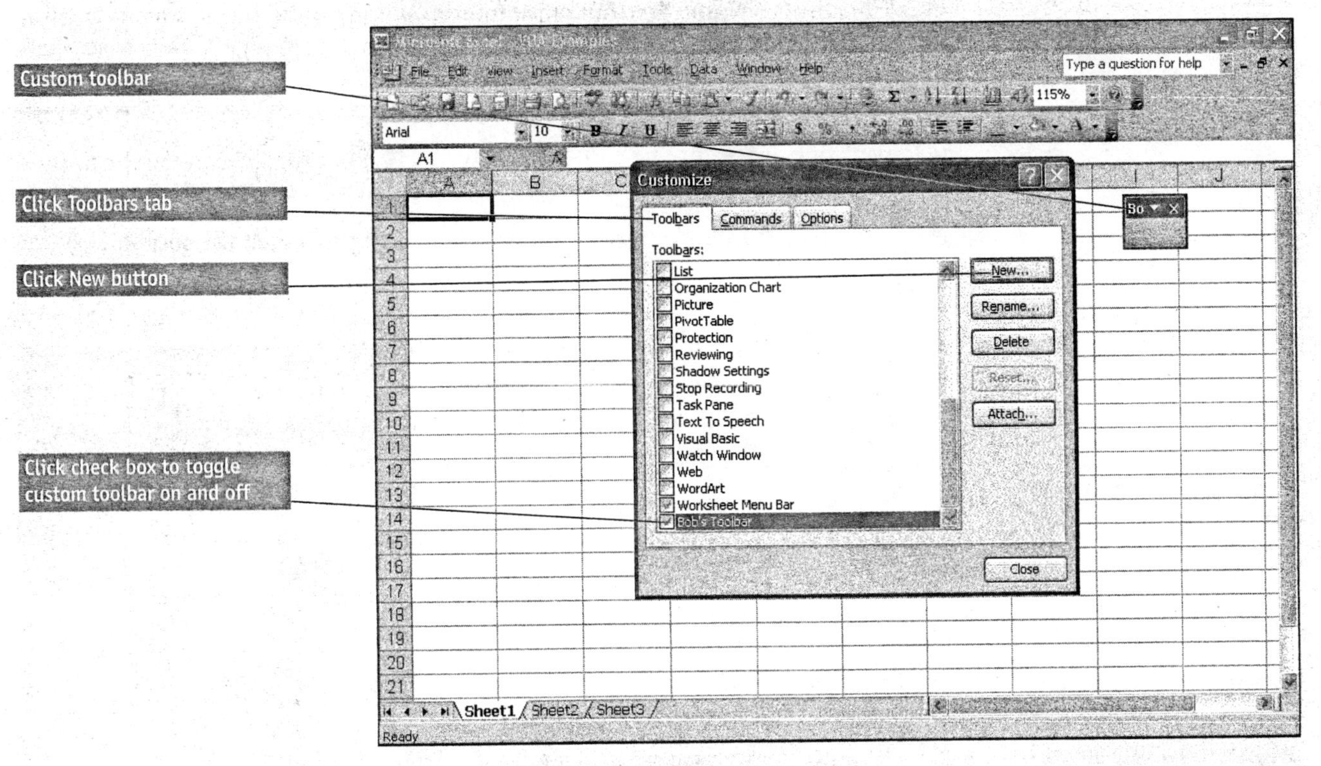

(f) Create a Custom Toolbar (step 6)

FIGURE 8 Hands-on Exercise 2 (*continued*)

FIXED VERSUS FLOATING TOOLBARS

A toolbar may be docked (fixed) along the edge of the application window, or it can be displayed as a floating toolbar anywhere within the window. You can switch back and forth by dragging the move handle of a docked toolbar to move the toolbar away from the edge. Conversely, you can drag the title bar of a floating toolbar to the edge of the window to dock the toolbar. You can also click and drag the border of a floating toolbar to change its size.

Step 7: Add Buttons to the Toolbar

- Click the **Commands tab** in the Customize dialog box, click the **down arrow** in the Categories list box, then scroll until you can select the **Macros category**. (If you are using Access and not Excel, you need to select the **File category**, then follow the steps as described in the boxed tip on the next page.)
- Click and drag the **Custom button** to your toolbar and release the mouse. A "happy face" button appears on the toolbar you just created. (You can remove a button from a toolbar by simply dragging the button from the toolbar.)
- Select the newly created button, then click the **Modify Selection command button** (or right click the button to display the context-sensitive menu) in Figure 8g. Change the button's properties as follows:
  - Click the **Assign Macro command** at the bottom of the menu to display the Assign Macro dialog box, then select the **IfThenElseExamples macro** (procedure) to assign it to the button. Click **OK**.
  - Click the **Modify Selection button** a second time.
  - Click in the **Name Textbox** and enter an appropriate name for the button, such as **IfThenElseExamples.**
  - Click the **Modify Selection button** a third time, then click **Text Only (Always)** to display text rather than an image.
- Repeat this procedure to add buttons to the toolbar for the MsgBoxExamples, InputBoxExamples, and CaseExample procedures that you created earlier.
- Close the Customize dialog box when you have completed the toolbar.
- Save the workbook.

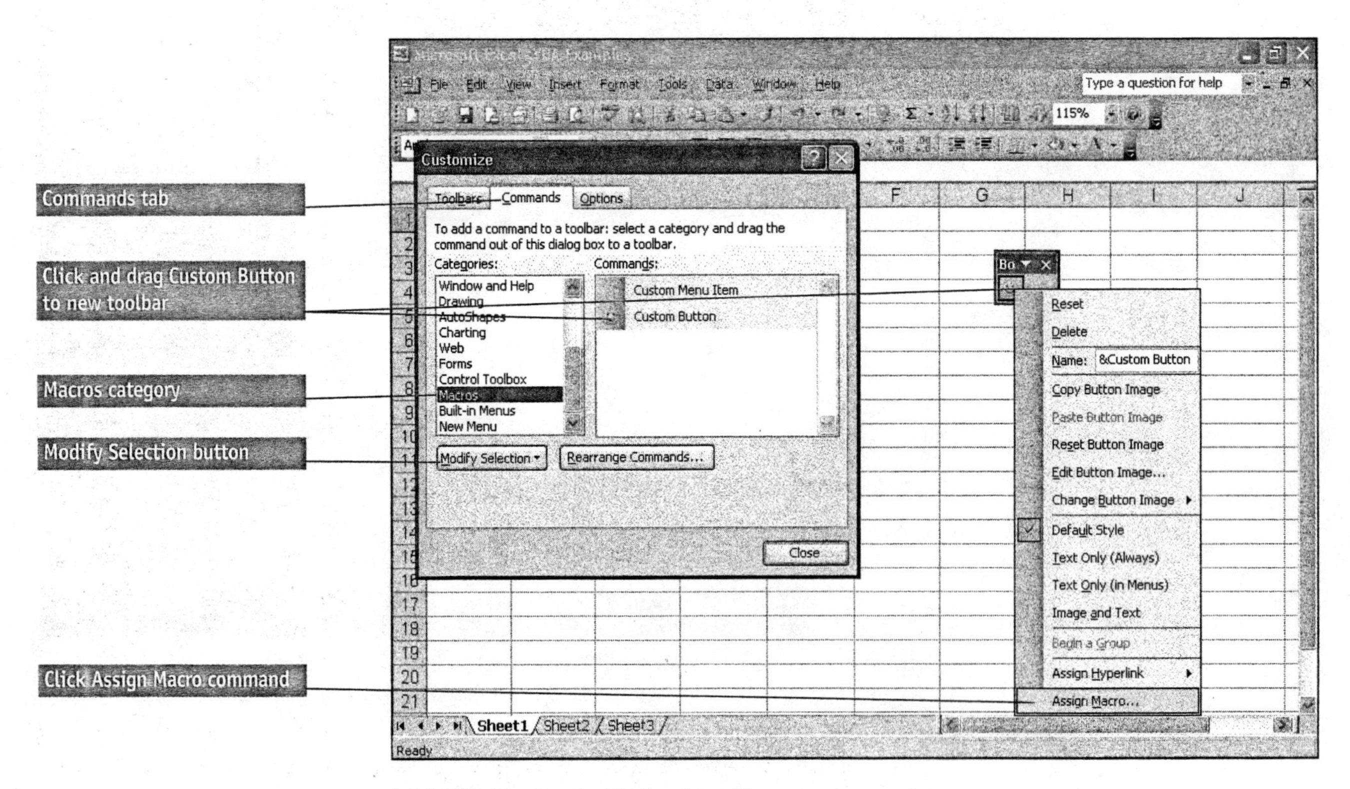

(g) Add Buttons to the Toolbar (step 7)

FIGURE 8 Hands-on Exercise 2 (*continued*)

Step 8: Test the Custom Toolbar

- Click any command on your toolbar as shown in Figure 8h. We clicked the **InputBoxExamples button**, which in turn executed the InputBoxExamples procedure that was created in the first exercise.
- Enter the appropriate information in any input boxes that are displayed. Click **OK**. Close your toolbar when you have completed testing it.
- If this is not your own machine, you should delete your toolbar as a courtesy to the next student. Pull down the **View menu**, click the **Toolbars command**, click **Customize** to display the Customize dialog box, then click the **Toolbars tab**. Select (highlight) the toolbar, then click the **Delete button** in the Customize dialog box. Click **OK** to delete the button. Close the dialog box.
- Exit Office if you do not want to continue with the next exercise.

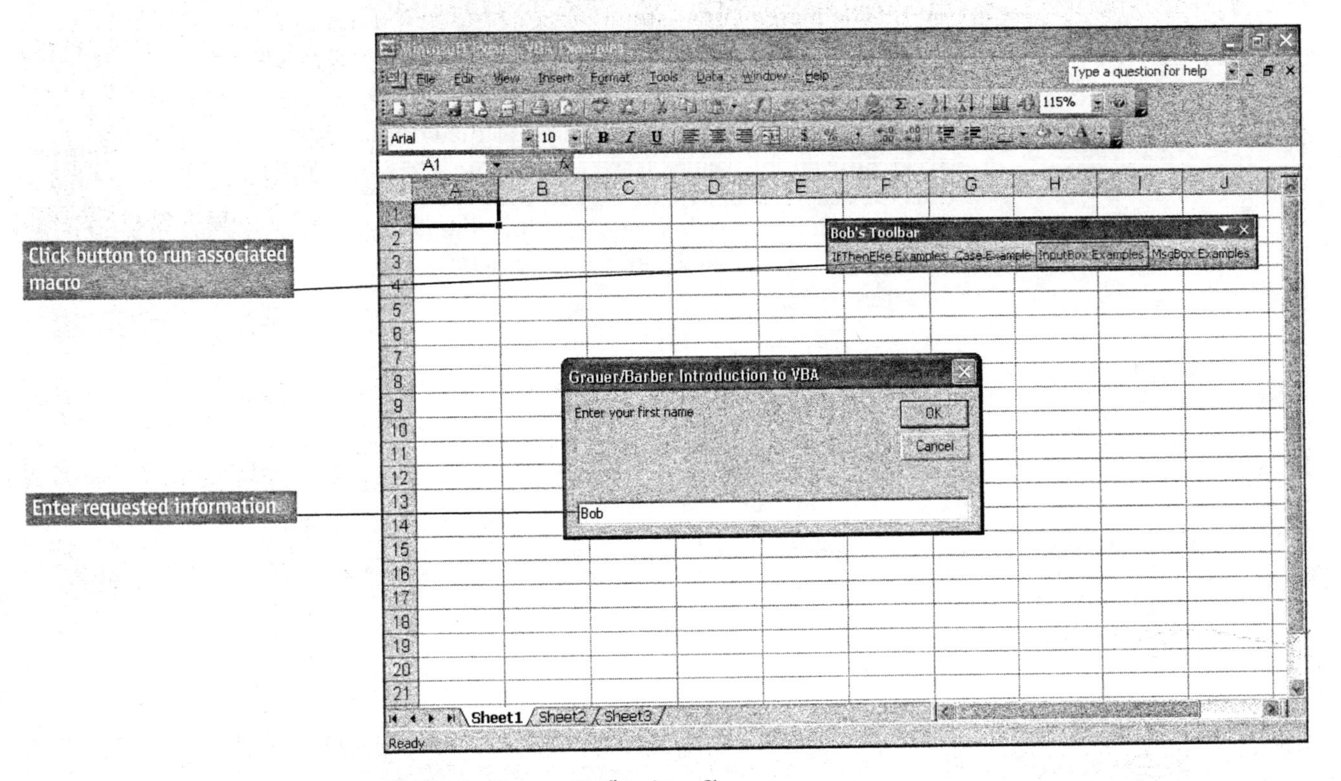

(h) Test the Custom Toolbar (step 8)

FIGURE 8 Hands-on Exercise 2 (*continued*)

ACCESS IS DIFFERENT

The procedure to add buttons to a custom toolbar in Access is different from the procedure in Excel. Pull down the View menu, click the Toolbars command, then click the Customize command. Select the File category within the Customize dialog box, then click and drag the Custom command to the newly created toolbar. Select the command on the toolbar, then click the Modify Selection command button in the dialog box. Click Properties, click the On Action text box, then type the name of the procedure you want to run in the format, =procedurename(). Close the dialog boxes, then press Alt+F11 to return to the VBA editor. Change the keyword "Sub" that identifies the procedure to "Function". Return to the database window, then test the newly created toolbar.

FOR . . . NEXT STATEMENT

The ***For . . . Next statement*** executes all statements between the words For and Next a specified number of times, using a counter to keep track of the number of times the statements are executed. The statement, For intCounter = 1 to N, executes the statements within the loop N times.

The procedure in Figure 9 contains two For . . . Next statements that sum the numbers from 1 to 10, counting by 1 and 2, respectively. The Dim statements at the beginning of the procedure declare two variables, intSumofNumbers to hold the sum and intCounter to hold the value of the counter. The sum is initialized to zero immediately before the first loop. The statements in the loop are then executed 10 times, each time incrementing the sum by the value of the counter. The result (the sum of the numbers from 1 to 10) is displayed after the loop in Figure 9b.

The second For . . . Next statement increments the counter by 2 rather than by 1. (The increment or step is assumed to be 1 unless a different value is specified.) The sum of the numbers is reset to zero prior to entering the second loop, the loop is entered, and the counter is initialized to the starting value of 1. Each subsequent time through the loop, however, the counter is incremented by 2. Each time the value of the counter is compared to the ending value, until it (the counter) exceeds the ending value, at which point the For . . . Next statement is complete. Thus the second loop will be executed for values of 1, 3, 5, 7, and 9. After the fifth time through the loop, the counter is incremented to 11, which is greater than the ending value of 10, and the loop is terminated.

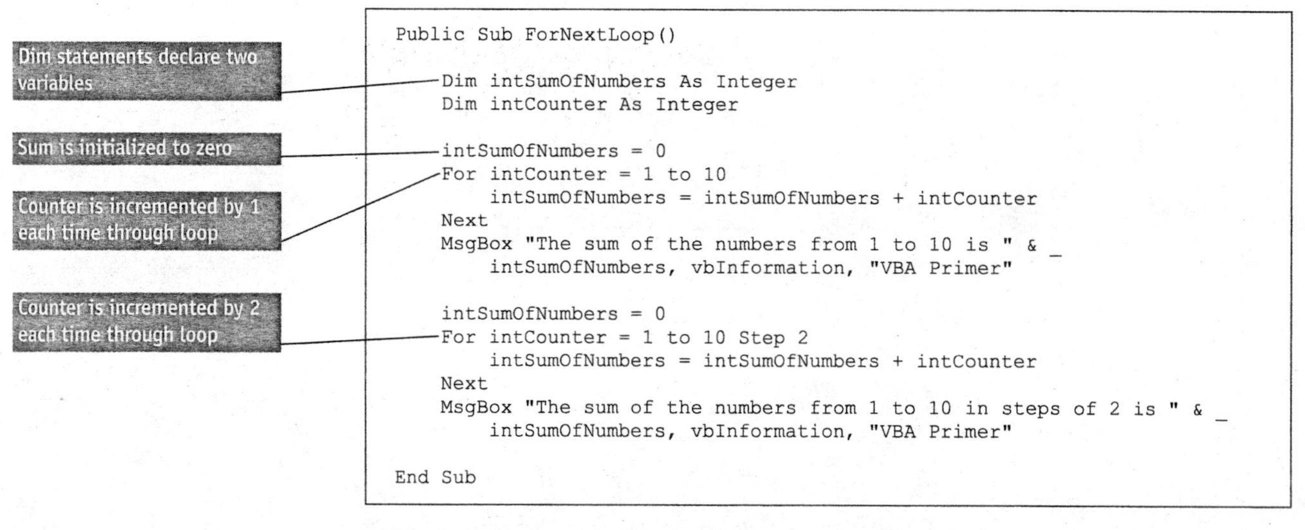


```
Public Sub ForNextLoop()

    Dim intSumOfNumbers As Integer
    Dim intCounter As Integer

    intSumOfNumbers = 0
    For intCounter = 1 to 10
        intSumOfNumbers = intSumOfNumbers + intCounter
    Next
    MsgBox "The sum of the numbers from 1 to 10 is " & _
        intSumOfNumbers, vbInformation, "VBA Primer"

    intSumOfNumbers = 0
    For intCounter = 1 to 10 Step 2
        intSumOfNumbers = intSumOfNumbers + intCounter
    Next
    MsgBox "The sum of the numbers from 1 to 10 in steps of 2 is " & _
        intSumOfNumbers, vbInformation, "VBA Primer"

End Sub
```

(a) VBA Code

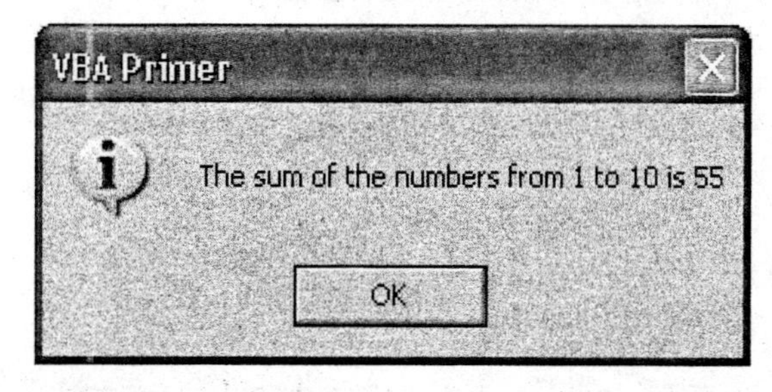

(b) In Increments of 1

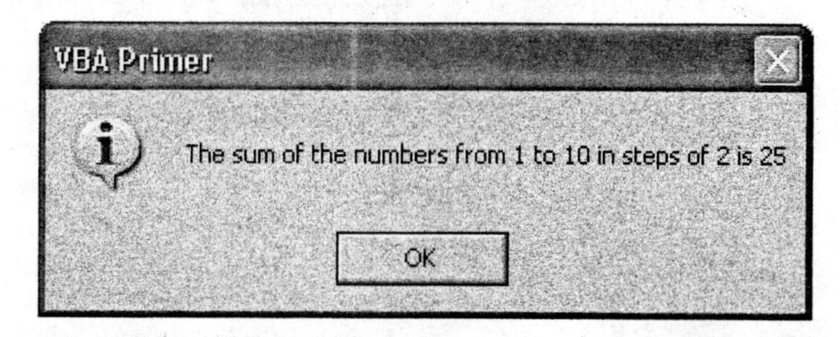

(c) In Increments of 2

FIGURE 9 For . . . Next Loops

DO LOOPS

The For . . . Next statement is ideal when you know in advance how many times you want to go through a loop. There are many instances, however, when the number of times through the loop is indeterminate. You could, for example, give a user multiple chances to enter a password or answer a question. This type of logic is implemented through a Do loop. You can repeat the loop as long as a condition is true (Do While), or until a condition becomes true (Do Until). The choice depends on how you want to state the condition.

Regardless of which keyword you choose, Do While or Do Until, two formats are available. The difference is subtle and depends on whether the keyword (While or Until) appears at the beginning or end of the loop. Our discussion will use the Do Until statement, but the Do While statement works in similar fashion.

Look closely at the procedure in Figure 10a, which contains two different loops. In the first example the Until condition appears at the end of the loop, which means the statements in the loop are executed, and then the condition is tested. This ensures that the statements in the loop will be executed at least once. The second loop, however, places the Until condition at the beginning of the loop, so that it (the condition) is tested prior to the loop being executed. Thus, if the condition is satisfied initially, the second loop will never be executed. In other words, there are two distinct statements ***Do . . . Loop Until*** and ***Do Until . . . Loop***. The first statement executes the loop, then tests the condition. The second statement tests the condition, then enters the loop.

```
Public Sub DoUntilLoop()

    Dim strCorrectAnswer As String, strUserAnswer As String

    strCorrectAnswer = "Earth"

    Do
        strUserAnswer = InputBox("What is the third planet from the sun?")
    Loop Until strUserAnswer = strCorrectAnswer
    MsgBox "You are correct, earthling!", vbExclamation

    strUserAnswer = InputBox("What is the third planet from the sun?")
    Do Until strUserAnswer = strCorrectAnswer
        strUserAnswer = InputBox("Your answer is incorrect. Try again.")
    Loop
    MsgBox "You are correct, earthling!", vbExclamation

End Sub
```

Until appears at end of loop

Until appears at beginning of loop

(a) (VBA Code)

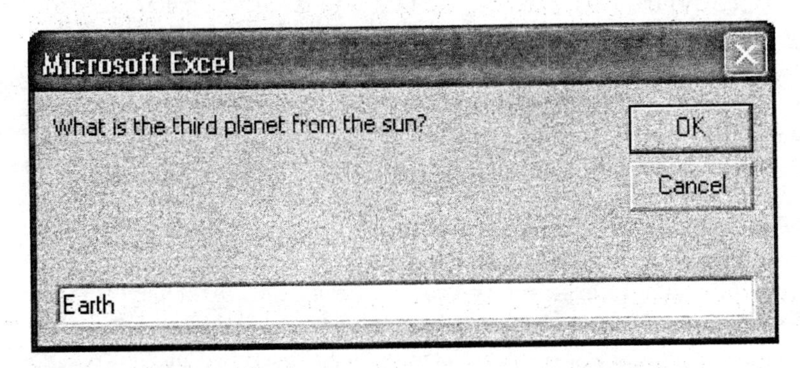

(b) Input the Answer

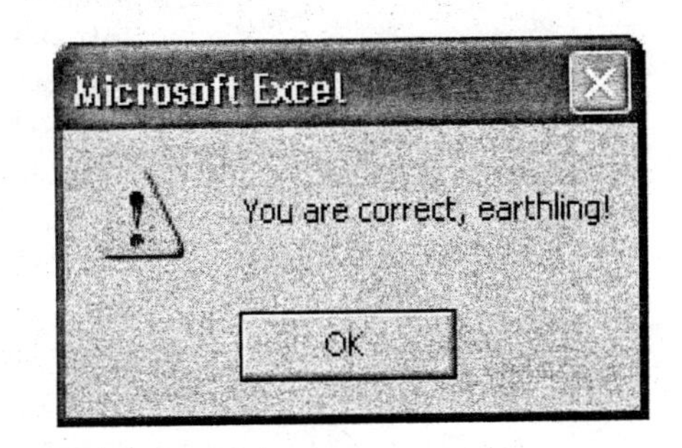

(c) Correct Response

FIGURE 10 Do Until Loops

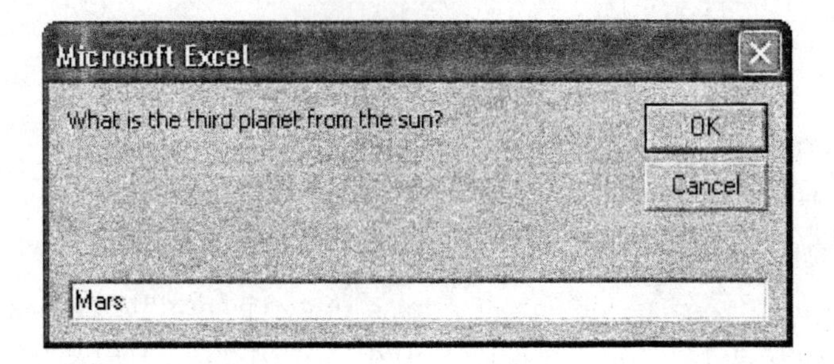

(d) Wrong Answer Initially

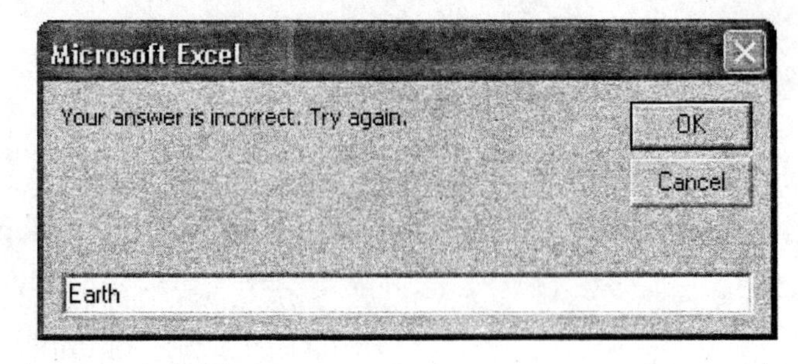

(e) Second Chance

FIGURE 10 Do Until Loops (*continued*)

It's tricky, but stay with us. In the first example the user is asked the question within the loop, and the loop is executed repeatedly until the user gives the correct answer. In the second example the user is asked the question outside of the loop, and the loop is bypassed if the user answers it correctly. The latter is the preferred logic because it enables us to phrase the question differently, before and during the loop. Look carefully at the difference between the InputBox statements and see how the question changes within the second loop.

DEBUGGING

As you learn more about VBA and develop more powerful procedures, you are more likely to make mistakes. The process of finding and correcting errors within a procedure is known as ***debugging*** and it is an integral part of programming. Do not be discouraged if you make mistakes. Everyone does. The important thing is how quickly you are able to find and correct the errors that invariably occur. We begin our discussion of debugging by describing two types of errors, ***compilation errors*** and ***execution*** (or ***run-time***) ***errors***.

A compilation error is simply an error in VBA syntax. (Compilation is the process of translating a VBA procedure into machine language, and thus a compilation error occurs when the VBA editor is unable to convert a statement to machine language.) Compilation errors occur for many reasons, such as misspelling a keyword, omitting a comma, and so on. VBA recognizes the error before the procedure is run and displays the invalid statement in red together with an associated error message. The programmer corrects the error and then reruns the procedure.

Execution errors are caused by errors in logic and are more difficult to detect because they occur without any error message. VBA, or for that matter any other programming language, does what you tell it to do, which is not necessarily what you want it to do. If, for example, you were to compute the sales tax of an item by multiplying the price by 60% rather than 6%, VBA will perform the calculation and simply display the wrong answer. It is up to you to realize that the results of the procedure are incorrect, and you will need to examine its statements and correct the mistake.

So how do you detect an execution error? In essence, you must decide what the expected output of your procedure should be, then you compare the actual result of the procedure to the intended result. If the results are different, an error has occurred, and you have to examine the logic in the procedure to find the error. You may see the mistake immediately (e.g., using 60% rather than 6% in the previous example), or you may have to examine the code more closely. And as you might expect, VBA has a variety of tools to help you in the debugging process. These tools are accessed from the ***Debug toolbar*** or the ***Debug menu*** as shown in Figure 11 on the next page.

The procedure in Figure 11 is a simple For . . . Next loop to sum the integers from 1 to 10. The procedure is correct as written, but we have introduced several debugging techniques into the figure. The most basic technique is to step through the statements in the procedure one at a time to see the sequence in which the statements are executed. Click the ***Step Into button*** on the Debug toolbar to enter (step into) the procedure, then continue to click the button to move through the procedure. Each time you click the button, the statement that is about to be executed is highlighted.

Another useful technique is to display the values of selected variables as they change during execution. This is accomplished through the ***Debug.Print statement*** that displays the values in the ***Immediate window***. The Debug.Print statement is placed within the For . . . Next loop so that you can see how the counter and the associated sum change during execution.

As the figure now stands, we have gone through the loop nine times, and the sum of the numbers from 1 to 9 is 45. The Step Into button is in effect so that the statement to be executed next is highlighted. You can see that we are back at the top of the loop, where the counter has been incremented to 10, and further, that we are about to increment the sum.

The ***Locals window*** is similar in concept except that it displays only the current values of all the variables within the procedure. Unlike the Immediate window, which requires the insertion of Debug.Print statements into a procedure to have meaning, the Locals window displays its values automatically, without any effort on the part of the programmer, other than opening the window. All three techniques can be used individually, or in conjunction with one another, as the situation demands.

We believe that the best time to practice debugging is when you know there are no errors in your procedure. As you may have guessed, it's time for the next hands-on exercise.

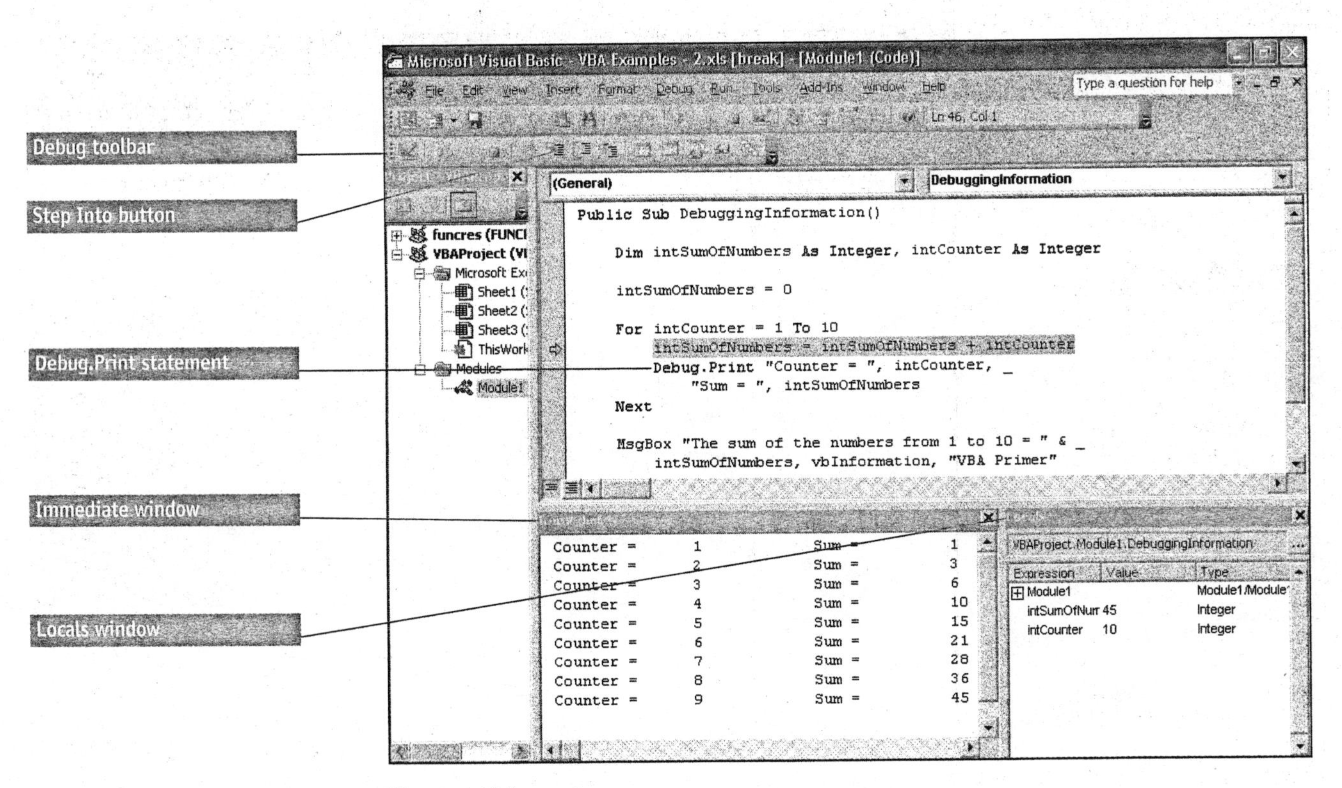

FIGURE 11 Debugging

hands-on exercise

3 Loops and Debugging

Objective To create a loop using the For . . . Next and Do Until statements; to open the Locals and Immediate windows and illustrate different techniques for debugging. Use Figure 12 as a guide in the exercise.

Step 1: Insert a New Procedure

- Open the **VBA Examples workbook** or the Access database from the previous exercise. Either way, pull down the **Tools menu**, click the **Macro command**, then click **Visual Basic editor** (or use the **Alt+F11** keyboard shortcut) to start the VBA editor.
- If necessary, double click **Module1** within the Project Explorer window to open this module. Pull down the **Insert menu** and click the **Procedure command** to display the Add Procedure dialog box.
- Click in the **Name** text box and enter **ForNextLoop** as the name of the procedure. Click the option buttons for a **Sub procedure** and for **Public scope**. Click **OK** to create the procedure.
- The Sub procedure should appear within the module and consist of the Sub and End Sub statements as shown in Figure 12a.
- Click the **Procedure View button** at the bottom of the window as shown in Figure 12a. The display changes to show just the current procedure, giving you more room in which to work.
- Save the module.

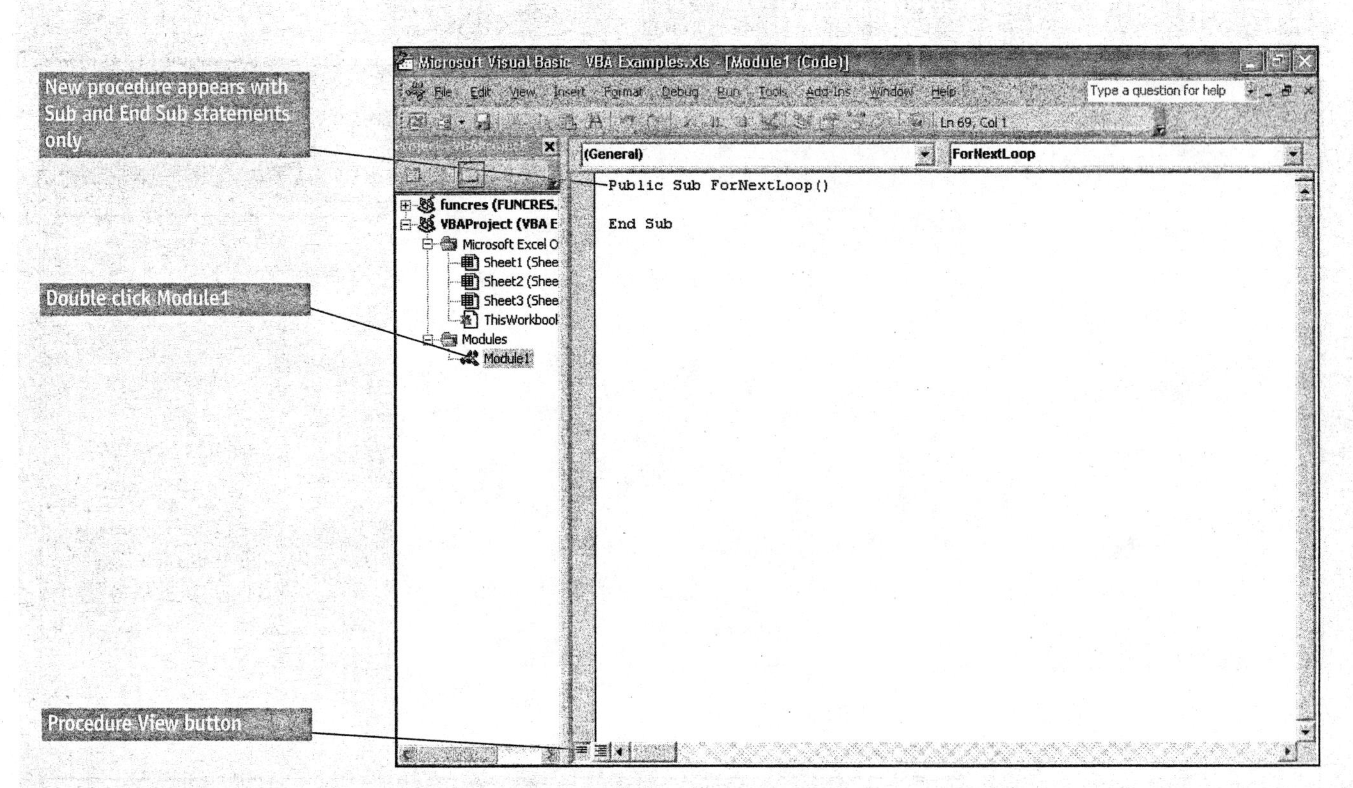

(a) Insert a New Procedure (step 1)

FIGURE 12 Hands-on Exercise 3

Step 2: Test the For . . . Next Procedure

- Enter the procedure exactly as it appears in Figure 12b. Note the following:
  - A comment is added at the beginning of the procedure to identify the author and the date.
  - Two variables are declared at the beginning of the procedure, one to hold the sum of the numbers and the other to serve as a counter.
  - The sum of the numbers is initialized to zero. The For . . . Next loop varies the counter from 1 to 10.
  - The statement within the For . . . Next loop increments the sum of the numbers by the current value of the counter. The equal sign is really a replacement operator; that is, replace the variable on the left (the sum of the numbers) by the expression on the right (the sum of the numbers plus the value of the counter.
  - Indentation and spacing within a procedure are for the convenience of the programmer and not a requirement of VBA. We align the For and Next statements at the beginning and end of a loop, then indent all statements within a loop.
  - The MsgBox statement displays the result and is continued over two lines as per the underscore at the end of the first line.
  - The ampersand concatenates (joins together) the text and the number within the message box.
- Click the **Save button** to save the module. Right click the **Excel button** on the Windows taskbar to display a context-sensitive menu, then click the **Minimize command**.
- Click the **Run Sub button** to test the procedure, which should display the MsgBox statement in Figure 12b. Correct any errors that may occur.

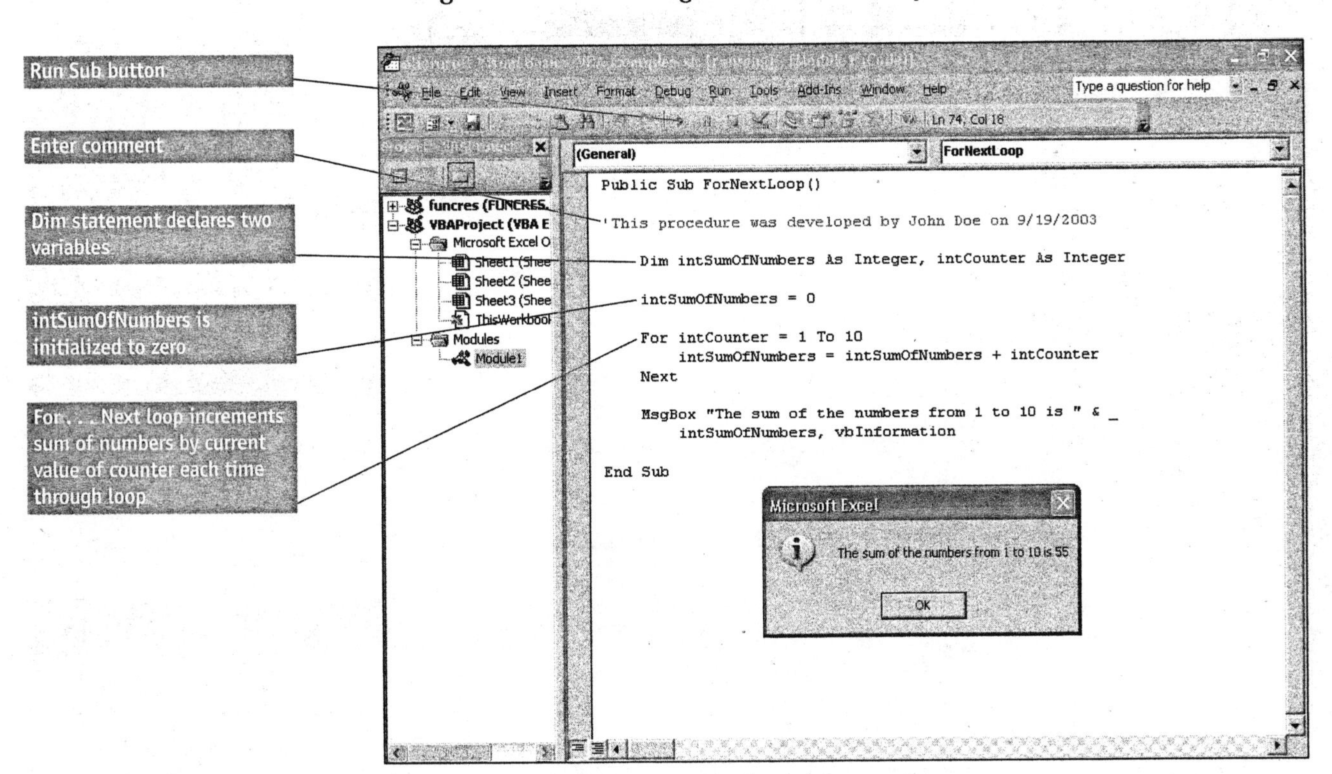

(b) Test the For . . . Next Procedure (step 2)

FIGURE 12 Hands-on Exercise 3 (*continued*)

Step 3: Compilation Errors

- The best time to practice debugging is when you know that the procedure is working properly. Accordingly, we will make some deliberate errors in our procedure to illustrate different debugging techniques.
- Pull down the **View menu**, click the **Toolbars command**, and (if necessary) toggle the Debug toolbar on, then dock it under the Standard toolbar.
- Click on the statement that initializes intSumOfNumbers to zero and delete the "s" at the end of the variable name. Click the **Run Sub button**.
- You will see the message in Figure 12c. Click **OK** to acknowledge the error, then click the **Undo button** to correct the error.
- The procedure header is highlighted, indicating that execution is temporarily suspended and that additional action is required from you to continue testing. Click the **Run Sub button** to retest the procedure.
- This time the procedure executes correctly and you see the MsgBox statement indicating that the sum of the numbers from 1 to 10 is 55. Click **OK**.

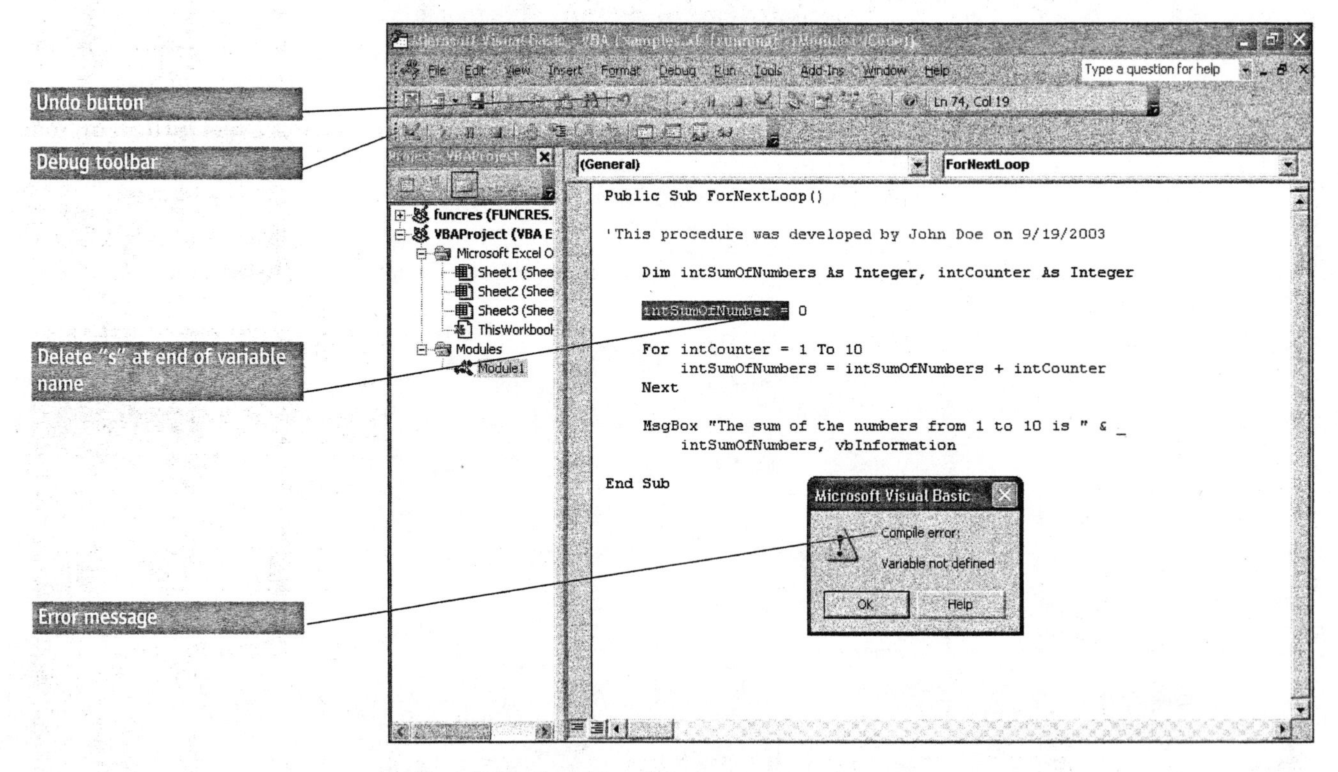

(c) Compilation Error (step 3)

FIGURE 12 Hands-on Exercise 3 (*continued*)

USE HELP AS NECESSARY

Pull down the Help menu at any time (or press the F1 key) to access the VBA Help facility to explore at your leisure. Use the Print command to create hard copy. (You can also copy the help text into a Word document to create your own reference manual.) The answers to virtually all of your questions are readily available if only you take the time to look.

Step 4: Step through a Procedure

- Pull down the **View menu** a second time and click the **Locals Window command** (or click the **Locals Window button** on the Debug toolbar).
- If necessary, click and drag the top border of the Locals window to size the window appropriately as shown in Figure 12d.
- Click anywhere within the procedure. Pull down the **Debug menu** and click the **Step Into command** (or click the **Step Into button** on the Debug toolbar). The first statement (the procedure header) is highlighted, indicating that you are about to enter the procedure.
- Click the **Step Into button** (or use the **F8** keyboard shortcut) to step into the procedure and advance to the next executable statement. The statement that initializes intSumOfNumbers to zero is highlighted, indicating that this statement is about to be executed.
- Continue to press the **F8 key** to step through the procedure. Each time you execute a statement, you can see the values of intSumOfNumbers and intCounter change within the Locals window. (You can click the **Step Out button** at any time to end the procedure.)
- Correct errors as they occur. Click the **Reset button** on the Standard or Debug toolbars at any time to begin executing the procedure from the beginning.
- Eventually you exit from the loop, and the sum of the numbers (from 1 to 10) is displayed within a message box.
- Click **OK** to close the message box. Press the **F8 key** a final time, then close the Locals window.
- Do you see how stepping through a procedure helps you to understand how it works?

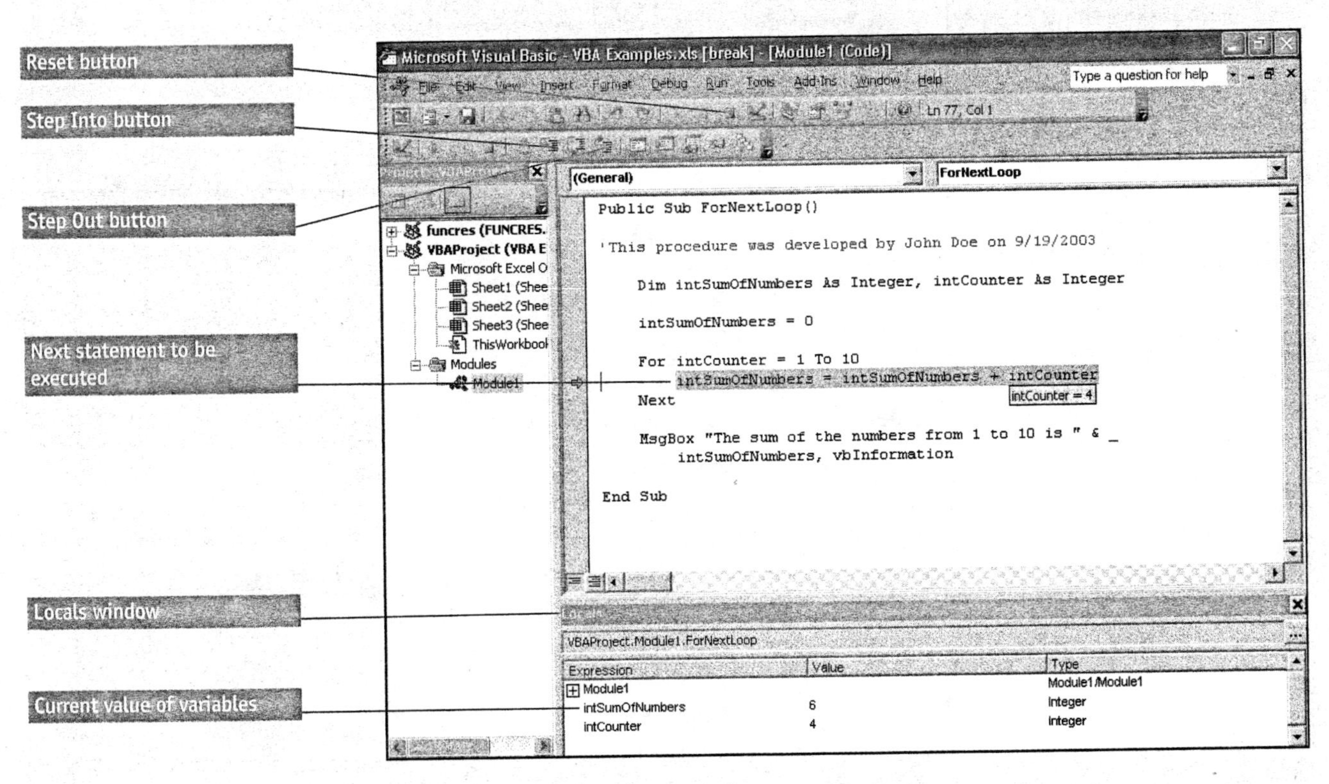

(d) Step through a Procedure (step 4)

FIGURE 12 Hands-on Exercise 3 (*continued*)

Step 5: The Immediate Window

- You should be back in the VBA window. Click immediately to the left of the Next statement and press **Enter** to insert a blank line. Type the **Debug.Print** statement exactly as shown in Figure 12e. (Click **OK** if you see a message indicating that the procedure will be reset.)
- Pull down the **View menu** and click the **Immediate Window command** (or click the **Immediate Window button** on the Debug toolbar). The Immediate window should be empty, but if not, you can click and drag to select the contents, then press the **Del key** to clear the window.
- Click anywhere within the For . . . Next procedure, then click the **Run Sub button** to execute the procedure. You will see the familiar message box indicating that the sum of the numbers is 55. Click **OK**.
- You should see 10 lines within the Immediate window as shown in Figure 12e, corresponding to the values displayed by the Debug.Print statement as it was executed within the loop.
- Close the Immediate window. Do you see how displaying the intermediate results of a procedure helps you to understand how it works?

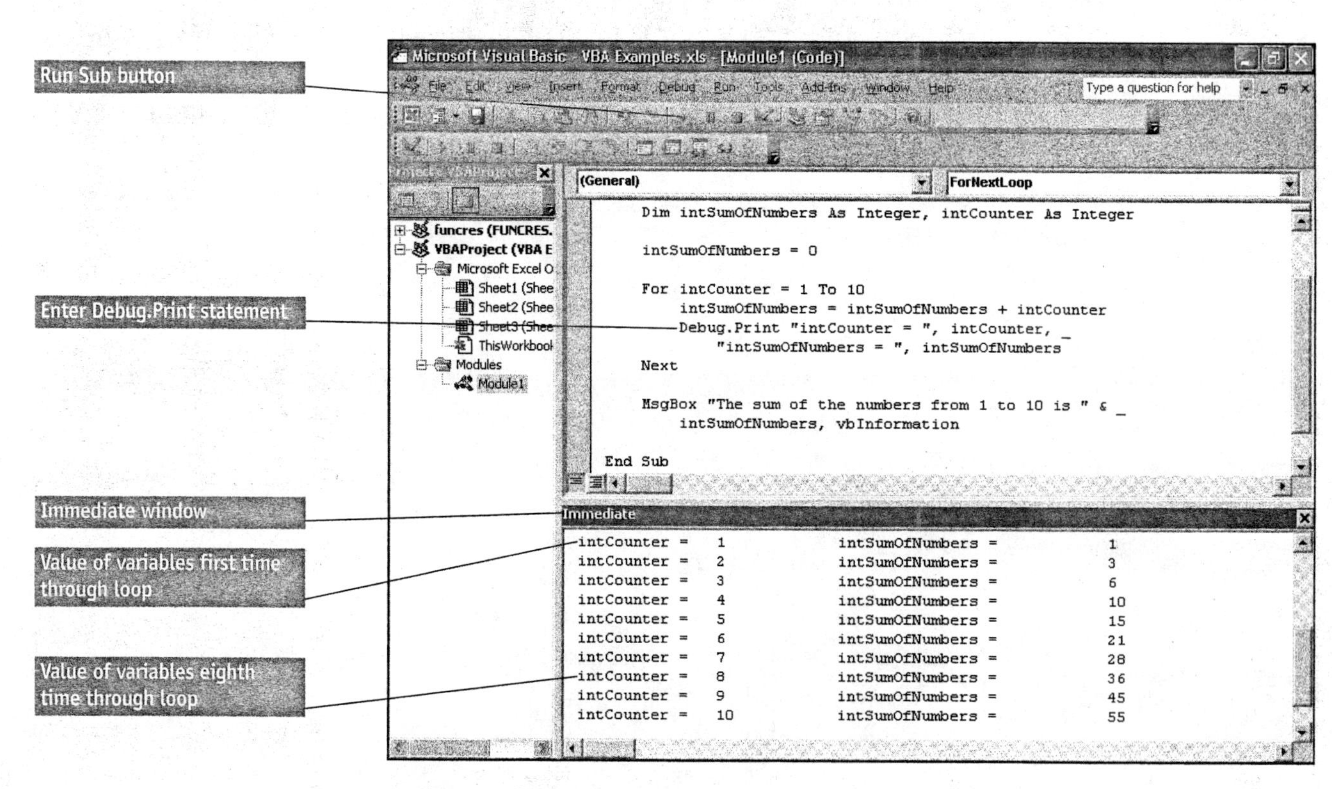

(e) The Immediate Window (step 5)

FIGURE 12 Hands-on Exercise 3 (*continued*)

INSTANT CALCULATOR

Use the Print method (action) in the Immediate window to use VBA as a calculator. Press Ctrl+G at any time to display the Immediate window. Click in the window, then type the statement Debug.Print, followed by your calculation, for example, Debug.Print 2+2, and press Enter. The answer is displayed on the next line in the Immediate window.

Step 6: A More General Procedure

- Modify the existing procedure to make it more general—for example, to sum the values from any starting value to any ending value:
  - Click at the end of the existing Dim statement to position the insertion point, press **Enter** to create a new line, then add the second **Dim statement** as shown in Figure 12f.
  - Click before the For statement, press **Enter** to create a blank line, press **Enter** a second time, then enter the two **InputBox statements** to ask the user for the beginning and ending values.
  - Modify the For statement to execute from **intStart** to **intEnd** rather than from 1 to 10.
  - Change the MsgBox statement to reflect the values of intStart and intEnd, and a customized title bar. Note the use of the ampersand and the underscore, to indicate concatenation and continuation, respectively.
- Click the **Save button** to save the module.

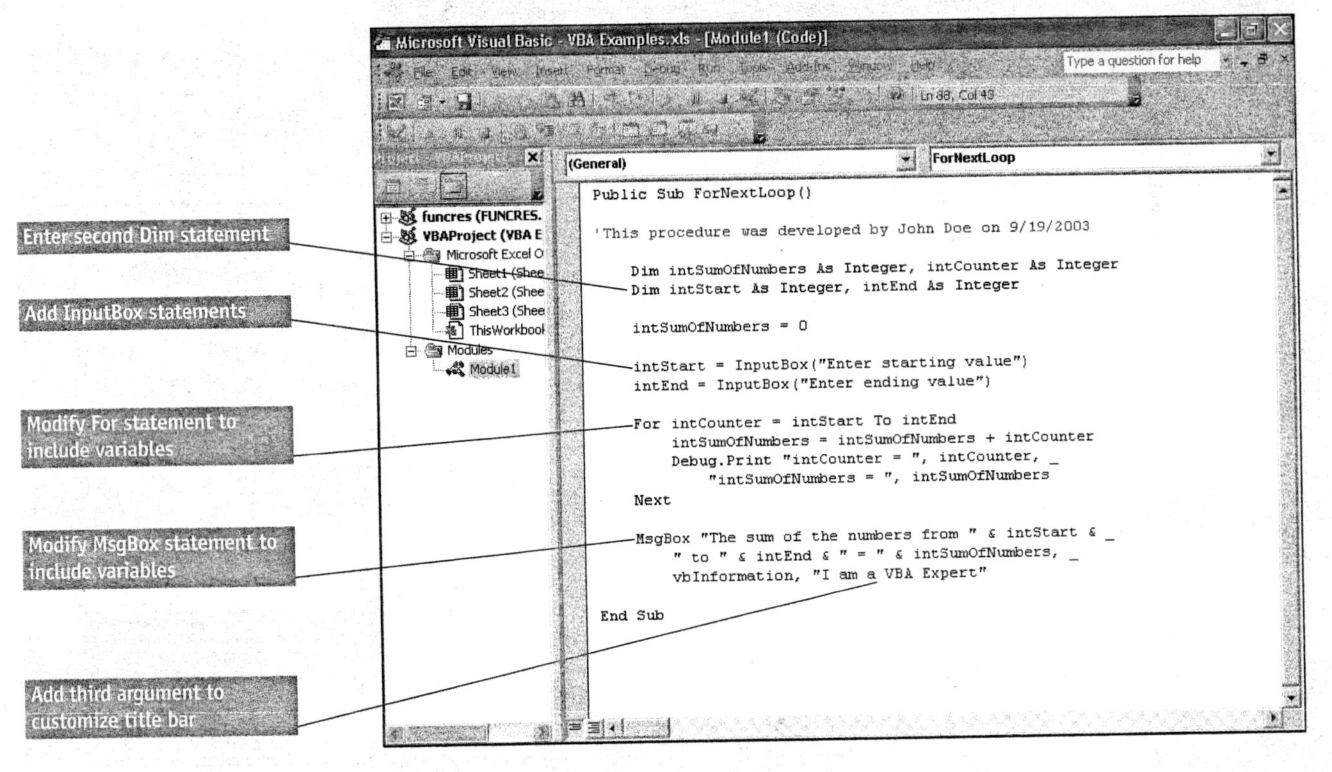

(f) A More General Procedure (step 6)

FIGURE 12 Hands-on Exercise 3 (*continued*)

USE WHAT YOU KNOW

Use the techniques acquired from other applications such as Microsoft Word to facilitate editing within the VBA window. Press the Ins key to toggle between the insert and overtype modes as you modify the statements within a VBA procedure. You can also cut, copy, and paste statements (or parts of statements) within a procedure and from one procedure to another. The Find and Replace commands are also useful.

Step 7: Test the Procedure

- Click the **Run Sub button** to test the procedure. You should be prompted for a beginning and an ending value. Enter any numbers you like, such as 10 and 20, respectively, to match the result in Figure 12g.
- The value displayed in the MsgBox statement should reflect the numbers you entered. For example, you will see a sum of 165 if you entered 10 and 20 as the starting and ending values.
- Look carefully at the message box that is displayed in Figure 12g. Its title bar displays the literal "I am a VBA expert", corresponding to the last argument in the MsgBox statement.
- Note, too, the spacing that appears within the message box, which includes spaces before and after each number. Look at your results and, if necessary, modify the MsgBox statement so that you have the same output. Click **OK**.
- Save the procedure.

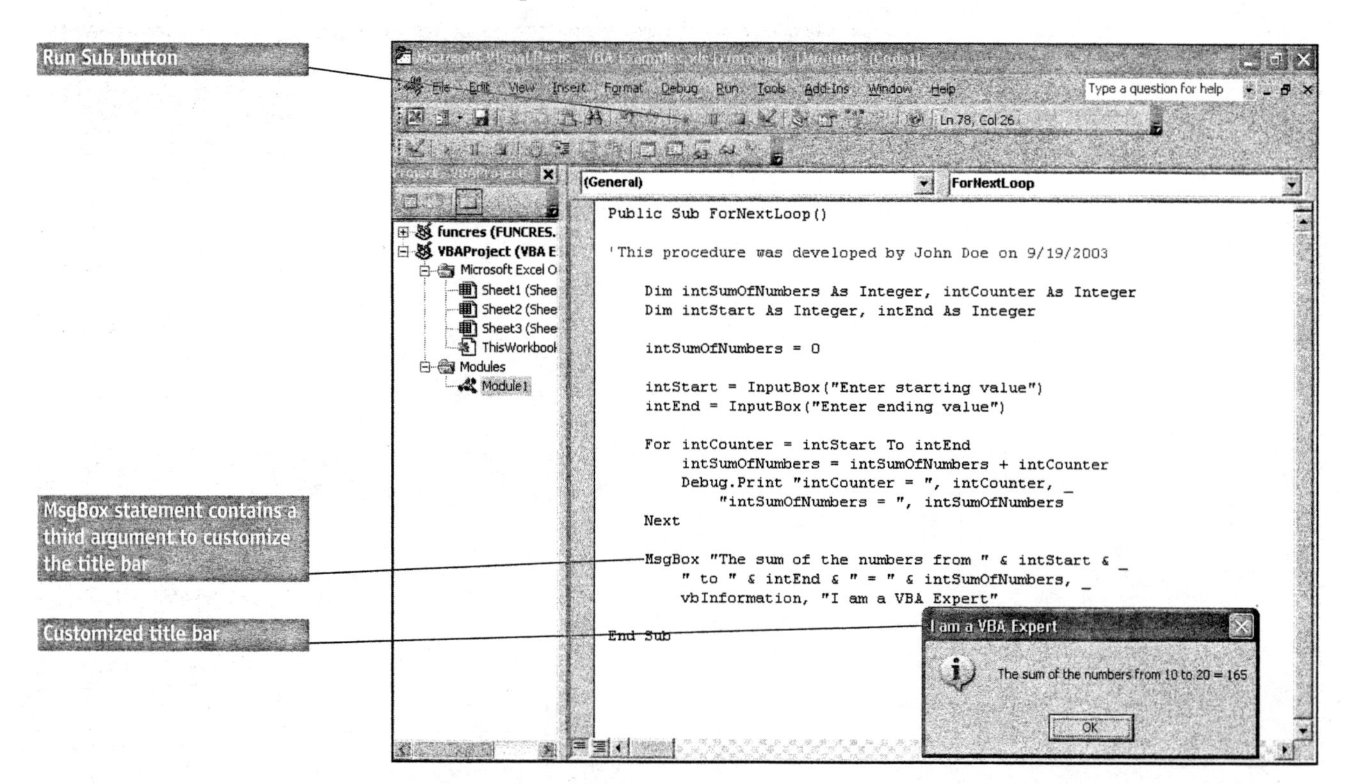

(g) Test the Procedure (step 7)

FIGURE 12 Hands-on Exercise 3 (*continued*)

CHANGE THE INCREMENT

The For . . . Next statement can be made more general by supplying an increment within the For statement. Try For intCount = 1 To 10 Step 2, or more generally, For intCount = intStart to intEnd Step intStepValue. "Step" is a Visual Basic keyword and must be entered that way. intCount, intEnd, and intStepValue are user-defined variables. The variables must be defined at the beginning of a procedure and can be initialized by requesting values from the user through the InputBox statement.

Step 8: Create a Do Until Loop

- Pull down the **Insert menu** and click the **Procedure command** to insert a new procedure called **DoUntilLoop**. Enter the procedure as it appears in Figure 12h. Note the following:
  - Two string variables are declared to hold the correct answer and the user's response, respectively.
  - The variable strCorrectAnswer is set to "Earth", which is the correct answer for our question.
  - The initial InputBox function prompts the user to enter his/her response to the question. A second InputBox function appears in the loop that is executed if and only if the user enters the wrong answer.
  - The Until condition appears at the beginning of the loop, so that the loop is entered only if the user answers incorrectly. The loop executes repeatedly until the correct answer is supplied.
  - A message to the user is displayed at the end of the procedure after the correct answer has been entered.
- Click the **Run Sub button** to test the procedure. Enter the correct answer on your first attempt, and you will see that the loop is never entered.
- Rerun the procedure, answer incorrectly, then note that a second input box appears, telling you that your answer was incorrect. Click **OK**.
- Once again you are prompted for the answer. Enter **Earth**. Click **OK**. The procedure ends.
- Save the procedure.

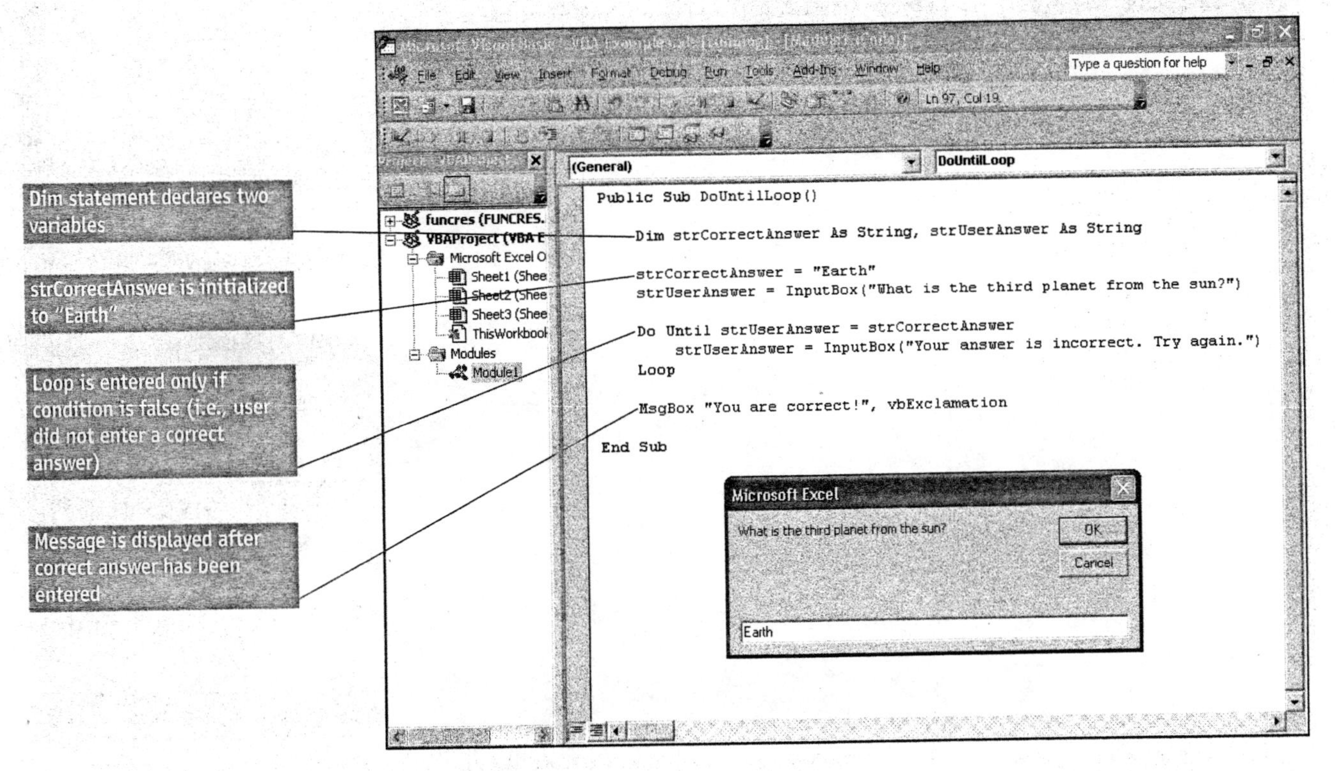

(h) Create a Do Until Loop (step 8)

FIGURE 12 Hands-on Exercise 3 (*continued*)

Step 9: A More Powerful Procedure

- Modify the procedure as shown in Figure 12i to include the statements to count and print the number of times the user takes to get the correct answer.
  - The variable intNumberOfAttempts is declared as an integer and is initialized to 1 after the user inputs his/her initial answer.
  - The Do loop is expanded to increment intNumberOfAttempts by 1 each time the loop is executed.
  - The MsgBox statement after the loop is expanded prints the number of attempts the user took to answer the question.
- Save the module, then click the **Run Sub button** to test the module. You should see a dialog box similar to the one in Figure 12i. Click **OK.** Do you see how this procedure improves on its predecessor?
- Pull down the **File menu** and click the **Print command** to display the Print dialog box. Click the option button to print the current module for your instructor. Click **OK.**
- Close the Debug toolbar. Exit Office if you do not want to continue with the next hands-on exercise at this time.

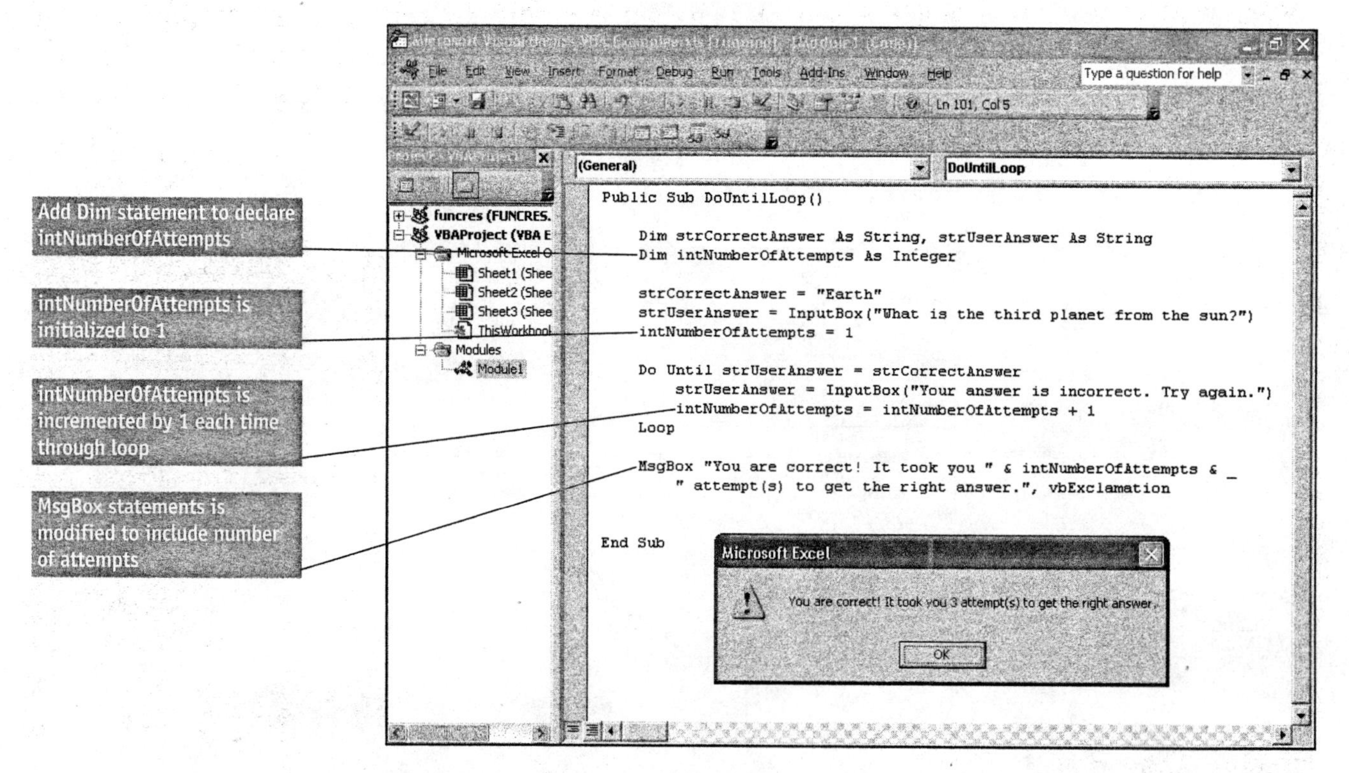

(i) A More Powerful Procedure (step 9)

FIGURE 12 Hands-on Exercise 3 (*continued*)

IT'S NOT EQUAL, BUT REPLACE

All programming languages use statements of the form N = N + 1, in which the equal sign does not mean equal in the literal sense; that is, N cannot equal N + 1. The equal sign is really a replacement operator. Thus, the expression on the right of the equal sign is evaluated, and that result replaces the value of the variable on the left. In other words, the statement N = N + 1 increments the value of N by 1.

Our approach thus far has focused on VBA as an independent entity that can be run without specific reference to the applications in Microsoft Office. We have covered several individual statements, explained how to use the VBA editor to create and run procedures, and how to debug those procedures, if necessary. We hope you have found the material to be interesting, but you may be asking yourself, "What does this have to do with Microsoft Office?" In other words, how can you use your knowledge of VBA to enhance your ability in Microsoft Excel or Access? The answer is to create ***event procedures*** that run automatically in response to events within an Office application.

VBA is different from traditional programming languages in that it is event-driven. An ***event*** is defined as any action that is recognized by an application such as Excel or Access. Opening or closing an Excel workbook or an Access database is an event. Selecting a worksheet within a workbook is also an event, as is clicking on a command button on an Access form. To use VBA within Microsoft Office, you decide which events are significant, and what is to happen when those events occur. Then you develop the appropriate event procedures.

Consider, for example, Figure 13, which displays the results of two event procedures in conjunction with opening and closing an Excel workbook. (If you are using Microsoft Access instead of Excel, you can skip this discussion and the associated exercise, and move to the parallel material for Access that appears after the next hands-on exercise.) The procedure associated with Figure 13a displays a message that appears automatically after the user executes the command to close the associated workbook. The procedure is almost trivial to write, and consists of a single MsgBox statement. The effect of the procedure is quite significant, however, as it reminds the user to back up his or her work after closing the workbook. Nor does it matter how the user closes the workbook—whether by pulling down the menu or using a keyboard shortcut—because the procedure runs automatically in response to the Close Workbook event, regardless of how that event occurs.

The dialog box in Figure 13b prompts the user for a password and appears automatically when the user opens the workbook. The logic here is more sophisticated in that the underlying procedure contains an InputBox statement to request the password, a Do Until loop that is executed until the user enters the correct password or exceeds the allotted number of attempts, then additional logic to display the worksheet or terminate the application if the user fails to enter the proper password. The procedure is not difficult, however, and it builds on the VBA statements that were covered earlier.

The next hands-on exercise has you create the two event procedures that are associated with Figure 13. As you do the exercise, you will gain additional experience with VBA and an appreciation for the potential event procedures within Microsoft Office.

HIDING AND UNHIDING A WORKSHEET

Look carefully at the workbooks in Figures 13a and 13b. Both figures reference the identical workbook, Financial Consultant, as can be seen from the title bar. Look at the worksheet tabs, however, and note that two worksheets are visible in Figure 13a, whereas the Calculations worksheet is hidden in Figure 13b. This was accomplished in the Open workbook procedure and was implemented to hide the calculations from the user until the correct password was entered.

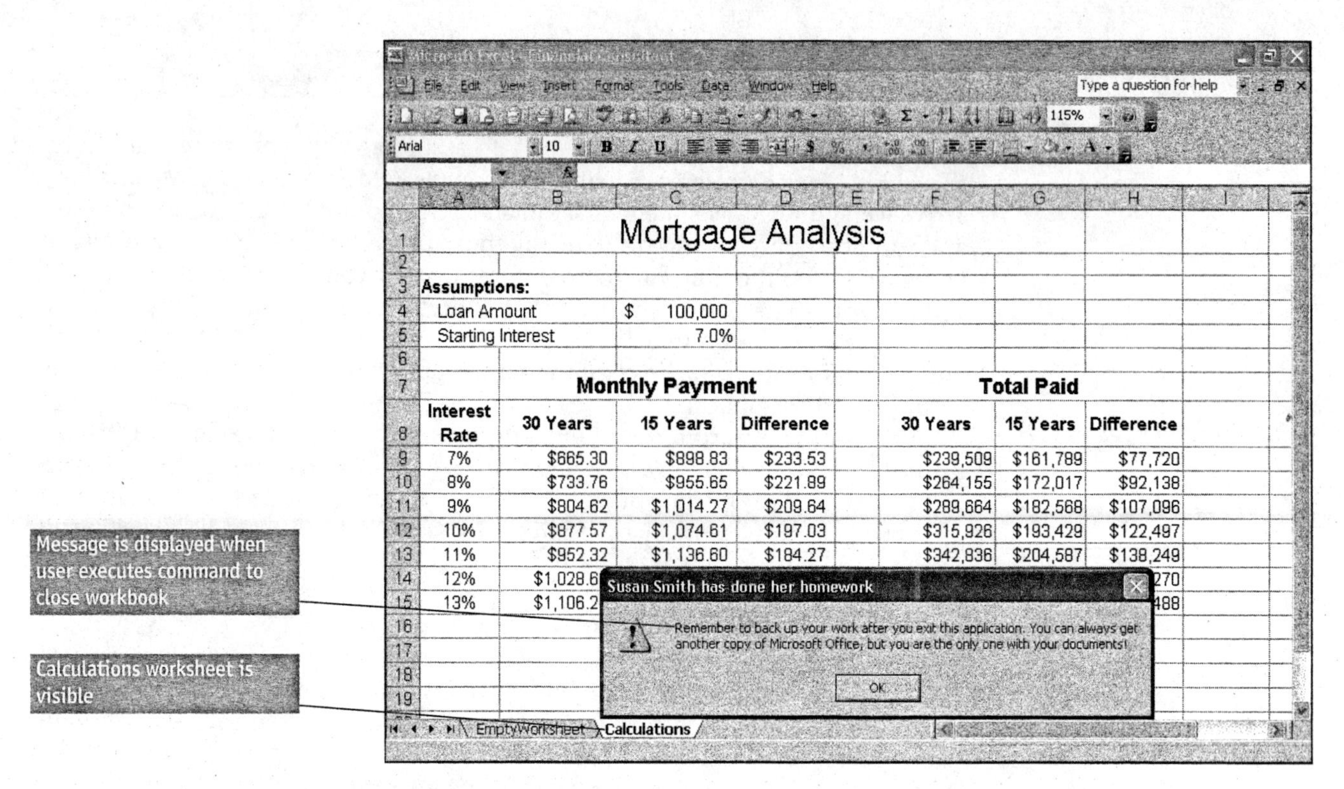

(a) Message to the User (Close Workbook event)

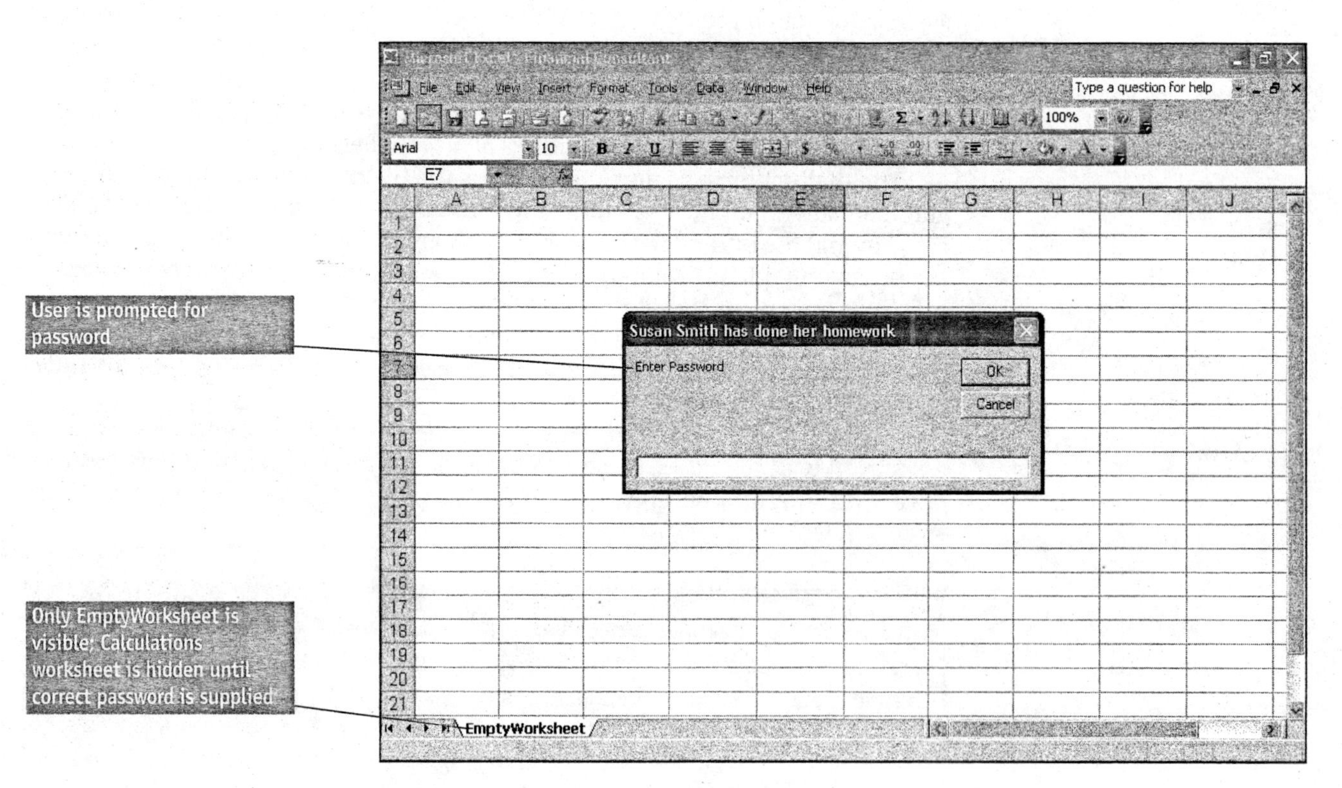

(b) Password Protection (Open Workbook event)

FIGURE 13 Event-Driven Programming

hands-on exercise

4 Event-Driven Programming (Microsoft Excel)

Objective To create an event procedure to implement password protection that is associated with opening an Excel workbook; to create a second event procedure that displays a message to the user upon closing the workbook. Use Figure 14 as a guide in the exercise.

Step 1: Create the Close Workbook Procedure

- Open the **VBA Examples workbook** you have used for the previous exercises and enable the macros. If you have been using Access rather than Excel, start Excel, open a new workbook, then save the workbook as **VBA Examples.**
- Pull down the **Tools menu**, click the **Macro command**, then click the **Visual Basic Editor command** (or use the **Alt+F11** keyboard shortcut).
- You should see the Project Explorer pane as shown in Figure 14a, but if not, pull down the **View menu** and click the **Project Explorer**. Double click **ThisWorkbook** to create a module for the workbook as a whole.
- Enter the **Option Explicit statement** if it is not there already, then press **Enter** to create a new line. Type the statement to declare the variable, **ApplicationTitle**, using your name instead of Susan Smith.
- Click the **down arrow** in the Object list box and select **Workbook**, then click the **down arrow** in the Procedure list box and select the **BeforeClose event** to create the associated procedure. (If you choose a different event by mistake, click and drag to select the associated statements, then press the **Del key** to delete the procedure.)
- Enter the comment and MsgBox statement as it appears in Figure 14a.
- Save the procedure.

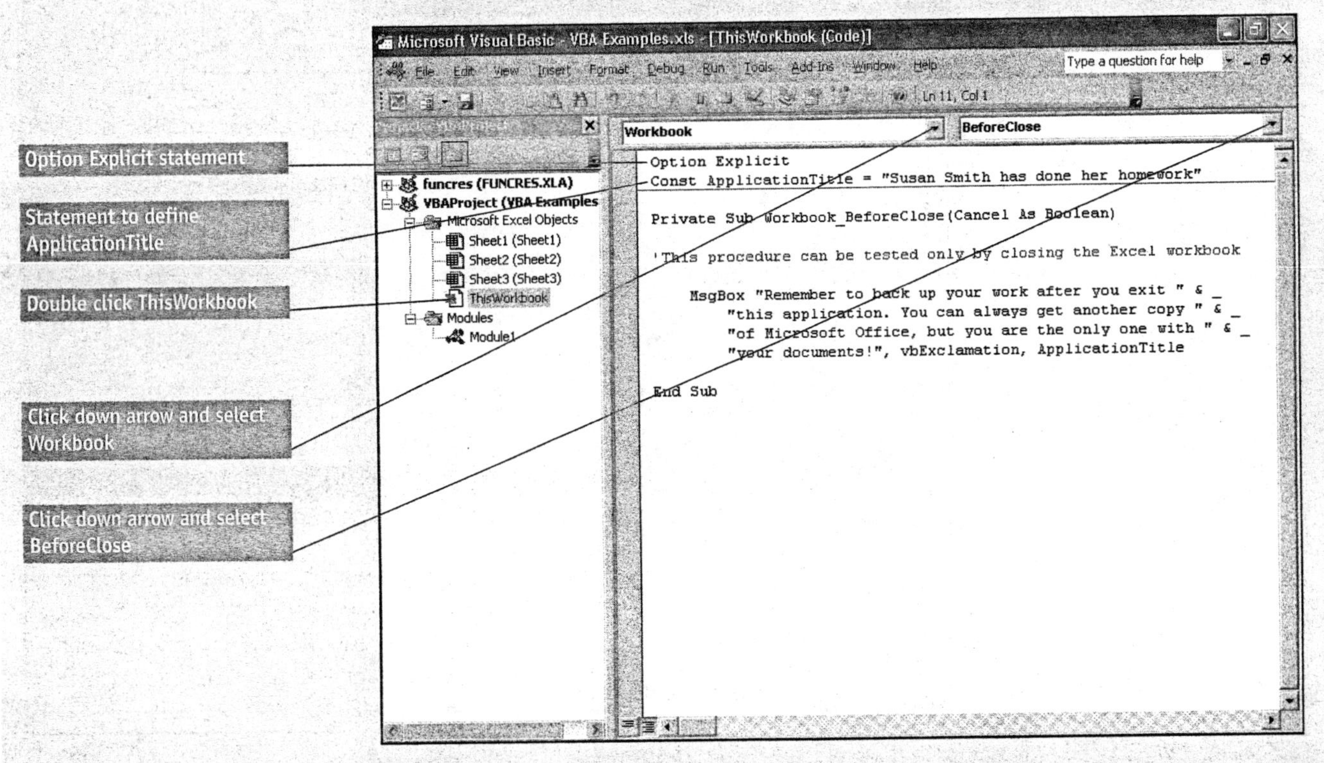

(a) Create the Close Workbook Procedure (step 1)

FIGURE 14 Hands-on Exercise 4

Step 2: Test the Close Workbook Procedure

- Click the **View Microsoft Excel button** on the Standard toolbar or on the Windows taskbar to view the Excel workbook. The workbook is not empty; that is, it does not contain any cell entries, but it does contain multiple VBA procedures.
- Pull down the **File menu** and click the **Close command**, which runs the procedure you just created and displays the dialog box in Figure 14b. Click **OK** after you have read the message, then click **Yes** if asked to save the workbook.
- Pull down the **File menu** and reopen the **VBA Examples workbook**, enabling the macros. Press **Alt+F11** to return to the VBA window to create an additional procedure.
- Double click **ThisWorkbook** from within the Project Explorer pane to return to the BeforeClose procedure and make the necessary corrections, if any.
- Save the procedure.

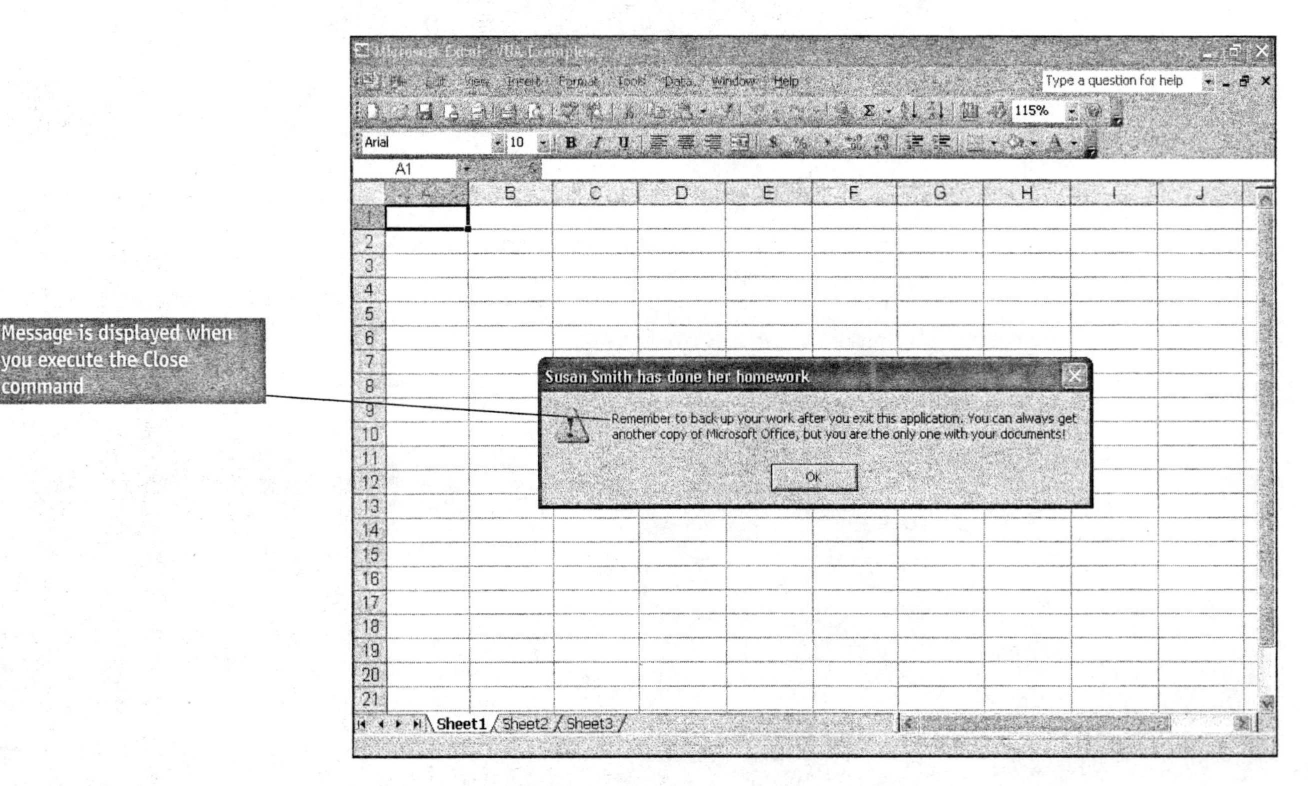

(b) Test the Close Workbook Procedure (step 2)

FIGURE 14 Hands-on Exercise 4 (*continued*)

THE MOST RECENTLY OPENED FILE LIST

One way to open a recently used workbook is to select the workbook directly from the File menu. Pull down the File menu, but instead of clicking the Open command, check to see if the workbook appears on the list of the most recently opened workbooks located at the bottom of the menu. If so, just click the workbook name, rather than having to make the appropriate selections through the Open dialog box.

Step 3: Start the Open Workbook Event Procedure

- Click within the Before Close procedure, then click the **Procedure View button** at the bottom of the Code window. Click the **down arrow** in the Procedure list box and select the **Open event** to create an event procedure.
- Enter the VBA statements as shown in Figure 14c. Note the following:
  - Three variables are required for this procedure—the correct password, the password entered by the user, and the number of attempts.
  - The user is prompted for the password, and the number of attempts is set to 1. The user is given two additional attempts, if necessary, to get the password correct. The loop is bypassed, however, if the user supplies the correct password on the first attempt.
- Minimize Excel. Save the procedure, then click the **Run Sub button** to test it. Try different combinations in your testing; that is, enter the correct password on the first, second, and third attempts. The password is case-sensitive.
- Correct errors as they occur. Click the **Reset button** at any time to begin executing the procedure from the beginning. Save the procedure.

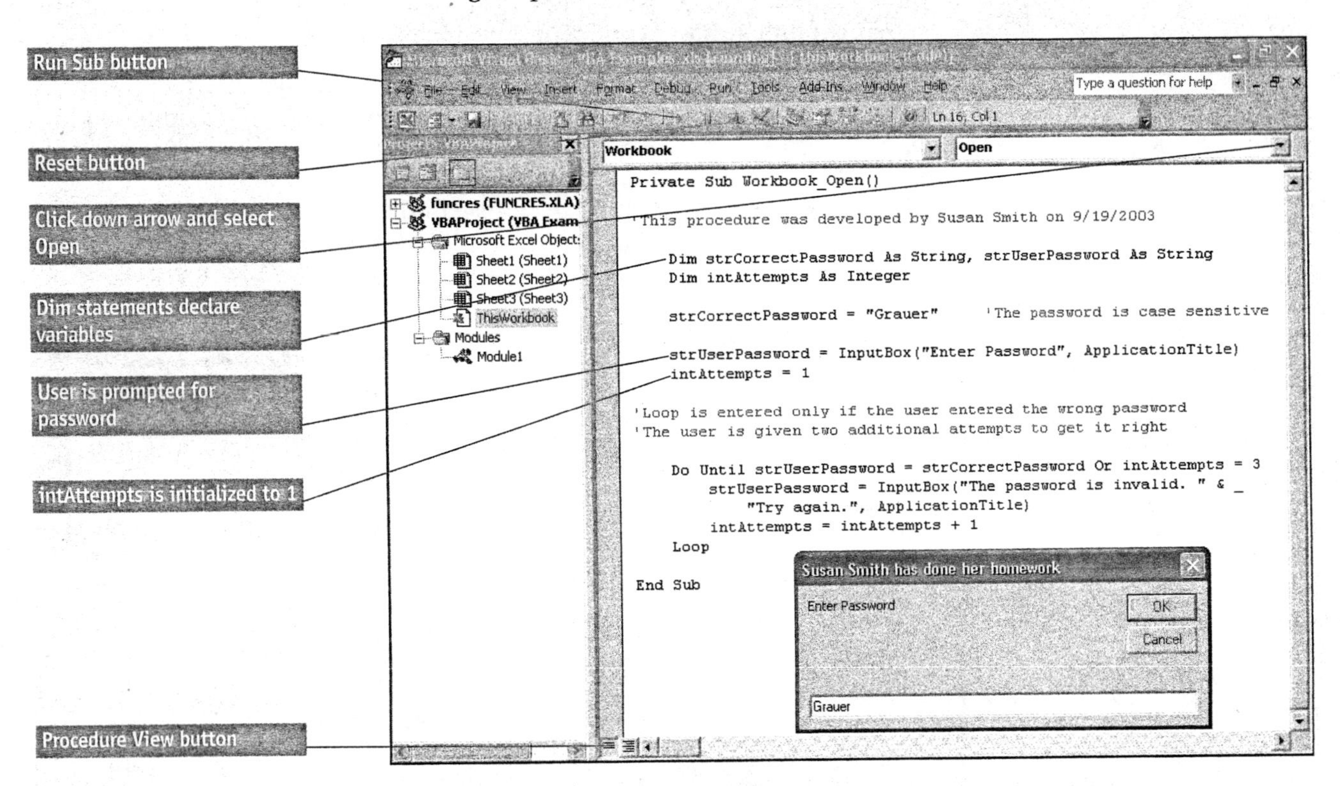

(c) Start the Open Workbook Event Procedure (step 3)

FIGURE 14 Hands-on Exercise 4 (*continued*)

THE OBJECT AND PROCEDURE BOXES

The Object box at the top of the code window displays the selected object such as an Excel workbook, whereas the Procedure box displays the name of the events appropriate to that object. Events that already have procedures appear in bold. Clicking an event that is not bold creates the procedure header and End Sub statements for that event.

Step 4: Complete the Open Workbook Event Procedure

- Enter the remaining statements in the procedure as shown in Figure 14d. Note the following:
  - The If statement determines whether the user has entered the correct password and, if so, displays the appropriate message.
  - If, however, the user fails to supply the correct password, a different message is displayed, and the workbook will close due to the **Workbooks.Close statement** within the procedure.
  - As a precaution, put an apostrophe in front of the Workbooks.Close statement so that it is a comment, and thus it is not executed. Once you are sure that you can enter the correct password, you can remove the apostrophe and implement the password protection.
- Save the procedure, then click the **Run Sub button** to test it. Be sure that you can enter the correct password (**Grauer**), and that you realize the password is case-sensitive.
- Delete the apostrophe in front of the Workbooks.Close statement. The text of the statement changes from green to black to indicate that it is an executable statement rather than a comment. Save the procedure.
- Click the **Run Sub button** a second time, then enter an incorrect password three times in a row. You will see the dialog box in Figure 14d, followed by a message reminding you to back up your workbook, and then the workbook will close.
- The first message makes sense, the second does not make sense in this context. Thus, we need to modify the Close Workbook procedure when an incorrect password is entered.

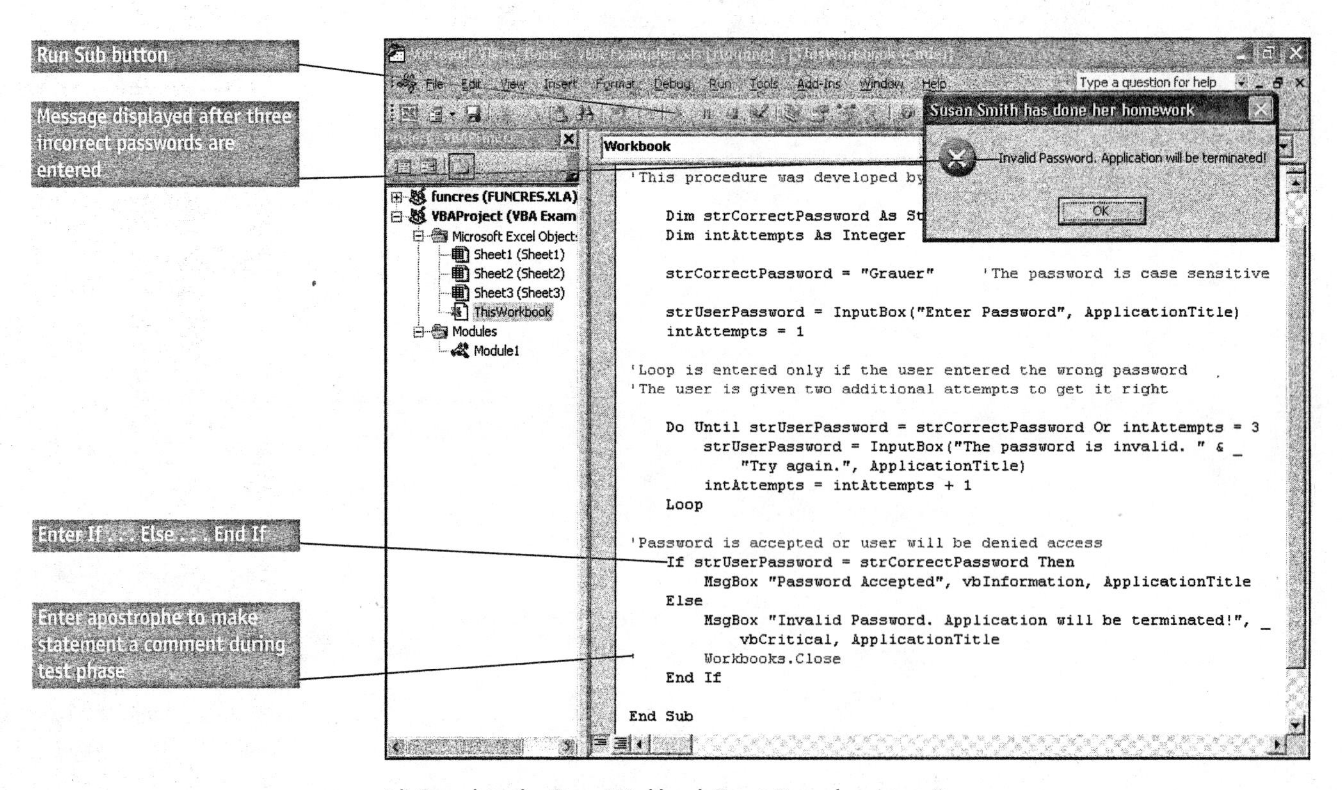

(d) Complete the Open Workbook Event Procedure (step 4)

FIGURE 14 Hands-on Exercise 4 (*continued*)

Step 5: **Modify the Before Close Event Procedure**

- Reopen the **VBA Examples workbook.** Click the button to **Enable Macros.**
- Enter the password, **Grauer** (the password is case-sensitive), press **Enter**, then click **OK** when the password has been accepted.
- Press **Alt+F11** to reopen the VBA editor, and (if necessary) double click **ThisWorkbook** within the list of Microsoft Excel objects.
- Click at the end of the line defining the ApplicationTitle constant, press **Enter**, then enter the statement to define the **binNormalExit** variable as shown in Figure 14e. (The statement appears initially below the line ending the General Declarations section, but moves above the line when you press Enter.)
- Modify the BeforeClose event procedure to include an If statement that tests the value of the binNormalExit variable as shown in Figure 14e. You must, however, set the value of this variable in the Open Workbook event procedure as described in step 6.
- Save the procedure.

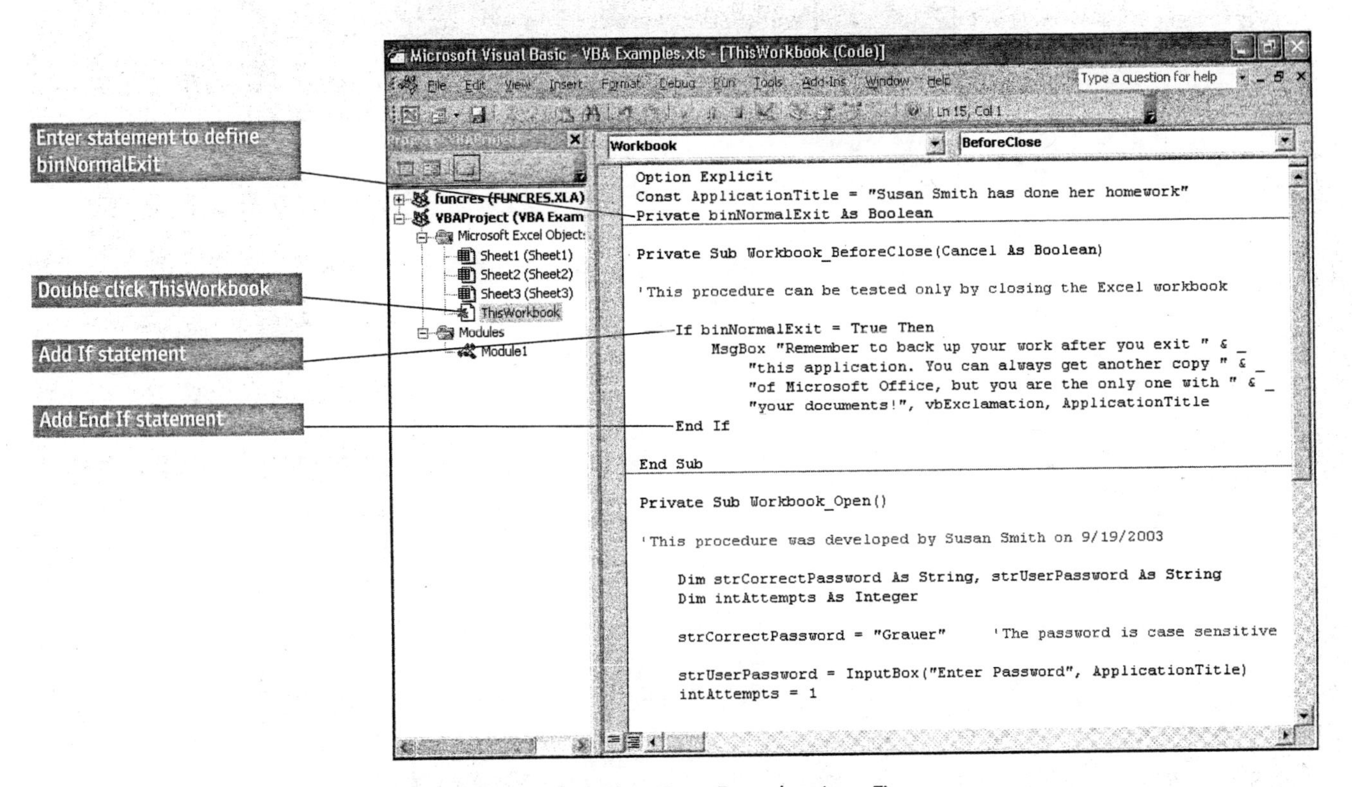

(e) Modify the Before Close Event Procedure (step 5)

FIGURE 14 Hands-on Exercise 4 (*continued*)

SETTING A SWITCH

The use of a switch (binNormalExit, in this example) to control an action within a procedure is a common programming technique. The switch is set to one of two values according to events that occur within the system, then the switch is subsequently tested and the appropriate action is taken. Here, the switch is set when the workbook is opened to indicate either a valid or invalid user. The switch is then tested prior to closing the workbook to determine whether to print the closing message.

Step 6: Modify the Open Workbook Event Procedure

- Scroll down to the Open Workbook event procedure, then modify the If statement to set the value of binNormalExit as shown in Figure 14f:
  - Take advantage of the Complete Word tool to enter the variable name. Type the first few letters, "**binN**", then press **Ctrl+Space**, and VBA will complete the variable name.
  - The indentation within the statement is not a requirement of VBA per se, but is used to make the code easier to read. Blank lines are also added for this purpose.
  - Comments appear throughout the procedure to explain its logic.
  - Save the modified procedure.
- Click the **Run Sub button**, then enter an incorrect password three times in a row. Once again, you will see the dialog box indicating an invalid password.
- Click **OK**. This time you will not see the message reminding you to back up your workbook. The workbook closes as before.

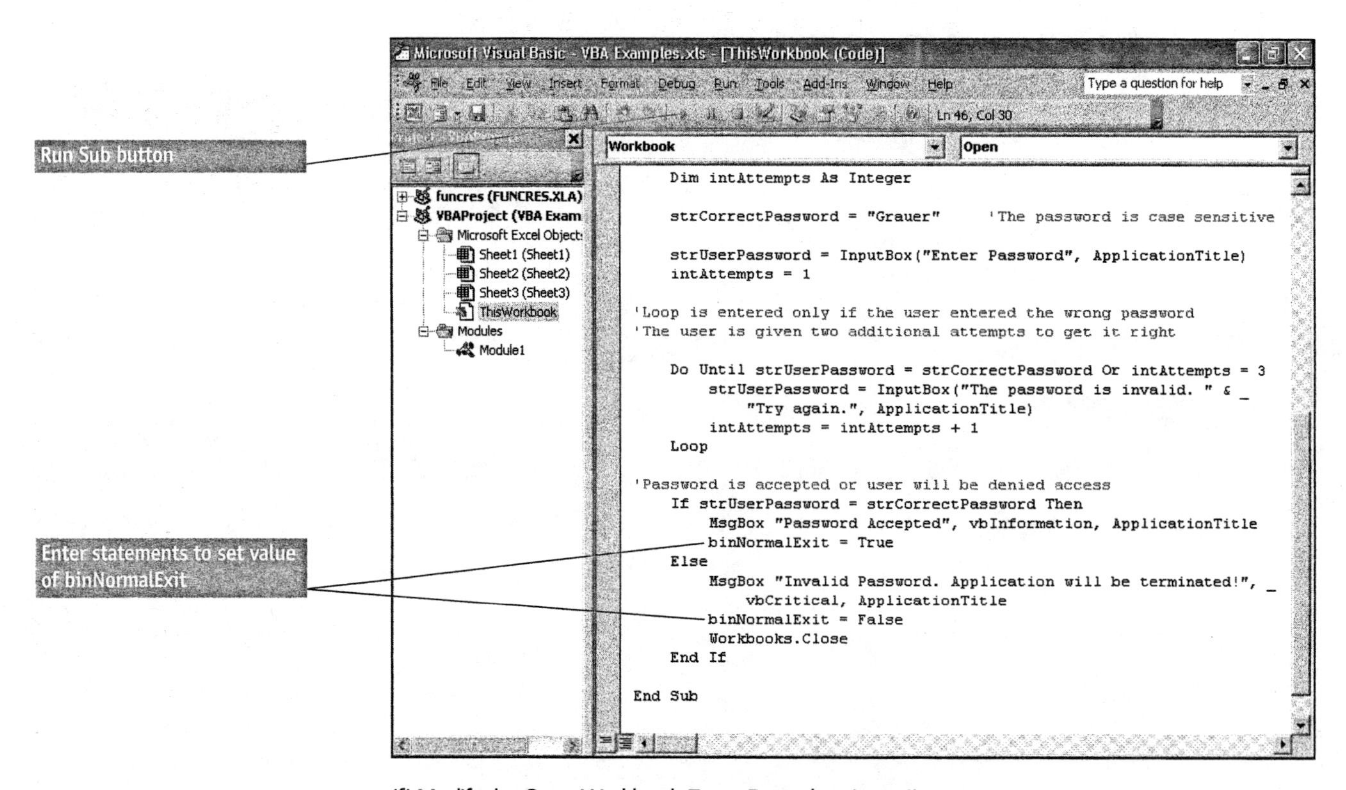

(f) Modify the Open Workbook Event Procedure (step 6)

FIGURE 14 Hands-on Exercise 4 (*continued*)

TEST UNDER ALL CONDITIONS

We cannot overemphasize the importance of thoroughly testing a procedure, and further, testing it under all conditions. VBA statements are powerful, but they are also complex, and a misplaced or omitted character can have dramatic consequences. Test every procedure completely at the time it is created, while the logic of the procedure is fresh in your mind.

Step 7: Open a Second Workbook

- Reopen the **VBA Examples workbook.** Click the button to **Enable Macros.**
- Enter the password, **Grauer**, then press **Enter**. Click **OK** when you see the second dialog box telling you that the password has been accepted.
- Pull down the **File menu** and click the **Open command** (or click the **Open button** on the Standard toolbar) and open a second workbook. We opened a workbook called **Financial Consultant**, but it does not matter which workbook you open.
- Pull down the **Window menu**, click the **Arrange command**, click the **Horizontal option button**, and click **OK** to tile the workbooks as shown in Figure 14g. The title bars show the names of the open workbooks.
- Pull down the **Tools menu**, click **Macro**, then click **Visual Basic editor.**

Financial Consultant workbook

VBA Examples workbook

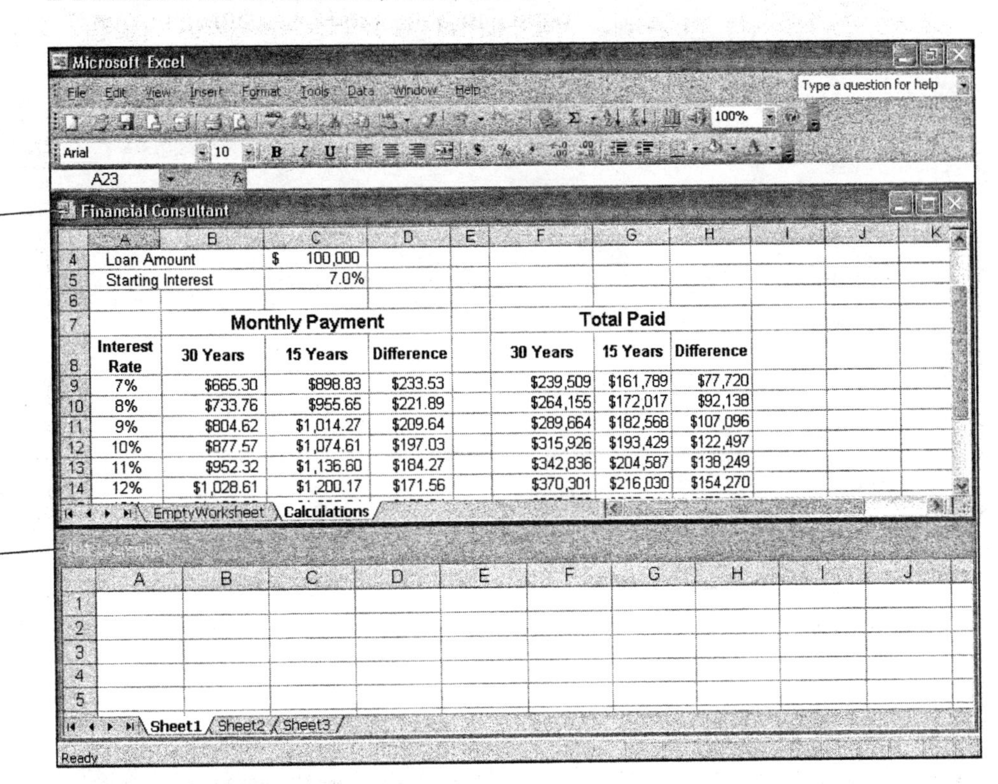

	B	C	D	E	F	G	H
4	Loan Amount	$ 100,000					
5	Starting Interest	7.0%					
6							
7		Monthly Payment			Total Paid		
8	Interest Rate	30 Years	15 Years	Difference	30 Years	15 Years	Difference
9	7%	$665.30	$898.83	$233.53	$239,509	$161,789	$77,720
10	8%	$733.76	$955.65	$221.89	$264,155	$172,017	$92,138
11	9%	$804.62	$1,014.27	$209.64	$289,664	$182,568	$107,096
12	10%	$877.57	$1,074.61	$197.03	$315,926	$193,429	$122,497
13	11%	$952.32	$1,136.60	$184.27	$342,836	$204,587	$138,249
14	12%	$1,028.61	$1,200.17	$171.56	$370,301	$216,030	$154,270

(g) Open a Second Workbook (step 7)

FIGURE 14 Hands-on Exercise 4 (*continued*)

THE COMPARISON IS CASE-SENSITIVE

Any literal comparison (e.g., strInstructorName = "Grauer") is case-sensitive, so that the user has to enter the correct name and case for the condition to be true. A response of "GRAUER" or "grauer", while containing the correct name, will be evaluated as false because the case does not match. You can, however, use the UCase (uppercase) function to convert the user's response to uppercase, and test accordingly. In other words, UCase(strInstructorName) = "GRAUER" will be evaluated as true if the user enters "Grauer" in any combination of upper- or lowercase letters.

Step 8: Copy the Procedure

- You should be back in the Visual Basic editor as shown in Figure 14h. Copy the procedures associated with the Open and Close Workbook events from the VBA Examples workbook to the other workbook, Financial Consultant.
  - ❑ Double click **ThisWorkbook** within the list of Microsoft Excel objects under the VBA Examples workbook.
  - ❑ Click and drag to select the definition of the ApplicationTitle constant in the General Declarations section, the binNormalExit definition, plus the two procedures (to open and close the workbook) in their entirety.
  - ❑ Click the **Copy button** on the Standard toolbar.
  - ❑ If necessary, expand the Financial Consultant VBA Project, then double click **ThisWorkbook** with the list of Excel objects under the Financial Consultant workbook. Click underneath the **Option Explicit command.**
  - ❑ Click the **Paste button** on the Standard toolbar. The VBA code should be copied into this module as shown in Figure 14h.
- Click the **Save button** to save the module.

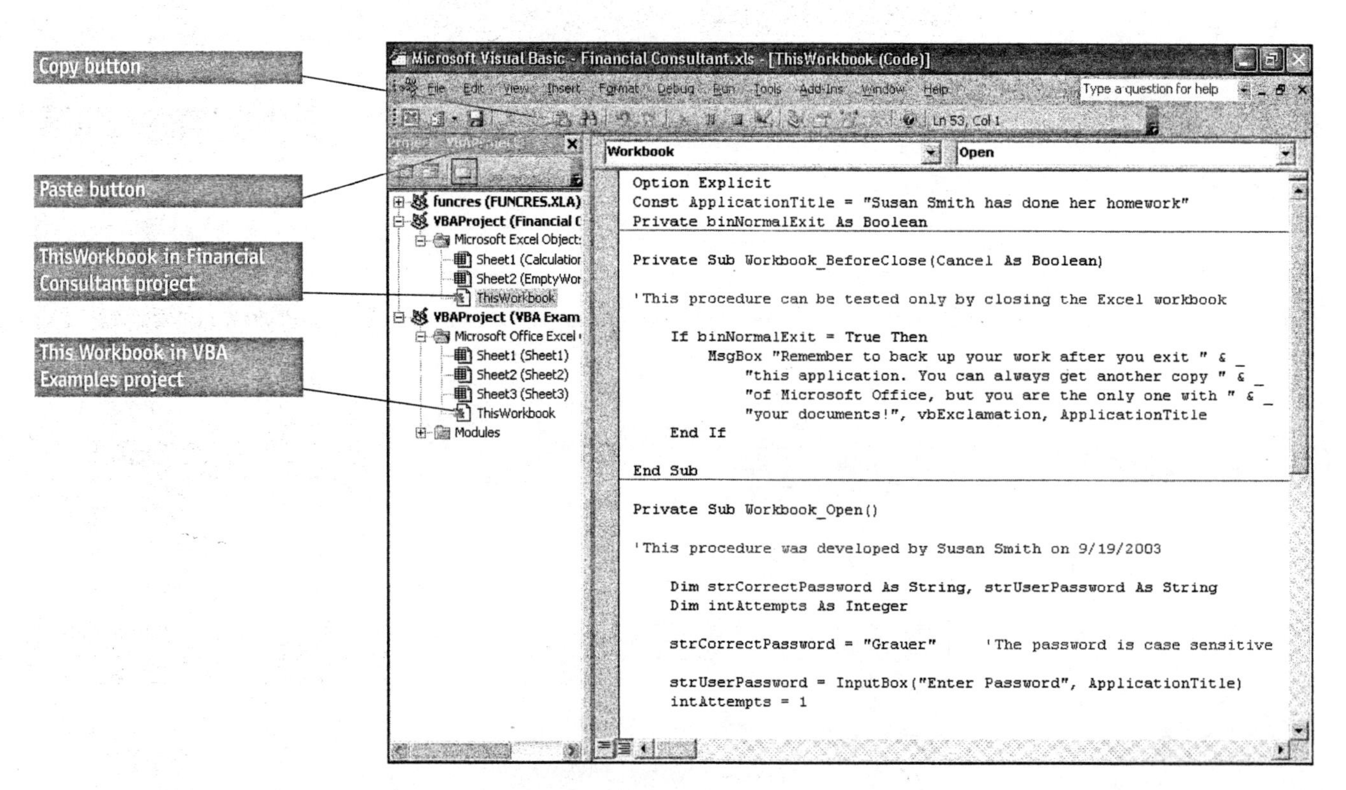

(h) Copy the Procedure (step 8)

FIGURE 14 Hands-on Exercise 4 (*continued*)

THE VISIBLE PROPERTY

The Calculations worksheet sheet should be hidden until the user enters the correct password. This is accomplished by setting the Visible property of the worksheet to false at the beginning of the Open Workbook event procedure, then setting it to true after the correct password has been entered. Click in the Open Workbook event procedure after the last Dim statement, press Enter, then enter the statement Sheet1.Visible = False to hide the Calculations worksheet. Scroll down in the procedure (below the MsgBox statement within the If statement that tests for the correct password), then enter the statement Sheet1.Visible = True followed by the statement Sheet1.Activate to select the worksheet.

Step 9: Test the Procedure

- Click the **View Microsoft Excel button** on the Standard toolbar within the VBA window (or click the **Excel button** on the Windows taskbar) to view the Excel workbook. Click in the window containing the Financial Consultant workbook (or whichever workbook you are using), then click the **Maximize button.**
- Pull down the **File menu** and click the **Close command.** (The dialog box in Figure 14i does not appear initially because the value of binNormalExit is not yet set; you have to open the workbook to set the switch.) Click **Yes** if asked whether to save the changes to the workbook.
- Pull down the **File menu** and reopen the workbook. Click the button to **Enable Macros,** then enter **Grauer** when prompted for the password. Click **OK** when the password has been accepted.
- Close this workbook, close the **VBA Examples workbook,** then pull down the **File menu** and click the **Exit command** to quit Excel.

Customized title bar

Message is displayed when you execute command to close workbook

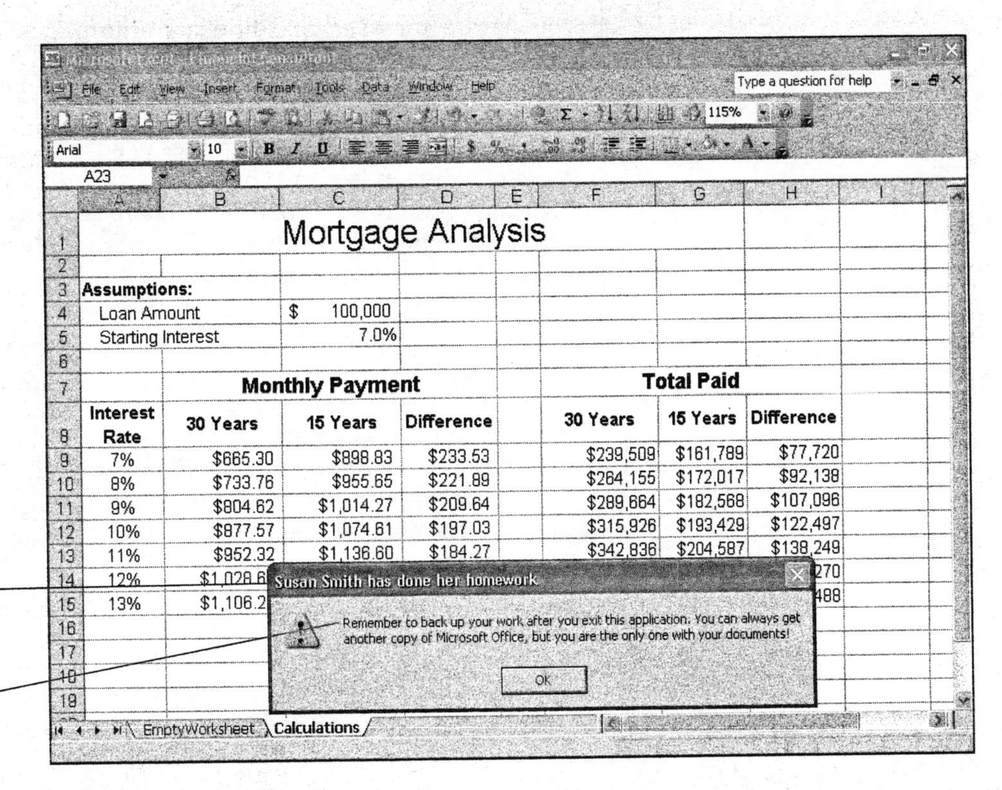

(i) Test the Procedure (step 9)

FIGURE 14 Hands-on Exercise 4 (*continued*)

SCREEN CAPTURE

Prove to your instructor that you have completed the hands-on exercise correctly by capturing a screen, then pasting the screen into a Word document. Do the exercise until you come to the screen that you want to capture, then press the PrintScreen key at the top of the keyboard. Click the Start button, start Word, and open a Word document, then pull down the Edit menu and click the Paste command to bring the captured screen into the Word document. Right click the screen within the Word document, click the Format Picture command, click the Layout tab, and select the Square layout. Click OK to close the dialog box. You can now move and size the screen within the document.

PUTTING VBA TO WORK (MICROSOFT ACCESS)

The same VBA procedure can be run from multiple applications in Microsoft Office, despite the fact that the applications are very different. The real power of VBA, however, is its ability to detect events that are unique to a specific application and to respond accordingly. An event is defined as any action that is recognized by an application. Opening or closing an Excel workbook or an Access database is an event. Selecting a worksheet within a workbook is also an event, as is clicking on a command button on an Access form. To use VBA within Microsoft Office, you decide which events are significant, and what is to happen when those events occur. Then you develop the appropriate ***event procedures*** that execute automatically when the event occurs.

Consider, for example, Figure 15, which displays the results of two event procedures in conjunction with opening and closing an Access database. (These are procedures similar to those we created in the preceding pages in conjunction with opening and closing an Excel workbook.) The procedure associated with Figure 15a displays a message that appears automatically after the user clicks the Switchboard button to exit the database. The procedure is almost trivial to write, and consists of a single MsgBox statement. The effect of the procedure is quite significant, however, as it reminds the user to back up his or her work. Indeed, you can never overemphasize the importance of adequate backup.

The dialog box in Figure 15b prompts the user for a password and appears automatically when the user opens the database. The logic here is more sophisticated in that the underlying procedure contains an InputBox statement to request the password, a Do Until loop that is executed until the user enters the correct password or exceeds the allotted number of attempts, then additional logic to display the switchboard or terminate the application if the user fails to enter the proper password. The procedure is not difficult, however, and it builds on the VBA statements that were covered earlier.

The next hands-on exercise has you create the event procedures that are associated with the database in Figure 15. The exercise references a switchboard, or user interface, that is created as a form within the database. The switchboard displays a menu that enables a nontechnical person to move easily from one object in the database (e.g., a form or report) to another.

The switchboard is created through a utility called the Switchboard Manager that prompts you for each item you want to add to the switchboard, and which action you want taken in conjunction with that menu item. You could do the exercise with any database, but we suggest you use the database we provide to access the switchboard that we created for you. The exercise begins, therefore, by having you download a data disk from our Web site.

EVENT-DRIVEN VERSUS TRADITIONAL PROGRAMMING

A traditional program is executed sequentially, beginning with the first line of code and continuing in order through the remainder of the program. It is the program, not the user, that determines the order in which the statements are executed. VBA, on the other hand, is event-driven, meaning that the order in which the procedures are executed depends on the events that occur. It is the user, rather than the program, that determines which events occur, and consequently which procedures are executed. Each application in Microsoft Office has a different set of objects and associated events that comprise the application's object model.

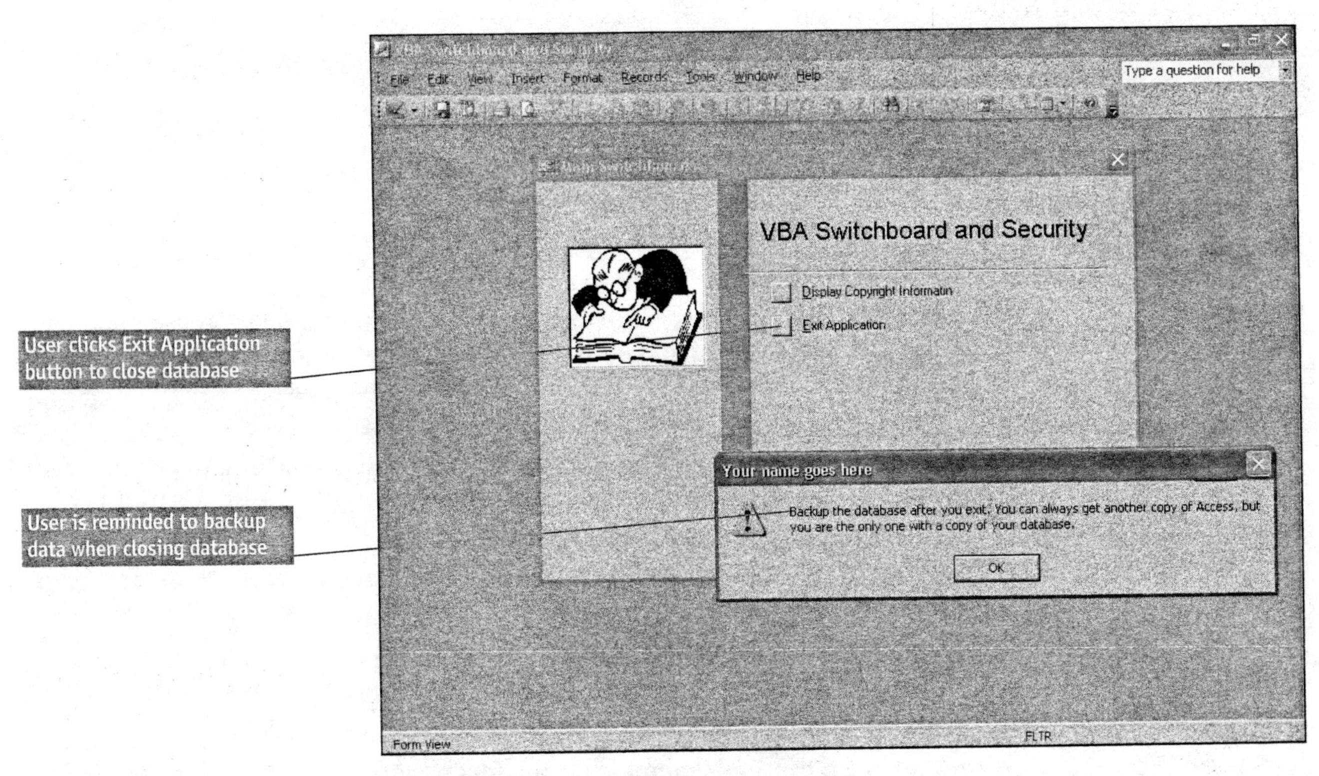

(a) Reminder to the User (Exit Application event)

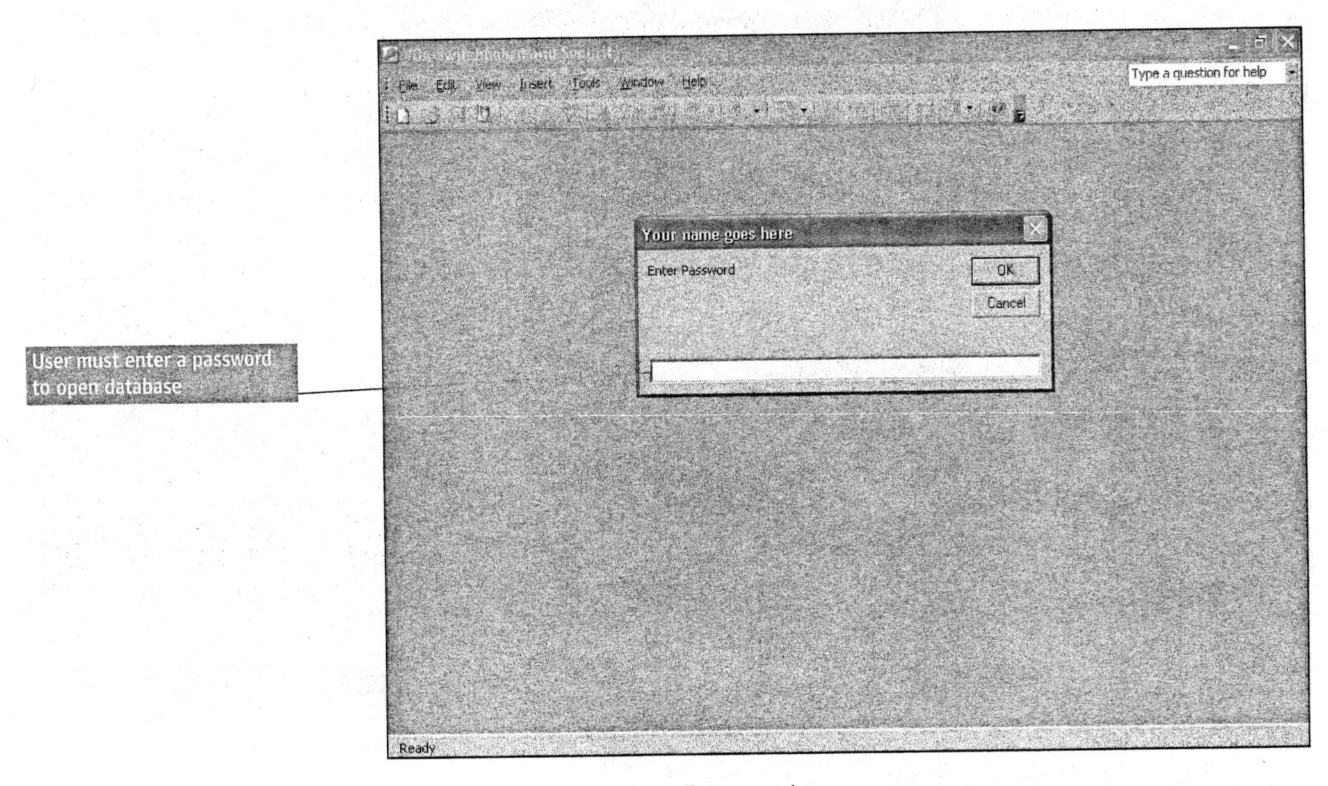

(b) Password Protection (Open Form event)

FIGURE 15 Event-Driven Programming (Microsoft Access)

hands-on exercise

5 Event-Driven Programming (Microsoft Access)

Objective To implement password protection for an Access database; to create a second event procedure that displays a message to the user upon closing the database. Use Figure 16 as a guide in the exercise.

Step 1: Open the Access Database

- You can do this exercise with any database, but we suggest you use the database we have provided. Go to **www.prenhall.com/grauer**, click the **Office 2003 book**, which takes you to the Office 2003 home page. Click the **Student Download tab** to go to the Student Download page.
- Scroll until you can click the link for **Getting Started with VBA**. You will see the File Download dialog box asking what you want to do. Click the **Save button** to display the Save As dialog box, then save the file on your desktop.
- Double click the file after it has been downloaded and follow the onscreen instructions to expand the self-extracting file that contains the database.
- Go to the newly created **Exploring VBA folder** and open the **VBA Switchboard and Security database**. Click the **Open button** when you see the security warning. You should see the Database window in Figure 16a.
- Pull down the **Tools menu**, click the **Macro command**, then click the **Visual Basic Editor command**. Maximize the VBA editor window.

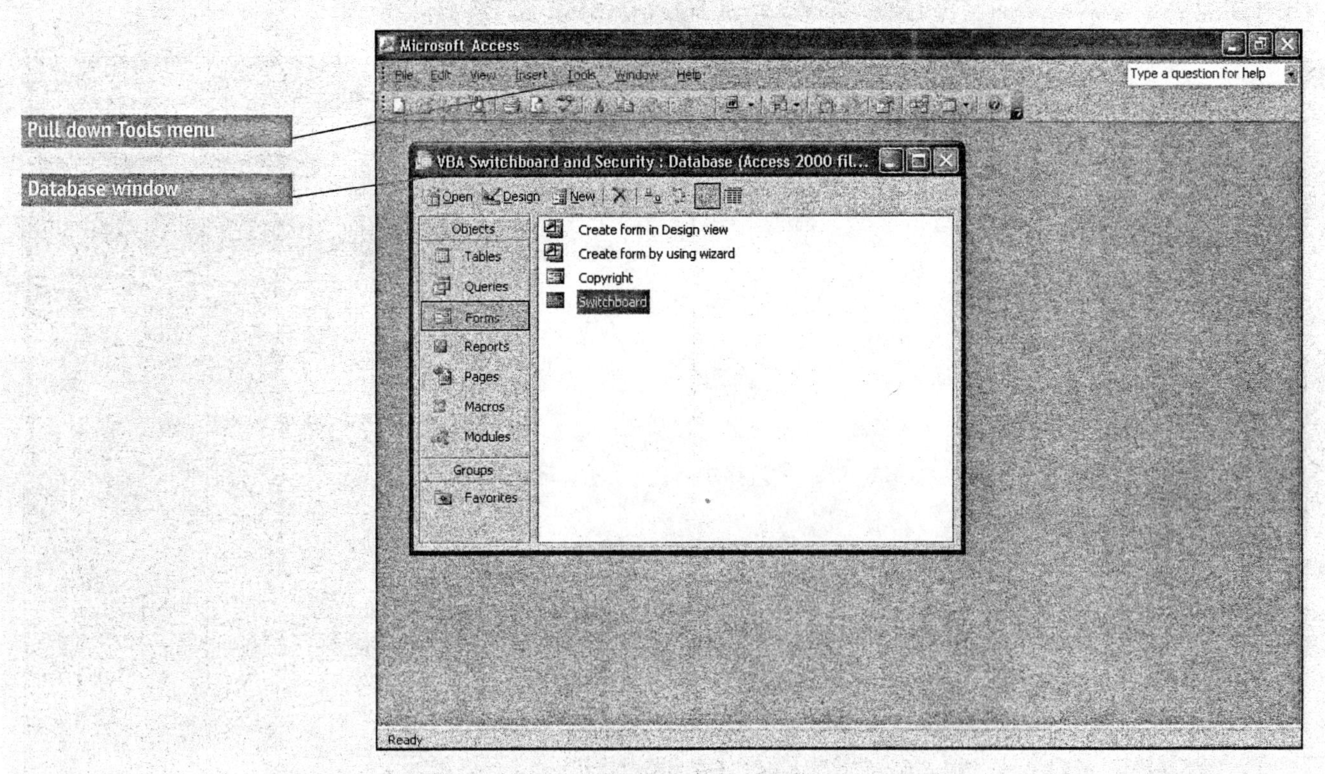

(a) Open the Access Database (step 1)

FIGURE 16 Hands-on Exercise 5

Step 2: Create the ExitDatabase Procedure

- Pull down the **Insert menu** and click **Module** to insert Module1. Complete the **General Declarations section** by adding the Option Explicit statement (if necessary) and the definition of the ApplicationTitle constant as shown in Figure 16b.
- Pull down the **Insert menu** and click **Procedure** to insert a new procedure called **ExitDatabase**. Click the option buttons for a **Sub procedure** and for **Public scope**. Click **OK**.
- Complete the ExitDatabase procedure by entering the **MsgBox** and **DoCmd.Quit** statements. The DoCmd.Quit statement will close Access, but it is entered initially as a comment by beginning the line with an apostrophe.
- Click anywhere in the procedure, then click the **Run Sub button** to test the procedure. Correct any errors that occur, then when the MsgBox displays correctly, **delete the apostrophe** in front of the DoCmd.Quit statement.
- Save the module. The next time you execute the procedure, you should see the message box you just created, and then Access will be terminated.

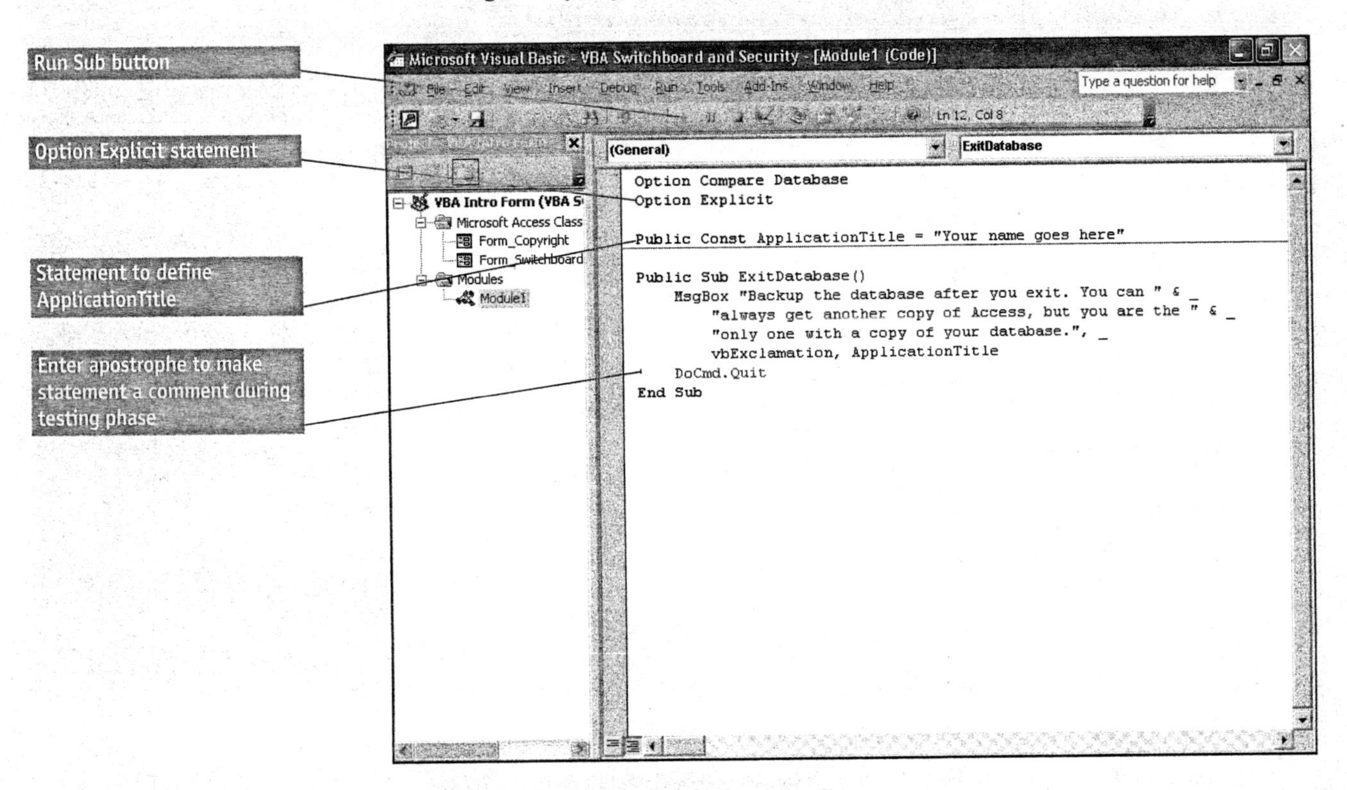

(b) Create the ExitDatabase Procedure (step 2)

FIGURE 16 Hands-on Exercise 5 (*continued*)

CREATE A PUBLIC CONSTANT

Give your application a customized look by adding your name or other identifying message to the title bar of the message and/or input boxes that you use. You can add the information individually to each statement, but it is easier to declare a public constant from within a general module. That way, you can change the value of the constant in one place and have the change reflected automatically throughout your application.

Step 3: Modify the Switchboard

- Click the **View Microsoft Access button** on the Standard toolbar within the VBA window to switch to the Database window (or use the **F11** keyboard shortcut).
- Pull down the **Tools menu**, click the **Database Utilities command**, then choose **Switchboard Manager** to display the Switchboard Manager dialog box in Figure 16c.
- Click the **Edit button** to edit the Main Switchboard and display the Edit Switchboard Page dialog box. Select the **&Exit Application command** and click its **Edit button** to display the Edit Switchboard Item dialog box.
- Change the command to **Run Code**. Enter **ExitDatabase** in the Function Name text box. Click **OK**, then close the two other dialog boxes.
- The switchboard has been modified so that clicking the Exit button will run the VBA procedure you just created.

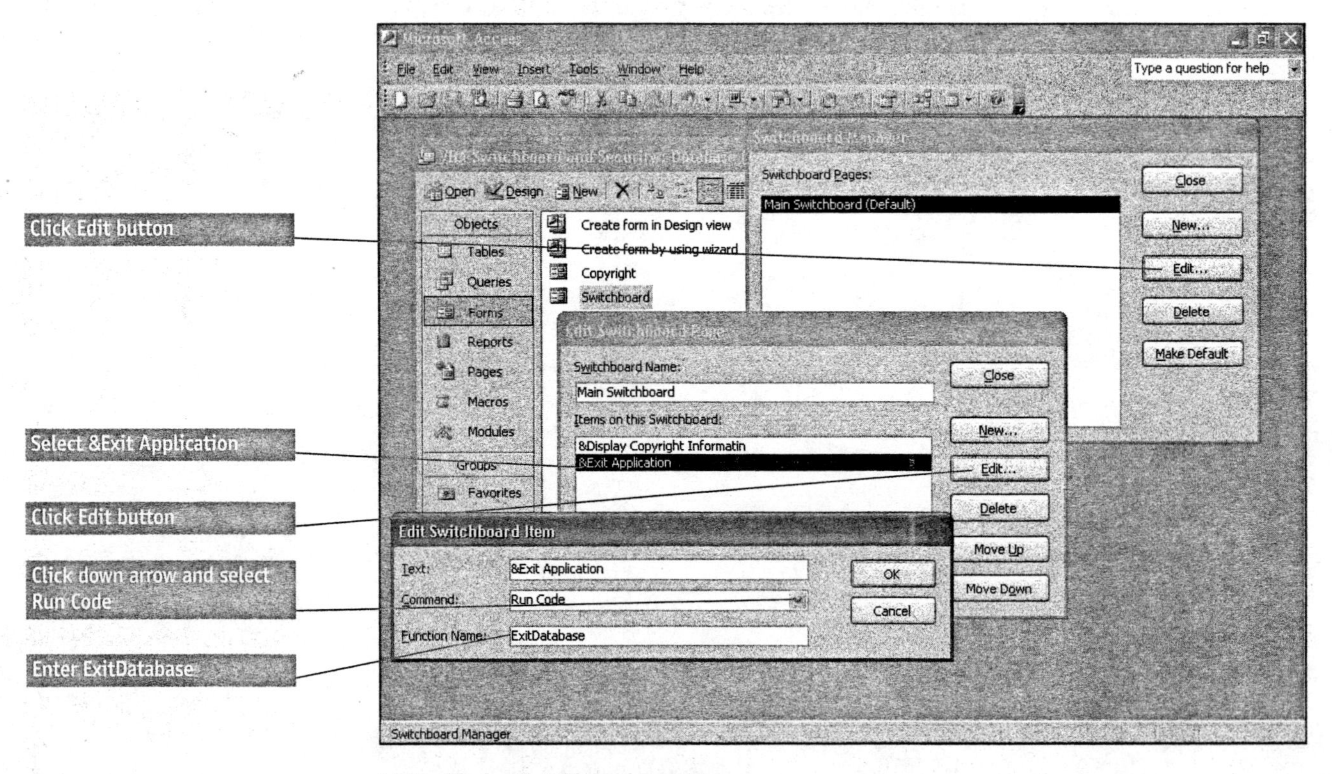

(c) Modify the Switchboard (step 3)

FIGURE 16 Hands-on Exercise 5 (*continued*)

CREATE A KEYBOARD SHORTCUT

The & has special significance when used within the name of an Access object because it creates a keyboard shortcut to that object. Enter "&Exit Application", for example, and the letter E (the letter immediately after the ampersand) will be underlined and appear as "Exit Application" on the switchboard. From there, you can execute the item by clicking its button, or you can use the Alt+E keyboard shortcut (where "E" is the underlined letter in the menu option).

Step 4: Test the Switchboard

- If necessary, click the **Forms button** in the Database window. Double click the **Switchboard form** to open the switchboard as shown in Figure 16d. The switchboard contains two commands.
- Click the **Display Copyright Information command** to display a form that we use with all our databases. (You can open this form in Design view and modify the text to include your name, rather than ours. If you do, be sure to save the modified form, then close it.)
- Click the **Exit Application command** (or use the **Alt+E** keyboard shortcut). You should see the dialog box in Figure 16d, corresponding to the MsgBox statement you created earlier. Click **OK** to close the dialog box.
- Access itself will terminate because of the DoCmd.Quit statement within the ExitDatabase procedure. (If this does not happen, return to the VBA editor and remove the apostrophe in front of the DoCmd statement.)

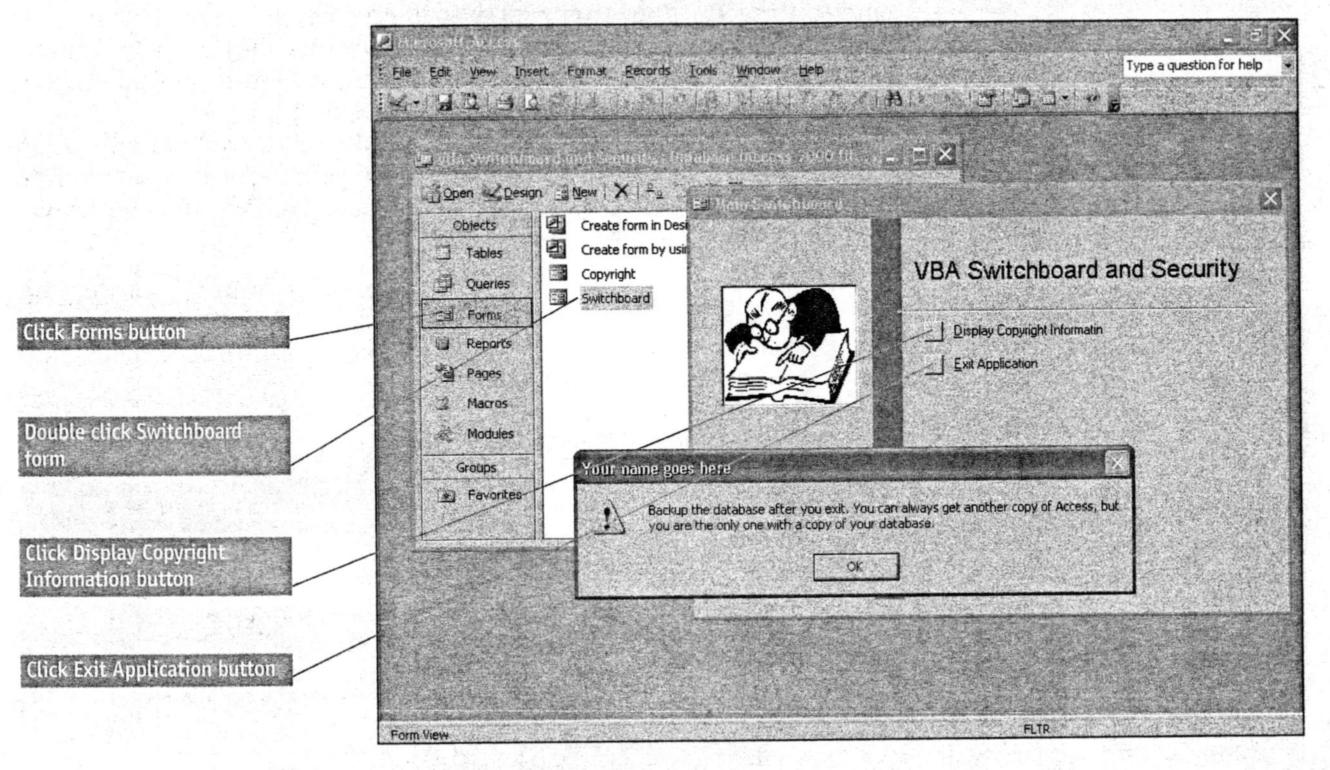

(d) Test the Switchboard (step 4)

FIGURE 16 Hands-on Exercise 5 (*continued*)

BACK UP IMPORTANT FILES

It's not a question of *if* it will happen, but *when*—hard disks die, files are lost, or viruses may infect a system. It has happened to us, and it will happen to you, but you can prepare for the inevitable by creating adequate backup before the problem occurs. The essence of a backup strategy is to decide which files to back up (your data), how often to do the backup (whenever it changes), and where to keep the backup (away from your computer). Do it!

Step 5: Complete the Open Form Event Procedure

- Start Access and reopen the **VBA Switchboard and Security database**. Press **Alt+F11** to start the VBA editor.
- Click the **plus sign** next to Microsoft Office Access Class objects, double click the module called **Form_Switchboard**, then look for the partially completed **Form_Open procedure** as shown in Figure 16e.
- The procedure was created automatically by the Switchboard Manager. You must, however, expand this procedure to include password protection. Note the following:
  - Three variables are required—the correct password, the password entered by the user, and the number of attempts.
  - The user is prompted for the password, and the number of attempts is set to 1. The user is given two additional attempts, if necessary, to get the correct password.
  - The If statement at the end of the loop determines whether the user has entered the correct password, and if so, it executes the original commands that are associated with the switchboard. If, however, the user fails to supply the correct password, an invalid password message is displayed and the **DoCmd.Quit** statement terminates the application.
  - We suggest you place an **apostrophe** in front of the statement initially so that it becomes a comment, and thus it is not executed. Once you are sure that you can enter the correct password, you can remove the apostrophe and implement the password protection.
- Save the procedure. You cannot test this procedure from within the VBA window; you must cause the event to happen (i.e., open the form) for the procedure to execute. Click the **View Microsoft Access button** on the Standard toolbar to return to the Database window.

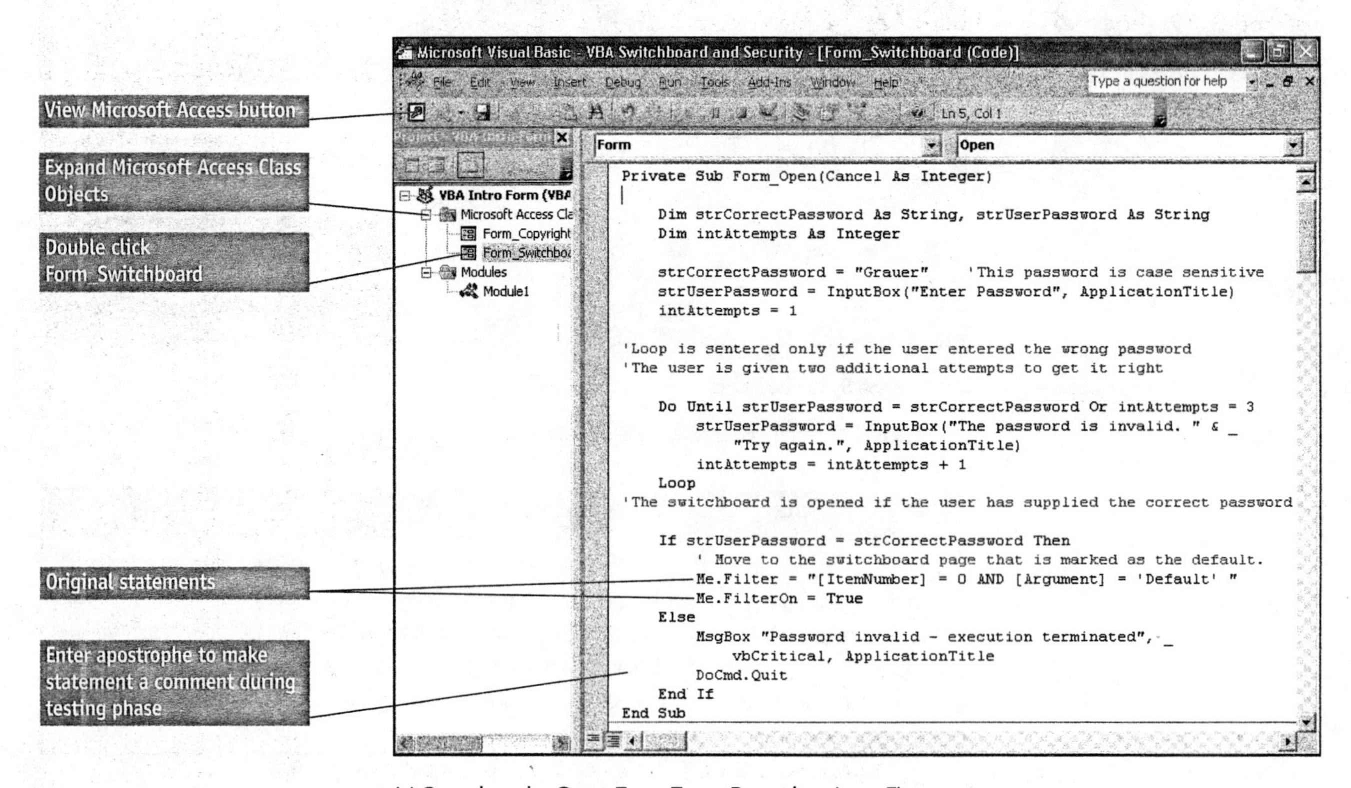

(e) Complete the Open Form Event Procedure (step 5)

FIGURE 16 Hands-on Exercise 5 (*continued*)

Step 6: **Test the Procedure**

- Close all open windows within the Access database except for the Database window. Click the **Forms button**, then double click the **Switchboard form**.
- You should be prompted for the password as shown in Figure 16f. The password (in our procedure) is **Grauer**.
- Test the procedure repeatedly to include all possibilities. Enter the correct password on the first, second, and third attempts to be sure that the procedure works as intended. Each time you enter the correct password, you will have to close the switchboard, then reopen it.
- Test the procedure one final time, by failing to enter the correct password. You will see a message box indicating that the password is invalid and that execution will be terminated. Termination will not take place, however, because the DoCmd.Quit statement is currently entered as a comment.
- Press **Alt+F11** to reopen the VBA editor. Open the **Microsoft Access Class Objects folder** and double click on **Form_Switchboard**. Delete the apostrophe in front of the DoCmd.Quit statement. The text of the statement changes from green to black to indicate that it is an executable statement. Save the procedure.

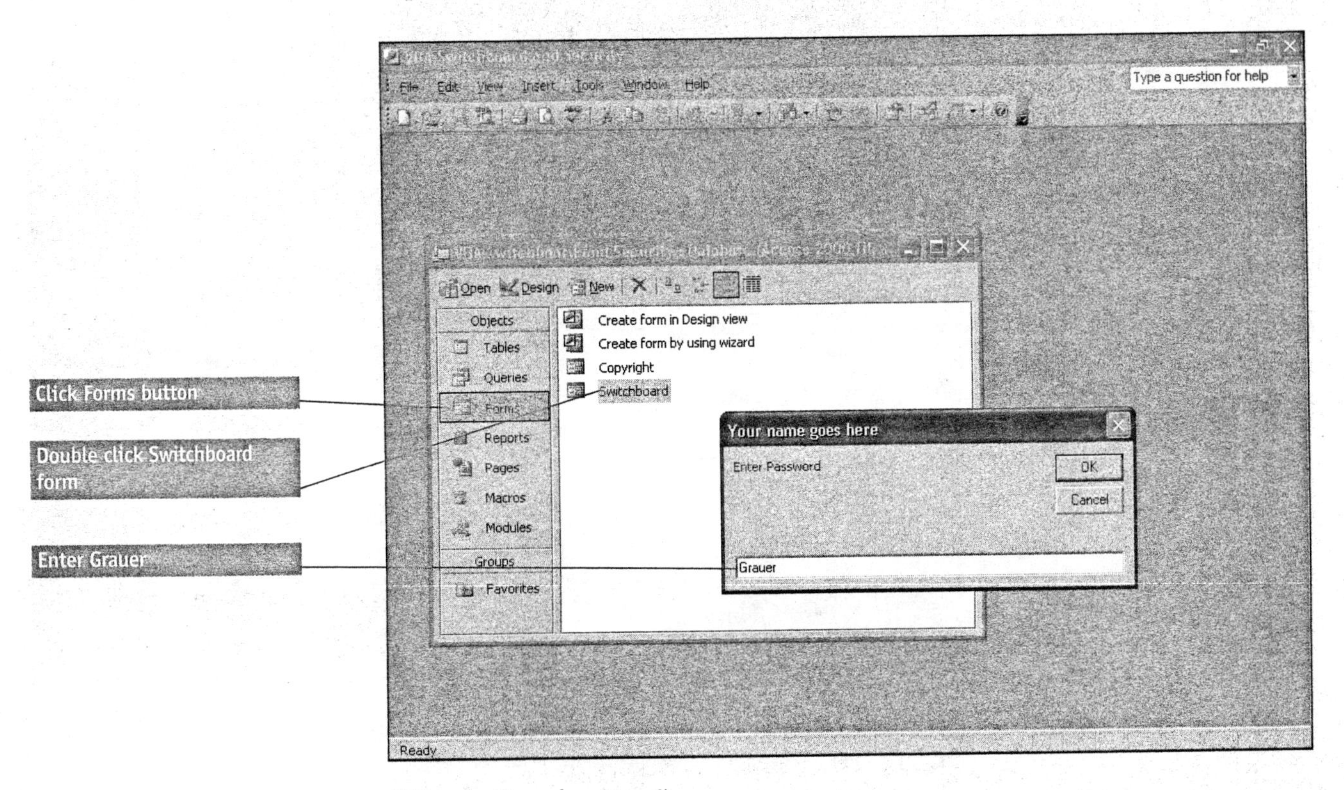

(f) Test the Procedure (step 6)

FIGURE 16 Hands-on Exercise 5 (*continued*)

TOGGLE COMMENTS ON AND OFF

Comments are used primarily to explain the purpose of VBA statements, but they can also be used to "comment out" code as distinct from deleting the statement altogether. Thus, you can add or remove the apostrophe in front of the statement, to toggle the comment on or off.

Step 7: Change the Startup Properties

- Click the **View Microsoft Access button** on the VBA Standard toolbar to return to the Database window.
- Close all open windows except the Database window. Pull down the **Tools menu** and click **Startup** to display the Startup dialog box as shown in Figure 16g.
- Click in the **Application Title** text box and enter the title of the application, **VBA Switchboard and Security** in this example.
- Click the **drop-down arrow** in the Display Form/Page list box and select the **Switchboard form** as the form that will open automatically in conjunction with opening the database.
- Clear the check box to display the Database window. Click **OK** to accept the settings and close the dialog box.
- The next time you open the database, the switchboard should open automatically, which in turn triggers the Open Form event procedure that will prompt the user to enter a password.

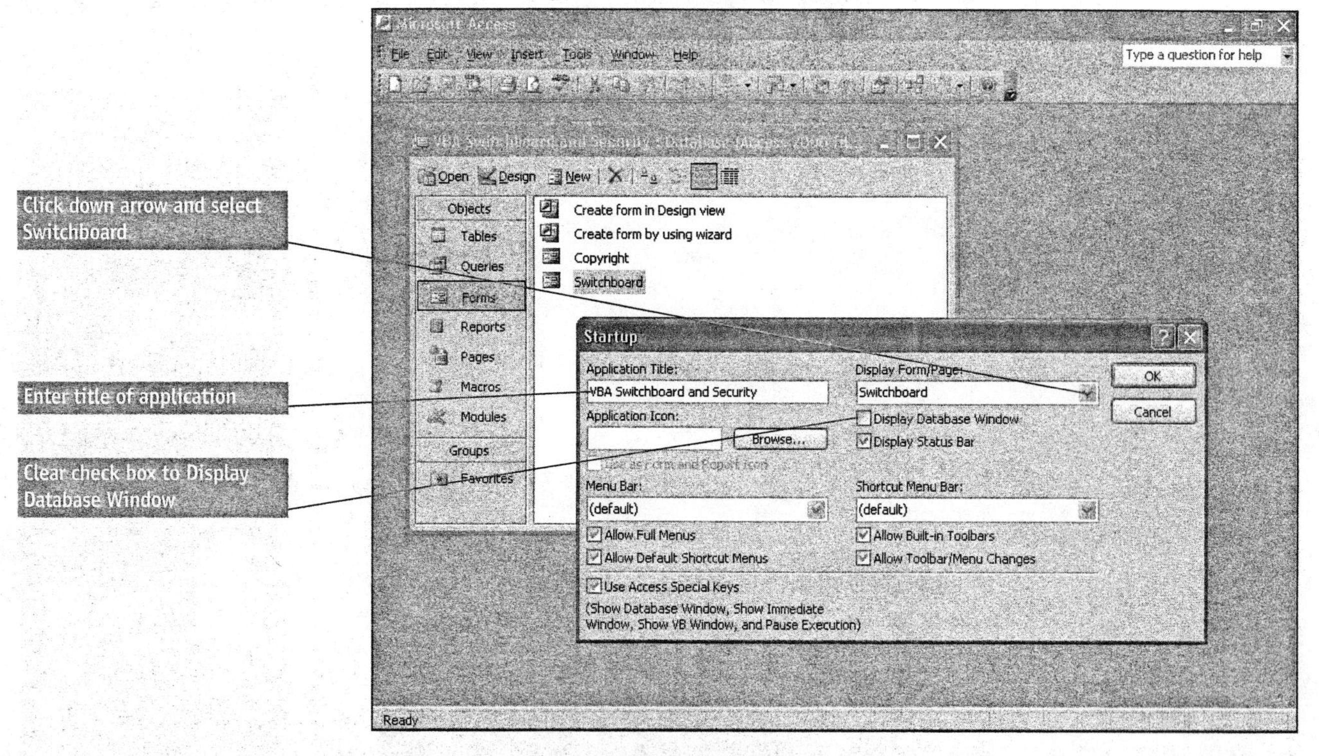

(g) Change the Startup Properties (step 7)

FIGURE 16 Hands-on Exercise 5 (*continued*)

HIDE THE DATABASE WINDOW

Use the Startup property to hide the Database window from the novice user. You avoid confusion and you may prevent the novice from accidentally deleting objects in the database. Of course, anyone with some knowledge of Access can restore the Database window by pulling down the Window menu, clicking the Unhide command, then selecting the Database window from the associated dialog box. Nevertheless, hiding the Database window is a good beginning.

Step 8: Test the Database

- Close the database, then reopen the database to test the procedures we have created in this exercise. The sequence of events is as follows:
  - The database is loaded and the switchboard is opened but is not yet visible. The Open Form procedure for the switchboard is executed, and you are prompted for the password as shown in Figure 16h.
  - The password is entered correctly and the switchboard is displayed. The Database window is hidden, however, because the Startup Properties have been modified.
- Click the **Exit Application command** (or use the **Alt+E** keyboard shortcut). You will see the message box reminding you to back up the system, after which the database is closed and Access is terminated.
- Reopen the database. This time, however, you are to enter the wrong password three times in a row. You should see a message indicating that the execution was terminated due to an invalid password.
- Testing is complete and you can go on to add the other objects to your Access database. Congratulations on a job well done.

Enter password

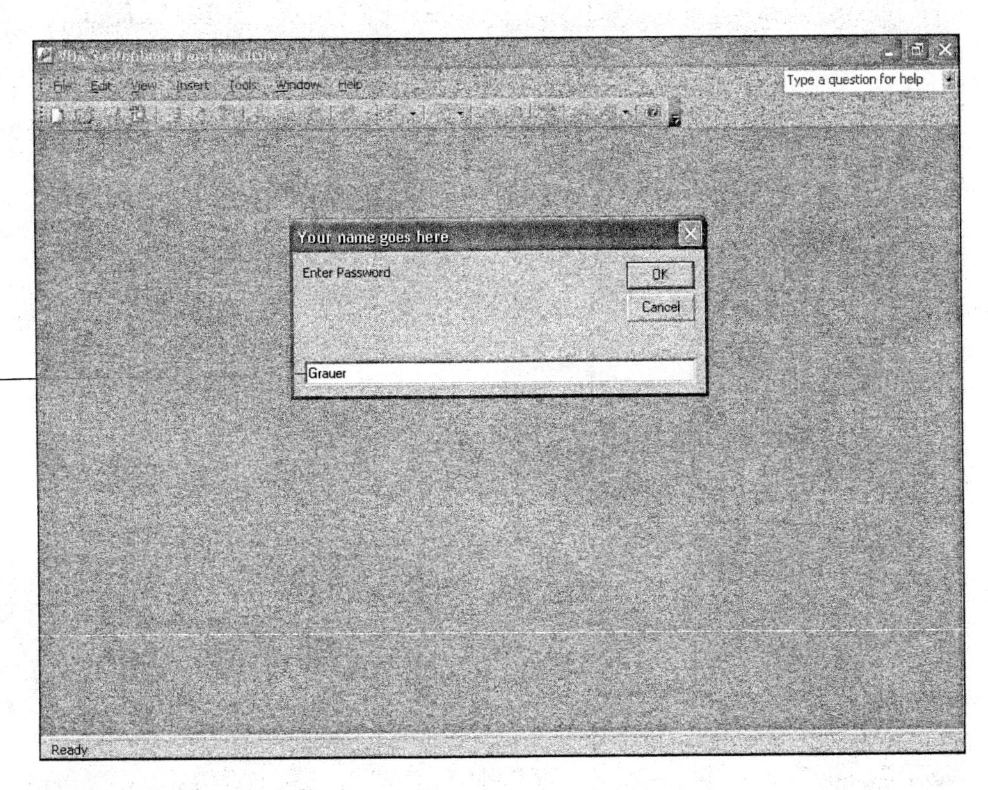

(h) Test the Database (step 8)

FIGURE 16 Hands-on Exercise 5 (*continued*)

RESTORING HIDDEN MENUS AND TOOLBARS

You can use the Startup property to hide menus and/or toolbars from the user by clearing the respective check boxes. A word of caution, however—once the menus are hidden, it is difficult to get them back. Start Access, pull down the File menu, and click Open to display the Open dialog box, select the database to open, then press and hold the Shift key when you click the Open button. This powerful technique is not widely known.

SUMMARY

Visual Basic for Applications (VBA) is a powerful programming language that is accessible from all major applications in Microsoft Office XP. A VBA statement accomplishes a specific task such as displaying a message to the user or accepting input from the user. Statements are grouped into procedures, and procedures in turn are grouped into modules. Every procedure is classified as either private or public.

The MsgBox statement displays information to the user. It has one required argument, which is the message (or prompt) that is displayed to the user. The other two arguments—the icon that is to be displayed in the dialog box and the text of the title bar—are optional. The InputBox function displays a prompt to the user requesting information, then it stores that information (the value returned by the user) for use later in the procedure.

Every variable must be declared (defined) before it can be used. This is accomplished through the Dim (short for Dimension) statement that appears at the beginning of a procedure. The Dim statement indicates the name of the variable and its type (for example, whether it will hold a character string or an integer number), which in turn reserves the appropriate amount of memory for that variable.

The ability to make decisions within a procedure, then branch to alternative sets of statements is implemented through the If . . . Then . . . Else or Case statements. The Else clause is optional, but may be repeated multiple times within an If statement. The Case statement is preferable to an If statement with multiple Else clauses.

The For . . . Next statement (or For . . . Next loop as it is also called) executes all statements between the words For and Next a specified number of times, using a counter to keep track of the number of times the loop is executed. The Do . . . Loop Until and/or Do Until . . . Loop statements are used when the number of times through the loop is not known in advance.

VBA is different from traditional programming languages in that it is event-driven. An event is defined as any action that is recognized by an application, such as Excel or Access. Opening or closing an Excel workbook or an Access database is an event. Selecting a worksheet within a workbook is also an event, as is clicking on a command button on an Access form. To use VBA within Microsoft Office, you decide which events are significant, and what is to happen when those events occur. Then you develop the appropriate event procedures.

KEY TERMS

Argument 2
Case statement 18
Character string 16
Comment 6
Compilation error 30
Complete Word tool 22
Concatenate 4
Custom toolbar 19
Debug menu 30
Debug toolbar 30
Debug.Print statement 31
Debugging 30
Declarations section 6
Dim statement 5
Do Loops 29
Else clause 16
End Sub statement 3
Event 41
Event procedure (Access) 52
Event procedure (Excel) 41
Execution error 30
For . . . Next Statement 28
Full Module view 21
If statement 16
Immediate window 31
InputBox function 4
Intrinsic constant 3
Literal 16
Locals window 31
Macro 2
Macro recorder 2
Module 2
MsgBox statement 3
Option Explicit 6
Private procedure 2
Procedure 2
Procedure header 3
Project Explorer 6
Public procedure 2
Run-time error 30
Statement 2
Step Into button 31
Syntax 2
Underscore 4
Variable 4
VBA 2
Visible property 50
Visual Basic editor 6
Visual Basic for Applications 2

MULTIPLE CHOICE

1. Which of the following applications in Office XP has access to VBA?
   (a) Word
   (b) Excel
   (c) Access
   (d) All of the above

2. Which of the following is a valid name for a VBA variable?
   (a) Public
   (b) Private
   (c) strUserFirstName
   (d) int Count Of Attempts

3. Which of the following is true about an If statement?
   (a) It evaluates a condition as either true or false, then executes the statement(s) following the keyword "Then" if the condition is true
   (b) It must contain the keyword Else
   (c) It must contain one or more ElseIf statements
   (d) All of the above

4. Which of the following lists the items from smallest to largest?
   (a) Module, procedure, statement
   (b) Statement, module, procedure
   (c) Statement, procedure, module
   (d) Procedure, module, statement

5. Given the statement, MsgBox "Welcome to VBA", "Bob was here", which of the following is true?
   (a) "Welcome to VBA" will be displayed within the resulting message box
   (b) "Welcome to VBA" will appear on the title bar of the displayed dialog box
   (c) The two adjacent commas will cause a compilation error
   (d) An informational icon will be displayed with the message

6. Where are the VBA procedures associated with an Office document stored?
   (a) In the same folder, but in a separate file
   (b) In the Office document itself
   (c) In a special VBA folder on drive C
   (d) In a special VBA folder on the local area network

7. The Debug.Print statement is associated with the:
   (a) Locals window
   (b) Immediate window
   (c) Project Explorer
   (d) Debug toolbar

8. Which of the following is the proper sequence of arguments for the MsgBox statement?
   (a) Text for the title bar, prompt, button
   (b) Prompt, button, text for the title bar
   (c) Prompt, text for the title bar, button
   (d) Button, prompt, text for the title bar

9. Which of the following is a true statement about Do loops?
   (a) Placing the Until clause at the beginning of the loop tests the condition prior to executing any statements in the loop
   (b) Placing the Until clause at the end of the loop executes the statements in the loop, then it tests the condition
   (c) Both (a) and (b)
   (d) Neither (a) nor (b)

10. Given the statement, For intCount = 1 to 10 Step 3, how many times will the statements in the loop be executed (assuming that there are no statements in the loop to terminate the execution)?
    (a) 10
    (b) 4
    (c) 3
    (d) Impossible to determine

... continued

multiple choice

11. Which of the following is a *false* statement?

(a) A dash at the end of a line indicates continuation
(b) An ampersand indicates concatenation
(c) An apostrophe at the beginning of a line signifies a comment
(d) A pair of quotation marks denotes a character string

12. What is the effect of deleting the apostrophe that appears at the beginning of a VBA statement?

(a) A compilation error will occur
(b) The statement is converted to a comment
(c) The color of the statement will change from black to green
(d) The statement is made executable

13. Which of the following If statements will display the indicated message if the user enters a response other than "Grauer" (assuming that "Grauer" is the correct password)?

(a) If strUserResponse <> "Grauer" Then MsgBox "Wrong password"
(b) If strUserResponse = "Grauer" Then MsgBox "Wrong password"
(c) If strUserResponse > "Grauer" Then MsgBox "Wrong password"
(d) If strUserResponse < "Grauer" Then MsgBox "Wrong password"

14. Which of the following will execute the statements in the loop at least once?

(a) Do . . . Loop Until
(b) Do Until ..... Loop
(c) Both (a) and (b)
(d) Neither (a) nor (b)

15. The copy and paste commands can be used to:

(a) Copy statements within a procedure
(b) Copy statements from a procedure in one module to a procedure in another module within the same document
(c) Copy statements from a module in an Excel workbook to a module in an Access database
(d) All of the above

16. Which of the following is true about indented text in a VBA procedure?

(a) The indented text is always executed first
(b) The indented text is always executed last
(c) The indented text is rendered a comment and is never executed
(d) None of the above

17. Which statement will prompt the user to enter his or her name and store the result in a variable called strUser?

(a) InputBox.strUser
(b) strUser = MsgBox("Enter your name")
(c) strUser = InputBox("Enter your name")
(d) InputBox("Enter strUser")

18. Given that strUser is currently set to "George", the expression "Good morning, strName" will return:

(a) Good morning, George
(b) Good morning, strName
(c) Good morning George
(d) Good morning strName

ANSWERS

1. d	**7.** b	**13.** a
2. c	**8.** b	**14.** a
3. a	**9.** c	**15.** d
4. c	**10.** b	**16.** d
5. a	**11.** a	**17.** c
6. b	**12.** d	**18.** b